3D PRINTING AND ADDITIVE MANUFACTURING

PRINCIPLES AND APPLICATIONS

(The 4th edition of Rapid Prototyping: Principles and Applications)

3D PRINTING AND ADDITIVE MANUFACTURING

PRINCIPLES AND APPLICATIONS

(The 4th edition of Rapid Prototyping: Principles and Applications)

Chee Kai Chua
Kah Fai Leong

Nanyang Technological University, Singapore

NEW JERSEY • LONDON • SINGAPORE • BEIJING • SHANGHAI • HONG KONG • TAIPEI • CHENNAI

Published by

World Scientific Publishing Co. Pte. Ltd.
5 Toh Tuck Link, Singapore 596224
USA office: 27 Warren Street, Suite 401-402, Hackensack, NJ 07601
UK office: 57 Shelton Street, Covent Garden, London WC2H 9HE

Library of Congress Cataloging-in-Publication Data
Chua, Chee Kai.
[Rapid prototyping]
3D printing and additive manufacturing : principles and applications / by Chee Kai Chua (Nanyang Technological University, Singapore) & Kah Fai Leong (Nanyang Technological University, Singapore). -- Fourth edition of Rapid prototyping (with companion CD-ROM).
pages cm
Includes bibliographical references.
ISBN 978-9814571401 (hardback : alk. paper) -- ISBN 978-9814571418 (pbk. : alk. paper)
1. Rapid prototyping. 2. CAD/CAM systems. 3. Three-dimensional printing. 4. Solid freeform fabrication. I. Leong, Kah Fai. II. Title. III. Title: Three dimensional printing and additive manufacturing.
TS155.6.C498 2014
620'.0042--dc23

2014017340

British Library Cataloguing-in-Publication Data
A catalogue record for this book is available from the British Library.

First published 2015
Reprinted 2016

In-house Editor: Amanda Yun

Printed in Singapore

To my wife, Wendy, and children, Cherie, Clement and Cavell, whose forbearance, support and motivation have made it possible for us to finish writing this book.

Chee Kai

To Soi Lin, for her patience and support, and Qian, who brings us cheer and joy.

Kah Fai

FOREWORD

Additive manufacturing and 3D printing (we use the two terms interchangeably) are moving quickly and in multiple directions. The level of activity is at an all-time high as researchers, investors, company management, and government agencies try to predict where it is headed. Many believe it is the next Big Thing. Organizations of all types and sizes are trying to understand the role they might play and when.

At the high end, companies are at work qualifying machines and materials for the direct manufacture of parts that go into final products. Aerospace companies such as Aerosud, Airbus, Boeing, and Honeywell Aerospace are qualifying AM processes and materials and certifying new designs for flight. GE Aviation announced that fuel nozzles for its next generation LEAP aircraft engine will be manufactured by AM. With 19 nozzles per engine and the production of 1,700 engines per year, the company expects to produce more than 32,000 very complex metal parts through AM annually.

When considering the advantages of AM, design and redesign are very important. With AM, it is possible to consolidate many individual parts of an assembly (as many as 20 or more) into a single, complex part. This is precisely what GE Aviation has done. This approach to design eliminates part numbers, inventory, assembly, labour, and inspection. With AM, it is possible to redesign parts with relatively thin skins that include internal lattice/mesh structures instead of solid material throughout, which can substantially reduce the amount of material, weight, and build time. It is also possible to redesign parts using topology optimisation—a method of letting mathematics decide where to put the material to optimise the strength to weight ratio. In some cases, the amount of material and weight has been reduced by more than 50% using these techniques.

Challenges abound, however. Amongst them are system reliability and process repeatability, especially when using AM for manufacturing. System manufacturers are addressing these challenges with real-time process monitoring and control software, but a lot of work remains. The current limitation in build speed and maximum part size are challenges, too. Manufacturers are developing systems with larger build volumes and methods that increase throughput.

Another challenge is the unprecedented amount of media attention and hype seen in the past year surrounding the technology. It seems as if some people believe 3D printing is a superior manufacturing method just because it is relatively new and different. Of course, this is not true. AM is a viable alternative to conventional methods in some cases, but not most at the present time. All opportunities for manufacturing must be carefully evaluated on a case-by-case basis.

Misconceptions about the technology are mentioned by many. For example, significant differences exist between industrial-grade systems and the sub-$3,000 3D printers that the mainstream media has discovered. Also, AM is relatively slow, and economies of scale apply to AM, but not in the same way as conventional methods of manufacturing. AM processes are generally more energy-intensive than conventional processes. AM is not a "push button" technology, and a lot of pre- and post-build work is required. AM is not inherently superior to subtractive or formative manufacturing, although it depends on the types of parts being built.

It is difficult to predict which industries will be most impacted or changed by additive manufacturing. The likely candidate industries in the foreseeable future are those that require low production volumes of high-value, highly complex parts. The aerospace, medical, and dental industries fall into this category. However, as system speeds increase, equipment and material costs decline, and more materials become available, new manufacturing applications will emerge in a range of industries. One example is the automotive industry, which has used AM for two decades for prototyping, but is not currently using it for most production applications. This could, and probably will, change in the future.

It is an exciting time for design and manufacturing. Never before have so many new options become available. The additive manufacturing industry has had 25 years to learn how to apply the technology to modelling and prototyping. For the most part, AM is well understood for these applications. The next frontier is applying AM to the production of parts for final products. This is where the largest opportunities lie and where most investments will be made in the future.

Terry Wohlers
Wohlers Associates, Inc.

PREFACE

The focus on productivity has been one of the main concerns of industries worldwide since the turn of new millennium. To increase productivity, industries have attempted to apply more computerised automation in manufacturing. Amongst the latest technologies to have significant stride over the past two and a half decades are the *Rapid Prototyping (RP) Technologies*, otherwise also known as *3D Printing* or, according to ASTM standards, now formally known as *Additive Manufacturing (AM)*.

The revolutionary change in factory production techniques and management requires a direct involvement of computer-controlled systems in the entire production process. Every operation in this factory, from product design, to manufacturing, to assembly and product inspection, is monitored and controlled by computers. CAD-CAM or Computer-Aided Design and Manufacturing has emerged since the 1960s to support product design. Up to the mid-1980s, it has never been easy to derive a physical prototype model, despite the existence of CNC (or Computer Numerical Controlled) machine tools. Early Rapid Prototyping Technologies provided that bridge from product conceptualisation to product realisation in a reasonably quick manner, without the fuss of NC programming, jigs and fixtures.

In recent years, there has been a growing trend in the development and use of AM. The Economist has forecasted that AM could be the 3rd industrial revolution due to its prospects of thriving into a new type of manufacturing industry. Manufacturing is an established and necessary sector in the spectrum of any world's economy. It makes significant contributions to a country's gross domestic production and national security through the constant supply of essential products and services.

From a local perspective of how the world adopts and adapts to AM, there is no reason for Singapore to stay stagnant. Indeed, there is a little internal pressure for Singapore to focus on AM development due to her relatively smaller market, but the external push is very real. First world countries, such as the United States, the United Kingdom and Australia have already established national centres or institutes on AM at various scales, setting an example for other advanced countries. In Asia, China is closely following this technological trend and has attempted to develop its own AM brands. However, the East is still far behind the West in general. Singapore, as a place where East meets West and a place where AM can be rooted back to the late 1980s, must take the leadership in developing new AM-based manufacturing technologies and training future AM engineer leaders, for Singapore as well as for the Asia Pacific region.

With this exciting promise, the industry and academia have internationally established research centres for Additive Manufacturing (AM), with the objectives of working in this leading edge and disruptive technology, as well as of educating and training more engineers in the field of AM. On that same premise, the Nanyang Technological University Additive Manufacturing Centre (NAMC) was founded in July 2013 to be the regional leader in Asia Pacific in the additive manufacturing technologies. Currently, there are seven AM processes categorised under the ASTM standard and each process has its unique advantages and challenges. All of these are covered in this book.

The NAMC focuses on the powder bed fusion process, which is one of the most accepted and (in the view of some, the) most important AM processes. However, much of this process is still under research today. The NAMC initiative aims to realise its full potential. NAMC will tap on the existing local AM experience and expertise to set up a large scale AM facility for research and education to generate new knowledge, create novel technologies and train manpower, especially in the area of Selective Laser Sintering, Selective Laser Melting and Electron Beam Melting. In addition, a conducive AM environment akin to the NAMC favours the development and building of AM capabilities and strengths.

Beyond that, these can be directly translated and extended into the industry for a variety of applications. By taking advantage of Singapore's position and manufacturing reputation in the aerospace and marine industries, coupled with anticipating infrastructures and an accessible pool of talents, NAMC offers an ideal setting in Asia to attract and support global businesses by focusing on three core sectors: Research, Education and Application.

In order to continue the focus on education in AM, this textbook, already in its fourth edition, is therefore needed as the basis for the development of a curriculum in AM. Like its previous three editions, the purpose of this updated book is to provide an introduction to the fundamental principles and application areas in AM. The book traces the development of AM in the arena of Advanced Manufacturing Technologies and explains the principles underlying each of the common AM techniques. Also covered are the detailed descriptions of the AM processes and their specifications. In this fourth edition, new AM techniques are introduced and existing ones updated, bringing the total number of AM techniques described to more than 30. To reflect the new terminology and for consistency, we have replaced "Rapid Prototyping" which was used from the first to third edition to "Additive Manufacturing". "3D Printing" will sometimes be used as a common term to refer to "Additive Manufacturing". The book would not be complete without emphasising the importance of AM applications in manufacturing and other industries. In addition to providing industrial examples for each vendor, an entire chapter is devoted to application areas. As AM has expanded its scope of applications, one whole chapter that focuses on the biomedical arena has been added. New applications include food printing, wearable fashion, garment printing, weapon printing and 3D printing in construction and building.

One key inclusion in this book is the use of multimedia to enhance the reader's understanding of the technique. In the accompanying Companion Media Pack, video and animation are used to demonstrate the working principles of major AM techniques such as Stereolithography, Polyjet, Selective Deposition Lamination, Fused

Deposition Modelling, Selective Laser Sintering and Colorjet Printing. In addition, the book focuses on some of the very important issues facing AM today, and these include, but are not limited to:

1. The problems with the *de facto* STL format.
2. The range of applications for tooling and manufacturing, including biomedical engineering.
3. The benchmarking methodology in selecting an appropriate AM technique.

The material in this book has been used, revised and updated several times for professional courses conducted for both academia and industry audiences since 1991. Certain materials were borne out of research conducted in the School of Mechanical and Aerospace Engineering at the Nanyang Technological University, Singapore.

To be used more effectively for graduate or final year (senior year) undergraduate students in Mechanical, Aerospace, Production or Manufacturing Engineering, problems have been included in this textbook. For university professors and other tertiary-level lecturers, the subject of AM can be combined easily with other topics such as: CAD, CAM, Machine Tool Technologies and Industrial Design.

Chua C. K.
Professor

Leong K. F.
Associate Professor

School of Mechanical and Aerospace Engineering
Nanyang Technological University
50 Nanyang Avenue
Singapore 639798

ACKNOWLEDGEMENTS

First, we would like to thank God for granting us His strength throughout the writing of this book. Secondly, we are especially grateful to our respective spouses, Wendy and Soi Lin, and our respective children, Cherie, Clement, Cavell Chua and Leong Qianyu for their patience, support and encouragement throughout the year it took to complete this edition.

We wish to thank the valuable support from the administration of Nanyang Technological University (NTU), especially the School of Mechanical and Aerospace Engineering (MAE) in particular the NTU Additive Manufacturing Centre (NAMC). In addition, we would like to thank our current and former students and research fellows, Dr. An Jia, Dr. Khalid Rafi Haludeen Kutty, Dr. Alex Liu, Dr. Zhang Danqing, Chan Lick Khor, Chong Fook Tien, Chow Lai May, Esther Chua Hui Shan, Derrick Ee, Foo Hui Ping, Witty Goh Wen Ti, Ho Ser Hui, Ketut Sulistyawati, Kwok Yew Heng, Angeline Lau Mei Ling, Jordan Liang Dingyuan, Liew Weng Sun, Lim Choon Eng, Liu Wei, Phung Wen Jia, Mahendra Suryadi, Mohamed Syahid Hassan, Anfee Tan Chor Kwang, Tan Leong Peng, Toh Choon Han, Wee Kuei Koon, Yap Kimm Ho, Yim Siew Yen, Yun Bao Ling, Zhang Jindong, and our colleague, Mr. Lee Kiam Gam for their valuable contributions that made the multimedia pack possible.

We would also like to express sincere appreciation to our special assistants Chu Shih Zoe, Lu Zhen, Vu Trong Thien, Ng Chyi Huey, Wang Yawei and Jesamine Lim for their selfless help and immense effort in the coordination and timely publication of this book.

Much of the research work which have been published in various journals and now incorporated into some chapters in the book can be attributed to our former students and colleagues, Dr. May Win Naing, Dr. Yeong Wai Yee, Dr. Florencia Edith Wiria, Dr. Novella Sudarmadji, Dr. Tan Jia Yong, Dr. Jolene Liu, Dr. Liu Dan, Dr, Zhang Danqing, Tong Mei, Micheal Ko, Ang Ker Chin, Tan Kwang Hui, Liew Chin Liong, Gui Wee Siong, Simon Cheong, William Ng, Chong Lee Lee, Lim Bee Hwa, Chua Ghim Siong, Ko Kian Seng, Chow Kin Yean, Verani, Evelyn Liu, Wong Yeow Kong, Althea Chua, Terry Chong, Tan Yew Kwang, Chiang Wei Meng, Ang Ker Ser, Lin Sin Chew, Ng Chee Chin and Melvin Ng Boon Keong, and colleagues Professor Robert Gay, Mr. Lee Han Boon, Dr. Jacob Gan, Dr. Du Zhaohui, Dr. M. Chandrasekeran, Dr. Cheah Chi Mun, Dr. Daniel Lim Chu Sing and Dr. Georg Thimm.

We would also like to extend our special appreciation to Mr. Terry Wohlers for his foreword, and Professor Richard Buswell of Loughborough University, Professor Michael C McAlpine of Princeton University, Associate Professor Hod Lipson of Cornell University, Dr. Ming Leu of the University of Missouri-Rolla, Dr. Amba D. Bhatt of the National Institute of Standards and Technology, Gaithersburg, MD, Ruston, and Prof. Fritz Prinz of Stanford University, USA for their assistance.

The acknowledgements would not be complete without the contributions of the following companies for supplying and helping us with the information about their products they develop, manufacture or represent:

1. 3D Systems Inc., USA
2. 3D-Micromac AG, Germany
3. Arcam AB, Sweden
4. Aeromet Corp., USA
5. Alpha Products and Systems Pte Ltd, Singapore
6. Blizzident, USA
7. CAM-LEM Inc., USA
8. Carl Zeiss Pte Ltd, Singapore

9. Champion Machine Tools Pte Ltd, Singapore
10. CMET Inc., Japan
11. Concept Laser GmbH, Germany
12 Creatz3d Pte Ltd, Singapore
13. Cubic Technologies Inc., USA
14. Cubital Ltd., Israel
15. Cybron Technology (S) Pte Ltd, Singapore
16. Defense Distributed, USA
17. EnvisionTec., Germany
18. Ennex Corporation, USA
19. EOS GmbH, Germany
20. Festo, Germany
21. Fraunhofer-Institute for Applied Materials Research, Germany
22. Fraunhofer-Institute for Manufacturing Engineering and Automation, Germany
23. Innomation Systems and Technologies Pte Ltd, Singapore
24. Kira Corporation, Japan
25. MCOR Technologics, Ireland
26. Meiko's RPS Co., Ltd., Japan
27. MIT, USA
28. Materialise, Belgium
29. Optomec Inc., USA
30. RegenHU Ltd, Switzerland
31. SLM Solutions, Germany
32. Solid Concepts, USA
33. Solidimension Ltd., Israel
34. Solidscape Inc., USA
35. Soligen Inc., USA
36. Stratasys Inc., USA
37. Teijin Seiki Co. Ltd, Japan
38. The ExOne Company, USA
39. ThreeASFOUR, USA
40. voxeljet AG, Germany
41. Zugo Technology Pte Ltd, Singapore

We would also like to express our special gratitude to the following individuals who have agreed to showcase their 3D printing work and creations in our book: Cody R. Wilson, Olaf Diegel, Amit Zoran, Neri Oxman, and Jeffrey Lipton.

Last but not least, we also wish to express our thanks and apologies to the many others not mentioned above for their suggestions, corrections and contributions to the success of the previous editions of the book. We would appreciate your comments and suggestions on this fourth edition book.

Chua C. K.
Professor

Leong K. F.
Associate Professor

ABOUT THE AUTHORS

CHUA Chee Kai is the Chair of the School of Mechanical and Aerospace Engineering, and Director of the NTU Additive Manufacturing Centre (NAMC) at Nanyang Technological University. Over the last 25 years, Prof. Chua has established a strong research group at NTU, pioneering and leading in computer-aided tissue engineering scaffold fabrication using various additive manufacturing techniques. He is internationally recognised for his significant contributions in the bio-material analysis and rapid prototyping process modelling and control for tissue engineering. His work has since extended further into additive manufacturing of metals and ceramics for defence applications.

Prof. Chua has published extensively with over 200 international journals and conferences, attracting over 3800 citations, and has a Hirsch index of 31 in the Web of Science. His book, *Rapid Prototyping: Principles and Applications*, now in its fourth edition, is widely used in American, European and Asian universities and is acknowledged by international academics as one of the best textbooks in the field. He is the World No. 1 Author for the area of Rapid Prototyping (or 3D Printing) in the Web of Science, and is the most 'Highly Cited Scientist' in the world for that topic. He is the Co-Editor-in-Chief of the *International Journal of Virtual & Physical Prototyping* and serves as an editorial board member of three other international journals. As a dedicated educator who is passionate in training the next generation, Prof. Chua is widely consulted on additive manufacturing (since 1990) and has conducted numerous professional development courses for mechanical engineers in Singapore and the region. In 2013, Prof. Chua was awarded the "Academic Career Award" for his contributions to Additive Manufacturing (or 3D Printing) at the 6th International Conference on Advanced Research in Virtual and

Rapid Prototyping (VRAP 2013), 1–5 October, 2013, at Leiria, Portugal. Dr. Chua can be contacted by email at mckchua@ntu.edu.sg.

LEONG Kah Fai is an Associate Professor at the School of Mechanical and Aerospace Engineering, Nanyang Technological University (NTU), Singapore. He has taught for more than 25 years in NTU in a variety of courses including, Creative Thinking and Design, Product Design and Development, Industrial Design, Additive Manufacturing, Collaborative Design and Engineering, amongst others. He graduated from the National University of Singapore and Stanford University for his undergraduate and graduate degrees respectively. He is actively involved in design, manufacturing and standards work. He has chaired several committees in the Singapore Institute of Standards and the Industrial Research and Productivity Standards Board, receiving Merit and Distinguished Awards for his services in 1994 and 1997 respectively. He has delivered keynote papers at several conferences on the application of RP in biomedical science and tissue engineering. He has authored over 140 publications, including books, book chapters, international journals, and conferences, and has one patent to his name. He has won several academic prizes and his publications have been cited more than 3500 times. His H-index is 26 based on the Science Citation Index. He is the Head of the Systems and Engineering Management Division. and was the Founding Director of the Design Research Centre at the University. He is also the Co-Director of the SIMTech-NTU Joint Laboratory on 3D Additive Manufacturing, and is the Deputy Director of the Institute for Sports Research. Leong Kah Fai can be contacted by email at mkfleong@ntu.edu.sg.

LIST OF ABBREVIATIONS

2D	= Two-Dimensional
3D	= Three-Dimensional
3DP	= Three-Dimensional Printing
ABS	= Acrylonitrile Butadiene Styrene
ACS	= Advanced Composite Structures
ASCII	= American Standard Code for Information Interchange
AIM	= ACES Injection Moulding
AM	= Additive Manufacturing
AOM	= Acoustic Optical Modulator
BDM	= Beam Delivery System
CAD	= Computer-Aided Design
CAE	= Computer-Aided Engineering
CAM	= Computer-Aided Manufacturing
CAM-LEM	= Computer-Aided Manufacturing of Laminated Engineering Materials
CBC	= Chemically Bonded Ceramics
CIM	= Computer-Integrated Manufacturing
CLI	= Common Layer Interface
CJP	= ColorJet Printing
CMM	= Coordinate Measuring Machine
CNC	= Computer Numerical Control
CSG	= Constructive Solid Geometry
CT	= Computerised Tomography
DLP	= Digital Light Processing
DSO	= Defence Science Organisation
DMD	= Direct Metal Deposition; Digital Mirror Device
DMLS	= Direct Metal Laser Sintering
DoD	= Drop on Demand

DPM = Digital Part Materialization
DSP = Digital Signal Processor
DSPC = Direct Shell Production Casting
EB = Electron Beam
EBM = Electron Beam Melting
EDM = Electric Discharge Machining
EMS = Engineering Modelling System
ERM = Enhanced Resolution Module
FDM = Fused Deposition Modelling
FEA = Finite Element Analysis
FEM = Finite Element Method
FRP = Fibre-Reinforced Polymer
GE = General Electrics
GIS = Geographic Information System
GPS = Global Positioning System
HPGL = Hewlett–Packard Graphics Language
HQ = High Quality
HR = High Resolution
HS = High Speed
IGES = Initial Graphics Exchange Specification
LAN = Local Area Network
LCD = Liquid Crystal Display
LEAF = Layer Exchange ASCII Format
LED = Light Emitting Diode
LENS = Laser Engineered Net Shaping
LMT = Laser Manufacturing Technologies
LOM = Laminated Object Manufacturing
LS = Laser Sintering
LWD = Linear Weld Density
M^3D = Maskless Mesoscale Material Deposition
M-RPM = Multi-Functional RPM
MEM = Melted Extrusion Modelling
MEMS = Micro-Electro-Mechanical Systems
MIG = Metal Inert Gas
MJM = Multi-Jet Modelling System
MJP = Multi-Jet Printing

MJS	= Multiphase Jet Solidification
MRI	= Magnetic Resonance Imaging
NASA	= National Aeronautical and Space Administration
NC	= Numerical Control
PC	= Personal Computer; Polycarbonate
PCB	= Printed Circuit Board
PDA	= Personal Digital Assistant
PLT	= Paper Lamination Technology
PPSF	= Polyphenylsulfone
PSL	= Plastic Sheet Lamination
RDM	= Resin Delivery Module
RFP	= Rapid Freeze Prototyping
RM&T	= Rapid Manufacturing and Tooling
RP	= Rapid Prototyping
RPI	= Rapid Prototyping Interface
RPM	= Rapid Prototyping and Manufacturing
RPS	= Rapid Prototyping Systems
RPT	= Rapid Prototyping Technologies
RSP	= Rapid Solidification Process
SAHP	= Selective Adhesive and Hot Press
SCS	= Solid Creation System
SDL	= Selective Deposition Lamination
SDM	= Shaped Deposition Manufacturing
SFF	= Solid Freeform Fabrication
SFM	= Solid Freeform Manufacturing
SGC	= Solid Ground Curing
SHR	= Single Head Replacement
SLA	= Stereolithography Apparatus
SLC	= Stereolithography Contour
SLM	= Selective Laser Melting
SLS	= Selective Laser Sintering
SMS	= Selective Mask Sintering
SSM	= Slicing Solid Manufacturing
SOUP	= Solid Object Ultraviolet-Laser Plotting
STL	= Stereolithography File
TPM	= Tripropylene Glycol Monomethyl Ether

TTL = Toyota Technical Centre
UC = Ultrasonic Consolidation
UV = Ultraviolet
UAS = Unmanned Aerial System

CONTENTS

Foreword		**vii**
Preface		**xi**
Acknowledgements		**xv**
About the Authors		**xix**
List of Abbreviations		**xxi**
Chapter 1	**Introduction**	**1**
1.1	Development of AM	2
1.2	Fundamentals of Additive Manufacturing	4
1.3	Classification of Additive Manufacturing Systems	6
1.4	Advantages of Additive Manufacturing	9
1.5	Standards on AM	14
1.6	Commonly Used Terms	15
	References	16
	Problems	17
Chapter 2	**Additive Manufacturing Process Chain**	**19**
2.1	Fundamental Automated Processes	19
2.2	Process Chain	20
2.3	3D Modelling	22
2.4	Data Conversion and Transmission	22
2.5	Checking and Preparing	24
2.6	Building	26
2.7	Postprocessing	26
	References	28
	Problems	29

Chapter 3 Liquid-Based Additive Manufacturing Systems 31

3.1 3D Systems' Sterolithography Apparatus (SLA) 31
3.2 Stratasys' Polyjet 48
3.3 3D Systems' Multi-Jet Printing System (MJP) 58
3.4 EnvisionTec's Perfactory 68
3.5 CMET's Solid Object Ultraviolet-Laser Printer 77
3.6 EnvisionTec's Bioplotter 81
3.7 RegenHU's 3D Bioprinting 87
3.8 Rapid Freeze Prototyping 92
3.9 Other Notable Liquid-Based AM Systems 100
References 122
Problems 124

Chapter 4 Solid-Based Additive Manufacturing Systems 127

4.1 Stratasys' Fused Deposition Modelling (FDM) 127
4.2 Solidscape's Benchtop System 140
4.3 MCOR Technologies' Selective Deposition Lamination (SDL) 145
4.4 Cubic Technologies' Laminated Object Manufacturing (LOM) 151
4.5 Ultrasonic Consolidation 163
4.6 Other Notable Solid-based AM Systems 165
References 189
Problems 191

Chapter 5 Powder-Based Additive Manufacturing Systems 193

5.1 3D Systems' Selective Laser Sintering (SLS) 193
5.2 3D Systems' ColorJet Printing (CJP) Technology 208
5.3 EOS's EOSINT Systems 218
5.4 Optomec's Laser Engineered Net Shaping (LENS) and Aerosol Jet System 232
5.5 Arcam's Electron Beam Melting (EBM) 244
5.6 Concept Laser's LaserCUSING® 250
5.7 SLM Solutions' Selective Laser Melting (SLM) 262

5.8	3D Systems' Phenix PX™ Series	270
5.9	3D-Micromac AG's Laser Structuring Systems	274
5.10	The ExOne's Digital Part Materialisation (DPM)	278
5.11	Voxeljet AG's VX System	283
5.12	Other Notable Powder-based AM systems	288
	References	296
	Problems	300
Chapter 6	**Additive Manufacturing Data Formats**	**303**
6.1	STL Format	303
6.2	STL File Problems	305
6.3	Consequences of Building a Valid and Invalid Tessellated Model	309
6.4	STL File Repair	311
6.5	Other Translators	340
6.6	Standards for Representing Additive Manufactured Objects	343
	References	351
	Problems	353
Chapter 7	**Applications and Examples**	**355**
7.1	Application – Material Relationship	355
7.2	Finishing Processes	357
7.3	Applications in Design	358
7.4	Applications in Engineering, Analysis and Planning	360
7.5	Applications in Manufacturing and Tooling	363
7.6	Aerospace Industry	379
7.7	Automotive Industry	383
7.8	Jewellery Industry	388
7.9	Coin Industry	389
7.10	Tableware Industry	392
7.11	Geographic Information System (GIS) Applications	395
7.12	Arts and Architecture	396

7.13	Construction	399
7.14	Fashion and Textile	401
7.15	Weapons	406
7.16	Musical Instruments	409
7.17	Food	412
7.18	Movies	414
	References	416
	Problems	421
Chapter 8	**Medical and Bioengineering Applications**	**423**
8.1	Planning and Simulation of Complex Surgery	423
8.2	Customised Implants and Prostheses	433
8.3	Design and Production of Medical Devices	444
8.4	Forensic Science and Anthropology	449
8.5	Visualisation of Bio-Molecules	452
8.6	Bionic Ear	453
8.7	Dentistry	455
	References	458
	Problems	464
Chapter 9	**Benchmarking and the Future of Additive Manufacturing**	**467**
9.1	Using Bureau Services	467
9.2	Technical Evaluation Through Benchmarking	468
9.3	Industrial Growth	486
9.4	Future Trends	488
	References	497
	Problems	501
Appendix	**List of AM Companies**	**503**
Companion Media Pack Attachment		**509**
Companion Media Pack User Guide		**511**

Chapter 1

INTRODUCTION

The competition in the world market for manufactured services and products has intensified tremendously in recent years. It has become important, if not vital, for new products to reach the market as early as possible, before the competitors [1, 2]. To bring products to the market swiftly, many of the processes involved in the design, test, manufacture, market and distribution of the product have been squeezed, both in terms of time and material resources. The efficient use of such valuable resources calls for more efficient tools and effective approaches in dealing with them, and many of these tools and approaches have evolved. They are almost entirely technology-driven, inevitably involving the computer, a result of the rapid development and advancement in such technologies over the last few decades.

Additive Manufacturing (AM), formerly known as Rapid Prototyping (RP), is one such technological development. AM is defined by the American Society for Testing and Materials (ASTM) as "a process of joining materials to make objects from 3D model data, usually layer upon layer, as opposed to subtractive manufacturing methodologies"[3]. Also commonly known as 3D Printing in public literature, this emerging technology is revolutionising the manufacturing industry with its ability to turn digital data into physical parts. Its distinct ability to manufacture complex shapes and structures has already made it invaluable for the production of prototypes such as engine manifolds for the automotive industry, and tools such as investment casting moulds in the jewellery and aeronautical industries. As the technology becomes more developed, additive manufacturing is poised to move on to the direct production of components and parts. The potential of AM has resulted in an increase in public interest in recent years, especially with the media coverage of novel and sometimes controversial applications of AM, such as the production of printable guns and food printing [4]. This book will provide a clear and comprehensive coverage of this exciting new technology.

1.1 Development of AM

In 1987, the first commercial AM system, SLA-1, was launched by 3D Systems in the United States [5]. It worked on the principle of stereolithography (SL) and for the first time enabled users to generate a physical object from digital data. The invention of AM technology was a "watershed event" [6] because of the tremendous time savings, especially for complicated and difficult to produce models. Since then, other new AM technologies have been commercialised, including Fused Deposition Modelling (see Chapter 4.1) and Laminated Object Manufacturing (see Chapter 4.4) in 1991 as well as Selective Laser Sintering (SLS) (see Chapter 5.1) in 1992.

New applications for AM systems have also been developed and expanded as users gained more experience with the machines. Improvements in materials, such as new resins with better mechanical properties when cured, and technologies like more accurate and faster builds have also made AM a feasible way to produce tooling. In 1993, Soligen brought Direct Shell Production Casting to the market and QuickCast (a method of producing indirect tooling from AM) from 3D Systems was also introduced as the potential savings in time and resources that AM can bring became apparent. Companies such as General Motors were early adopters of AM for rapid tooling, acquiring the SLA-250 (the model immediately following SLA-1) in 1991 and using it for rapid tooling and prototyping of parts such as cranking motor nose housings and connector feeder tracks [7].

New AM technologies continued to emerge and old ones became more refined as Kira Corp introduced a system that works on paper lamination in 1994; 3D printers based on technology similar to inkjet printers began to appear in the market in 1996. In 1998, Optomec sold its first Laser Engineered Net Shaping (LENS) metal powder system (see Chapter 5.4) as the LENS process was capable of producing fully dense metal parts. In 1999, selective laser-melting (SLM) system (see Chapter 5.7) was introduced by Fockele & Schwarze of Germany. As more AM systems capable of producing fully dense metal parts emerged, there was growing interest in refining the process of producing metal parts directly for the aerospace and automobile industries. While some companies such as Boeing were already using AM to produce parts such as electrical boxes, brackets and environmental control system ducting [5], the development of international standards was expected to further increase the use of AM for direct manufacturing. AM has developed from a tool, first for "visualisation", then to RP, before being applied to produce tooling and eventually becoming able to build parts with sufficient properties allowing it to be used for direct manufacturing.

New applications for AM and new capabilities are emerging all the time, including the application to bioengineering which led to the development of machines such as the Bioplotter from EnvisionTEC GmbH (see Chapter 3.6). Released in 2002, it produces scaffold using biocompatible materials. New machines with new functionalities such as multi-material, multi-colour printing are being released.

The expiration of older patents, such as the thermoplastic extrusion technology patent used by the Fused Deposition Modelling (FDM) (expired in 2009), has led to an explosion of development of an array of low-cost personal 3D printers such as machines by MakerBot, RepRap etc. These low-cost machines can even be self-assembled from spare parts or kits, and allow individuals to access AM technology at a relatively low price. These machines have also spawned online communities where AM design files are shared on open sources. All these hint at a future where AM will become an integral part of everyday life.

1.2 Fundamentals of Additive Manufacturing

Regardless of the different techniques used in the AM systems developed, they generally adopt the same basic approach, which can be described as follows:

(1) A model or component is modelled on a Computer-Aided Design-Computer-Aided Manufacturing (CAD-CAM) system. The model which describes the physical part to be built, must be represented as closed surfaces which unambiguously define an enclosed volume. This means that the data must specify the inside, the outside and the boundary of the model. This requirement will become redundant if the modelling technique used is based on solid modelling. This is by virtue that a valid solid model will automatically have an enclosed volume. This requirement ensures that all horizontal cross sections that are essential to AM are closed curves to create the solid object.

(2) The solid or surface model to be built is next converted into a format dubbed the "STL" (Stereolithography) file format which originated from 3D Systems. The STL file format approximates the surfaces of the model using the simplest of polygons, triangles. Highly curved surfaces must employ many triangles, and this means that STL files for curved parts can be very large. However, some additive manufacturing systems also accept data in the IGES (Initial Graphics Exchange Specifications) format, provided it is of the correct "flavour".

(3) A computer program analyses a .STL file that defines the model to be fabricated and "slices" the model into cross sections. The cross sections are systematically recreated through the solidification of either liquids or powders and then combined to form a 3D model. Another possibility is that the cross sections are already thin, solid laminations and these thin laminations are glued together with adhesives to form a 3D model. Other similar methods may also be employed to build the model.

Fundamentally, the development of AM can be described in four primary areas. The Additive Manufacturing Wheel in Figure 1.1 depicts these four key aspects of Additive Manufacturing. They are: input, method, material and applications.

1.2.1 *Input*

Input refers to the electronic information required to describe the object in 3D. There are two possible starting points - a computer model or a physical model or part. The computer model created by a computer-aided design (CAD) system can be either a surface model or a solid model. On the other hand, 3D data from the physical model is not so straightforward. It requires data acquisition through a method known as reverse engineering. In reverse engineering, a wide range of equipment, such as CMM (coordinate measuring machine) or a laser digitiser, can be used to capture data points of the physical model, usually in a raster format, and to then "reconstruct" it in a CAD system.

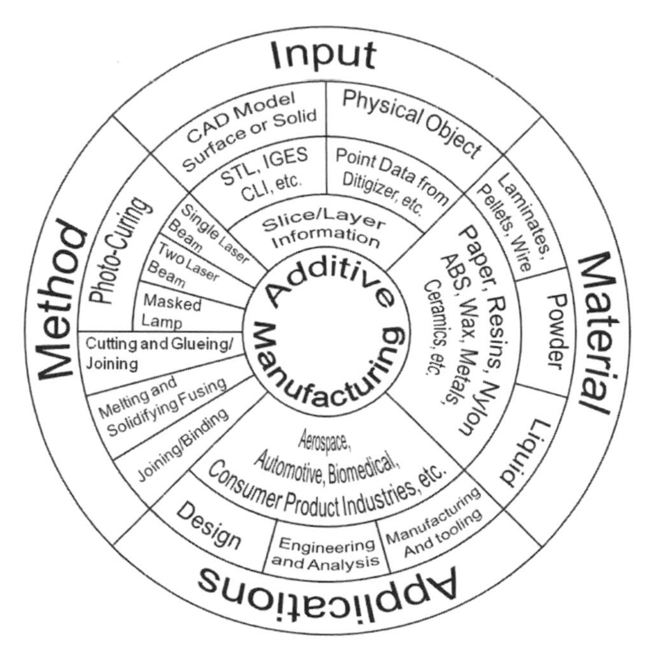

Fig. 1.1. The Additive Manufacturing Wheel depicting the four major aspects of AM.

1.2.2 *Method*

While they are currently more than 40 vendors for AM systems, the method employed by each vendor can be generally classified into the following categories: photo-curing; cutting and joining; melting and solidifying or fusing; and joining or binding. Photo-curing can be further divided into categories of single laser beam, double laser beams and masked lamp.

1.2.3 *Material*

The initial state of material can come in one of the following forms: solid, liquid or powder state. Solid materials come in various forms such as pellets, wire or laminates. The current range of materials includes paper, polymers, wax, resins, metals and ceramics.

1.2.4 *Applications*

Most of the AM parts are finished or touched up before they are used for their intended applications. Applications can be grouped into (1) Design (2) Engineering Analysis and Planning and (3) Manufacturing and Tooling. A wide range of industries can benefit from AM and these include, but are not limited to, aerospace, automotive, biomedical, consumer, electrical and electronic products.

1.3 Classification of Additive Manufacturing Systems

While there are many ways in which one can classify the numerous AM systems in the market, one of the better ways is to classify AM systems broadly by the initial form of its material, i.e. the material that the prototype or part is built with. In this manner, all AM systems can be easily categorised into (1) liquid-based (2) solid-based and (3) powder-based.

1.3.1 *Liquid-based*

The initial form of liquid-based AM systems' building material is the liquid state. Through a process commonly known as curing, the liquid is converted to the solid state. The following AM systems fall into this category:

(1) 3D Systems' Stereolithography Apparatus (SLA)
(2) Stratasys' Polyjet
(3) 3D Systems' Multijet Printing (MJP)
(4) EnvisionTEC's Perfactory
(5) CMET's Solid Object Ultraviolet-Laser Printer (SOUP)
(6) EnvisionTEC's Bioplotter
(7) RegenHU's 3D Bioprinting
(8) Rapid Freeze Prototyping

As illustrated in the AM Wheel in Figure 1.1, three methods are possible under the *"Photo-curing"* method. The *single laser beam* method is a widely used method and includes two of the above AM systems: (1) and (5). (2) and (3) both use UV lamps for curing after deposition of photocurable polymers via jetting heads. (4) uses an imaging system called Digital Light Processing (DLP) and both (6) and (7) use the extrusion method in a liquid medium for creating the objects. (8) involves the freezing of water droplets and deposits in a manner much like FDM (see below) to create the prototype. These will be described in more details in Chapter 3.

1.3.2 *Solid-based*

Except for powder, solid-based AM systems are meant to encompass all forms of material in the solid state. In this context, the solid form can include the shape in the form of wires, rolls, laminates and pellets. The following AM systems fall into this definition:

(1) Stratasys' Fused Deposition Modelling (FDM)
(2) Solidscape's Benchtop System

(3) MCOR Technologies' Selective Deposition Lamination (SDL)
(4) Cubic Technologies' Laminated Object Manufacturing (LOM)
(5) Ultrasonic Consolidation

Referring to the AM Wheel in Figure 1.1, two methods are possible for solid-based AM systems. AM systems (1) and (2) belong to the *Melting and Solidifying or Fusing* method, while the *Cutting and Joining* method is used for AM systems (3), (4) and (5). The various AM systems will be described in more detail in Chapter 4.

1.3.3 *Powder-based*

In a strict sense, powder is by-and-large in the solid state. However, it is intentionally created as a category outside the solid-based AM systems to mean powder in grain-like form. The following AM systems fall into this definition:

(1) 3D Systems' Selective Laser Sintering (SLS)
(2) 3D Systems' ColorJet Printing (CJP)
(3) EOS's EOSINT Systems
(4) Optomec's Laser Engineered Net Shaping (LENS)
(5) Arcam's Electron Beam Melting (EBM)
(6) Concept Laser GmbH's LaserCUSING®
(7) SLM Solutions GmbH's SLM®
(8) 3D Systems' Phenix PXTM
(9) 3D-Micromac AG's MicroSTRUCT
(10) The ExOne Company's ProMetal
(11) Voxeljet AG's VX System

All the above AM systems employ the *Joining/Binding* method. The method of joining/binding differs for the above systems in that some employ a laser while others use a binder/glue to achieve the joining effect. Similarly, the above AM systems will be described in more detail in Chapter 5.

1.4 Advantages of Additive Manufacturing

Today's automated, tool-less, pattern-less AM systems can directly produce functional parts in small production quantities. Parts produced in this way usually have an accuracy and surface finish inferior to those made by machining. However, some advanced systems are able to produce near tooling quality parts that are close to or are in the final shape. The parts produced, with appropriate postprocessing, have material qualities and properties close to the final product. More importantly, the time taken to produce any part – once the design data are available – is short, and can be a matter of hours.

The benefits of AM systems are immense and can be broadly categorised into direct and indirect benefits.

1.4.1 *Direct benefits*

There are many benefits for the company using AM systems. One would be the ability to experiment with physical models of any complexity in a relatively short time. In the last 40 years, products realised to the market place have become increasingly complex in shape and form [8]. For instance, compare the aesthetically beautiful car body of today with that of the 1970s. On a relative complexity scale of 1 to 3, as seen in Figure 1.2, it can be noted that from a base of 1 in 1970, this relative complexity index has increased to about 2 in 1980, approached 3 in the 1990s, and exceeded 3 after 2000. More interestingly and ironically, the relative project completion times have not correspondingly increased. It increased from an initial base of about 4 weeks' project completion time in 1970 to 16 weeks in 1980. However, with the use of CAD-CAM and computer numerical control (CNC) technologies, project completion time was reduced to 8 weeks. Eventually, AM systems allowed the project manager to further cut the completion time to less than 2 weeks in 2010.

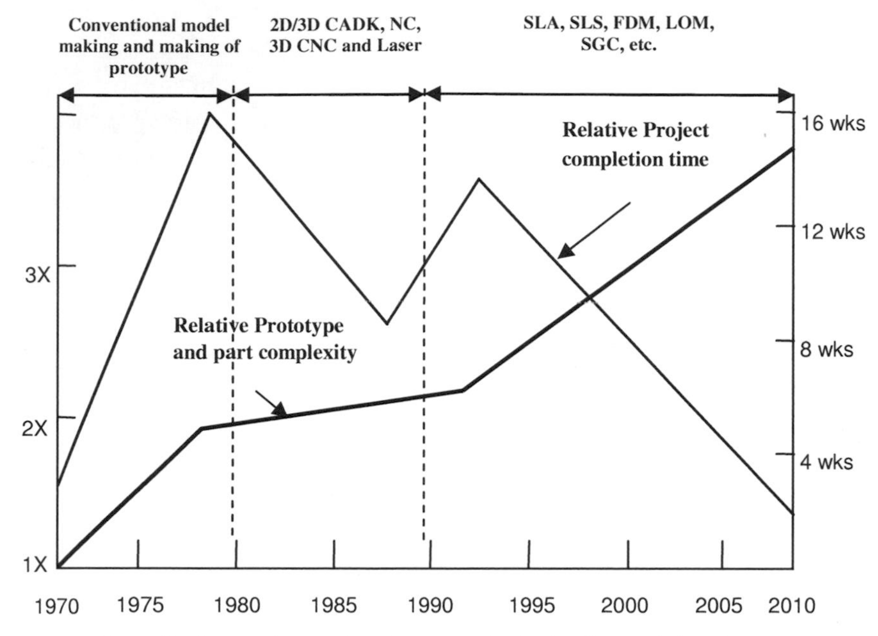

Fig. 1.2. Project time and product complexity in 40 years' time frame.

To the individual in the company, the benefits can be varied and have different impacts. It depends on the role they play in the company. The full production of any product encompasses a wide spectrum of activities. Kochan and Chua [9] describe the impact of AM technologies on the entire spectrum of product development and process realisation. In Figure 1.3, the activities required for full production in a conventional model are depicted at the top. The bottom of Figure 1.3 shows the AM model. Depending on the size of production, savings on time and cost could range from 50% to 90%.

1.4.1.1 *Benefits to product designers*

Product designers can increase part complexity with little effect on lead time and cost. More organic, sculptured and complex shapes for functional or aesthetic features can be accommodated. They can optimise part design to meet customer requirements, with little restrictions imposed by manufacturing. In addition, they can reduce parts count by combining features into single-piece parts that were previously made from several because of poor tool accessibility or the need to minimise machining and waste. With fewer parts, time spent on tolerance analysis, selecting fasteners, detailing screw holes and assembly drawings is greatly reduced.

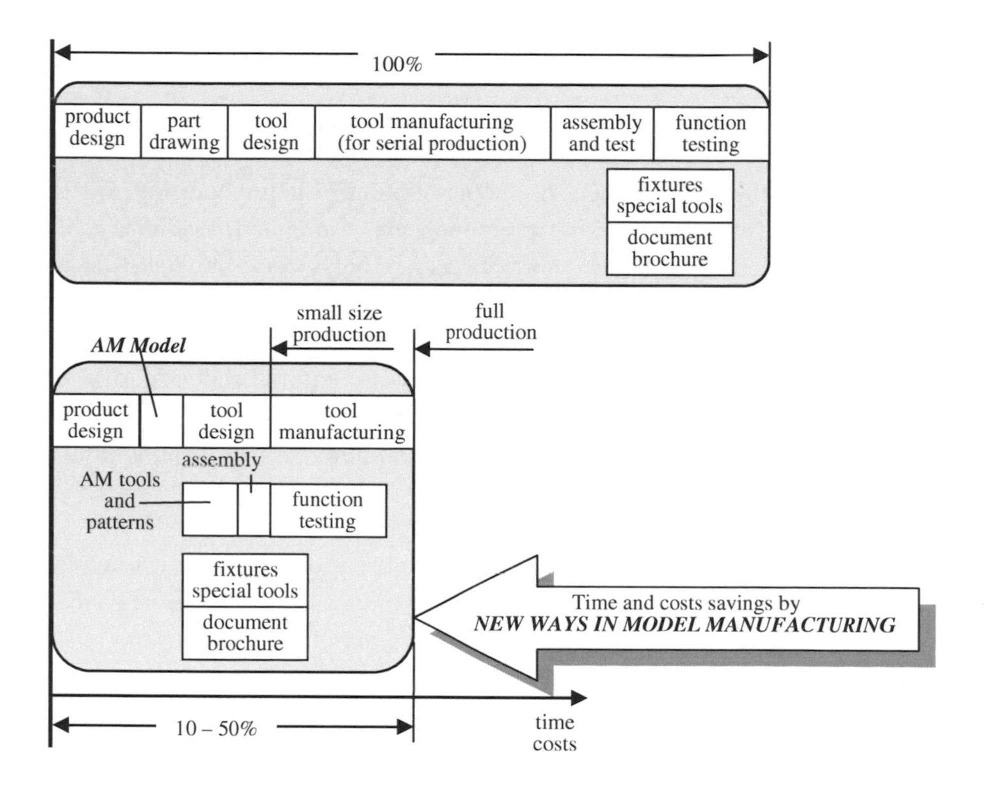

Fig. 1.3. Results of the integration of AM technologies.

There will also be fewer constraints in that parts can be designed without regard to draft angles, parting lines or other such constraints. Parts which cannot easily be set up for machining, or have accurate, large, thin walls, or do not use stock shapes to minimise machining and waste can now be designed. They can minimise the use of material and optimise strength/weight ratios without regard to the cost of machining. Finally, they can minimise time-consuming discussions and evaluations of manufacturing possibilities.

1.4.1.2 *Benefits to tooling and manufacturing engineers*

The main savings are in costs. The manufacturing engineer can minimise design, manufacturing and verification of tooling. He can realise profit earlier on new products, since fixed costs are lower. He can also reduce parts count and, therefore, assembly, purchasing and inventory expenses.

The manufacturer can reduce the labour content of manufacturing, since part-specific setting up and programming are eliminated, machining or casting labour is reduced, and inspection and assembly are also consequently minimised as well. Reducing material waste, waste disposal costs, material transportation costs, inventory cost for raw stock and finished parts (producing only as many as required reduces storage requirements) can contribute to lower overheads. Fewer inventories are scrapped because of fewer design changes, and the risks of disappointing sales are reduced.

In addition, the manufacturer can simplify purchasing since unit price is almost independent of quantity, therefore, only as many as are needed short-term need to be ordered. Quotations vary little among suppliers, since fabrication is automatic and standardised. One can purchase one general purpose machine rather than many specialised machines and therefore, reduce capital equipment and maintenance expenses, and minimise the need for specialised operators and training. A smaller production facility will also make scheduling production easier. Furthermore, one can reduce the inspection reject rate since the number of tight tolerances required when parts must mate can be reduced. One can avoid design misinterpretations (instead, "what you design is what you get"), quickly change design dimensions to deal with tighter tolerances and achieve higher part repeatability, since tool wear is eliminated. Lastly, one can reduce spare parts inventories (produce spares on demand, even for obsolete products).

1.4.2 *Indirect benefits*

Outside the design and production departments, indirect benefits can also be derived. Marketers as well as consumers will benefit from the utilisation of AM technologies.

1.4.2.1 *Benefits to marketing and supply chain*

To the market, it presents new capabilities and opportunities. It can greatly reduce time-to-market, resulting in (1) reduced risk as there is no need to project customer needs and market dynamics several years into the future, (2) products which better satisfy customers' needs, (3) products offering the price/performance of the latest technology, (4) new products being test-marketed economically.

AM can change production capacity according to market demand, possibly in real time, with little or no impact on manufacturing facilities. One can increase the diversity of product offerings and pursue market niches which are currently too small to justify due to tooling cost (including custom and semi-custom production). One can easily expand distribution and quickly enter new markets.

AM can also potentially simplify the supply chain for certain types of businesses, resulting in lower logistic costs. Unlike traditional manufacturing practices, multiple AM machines can be operated by a single person, reducing labour costs. Its ability to produce complex parts at no extra cost may lead to new designs with fewer separate components and hence, reduce costs incurred from product assembly. Moreover, the AM machine may be located close to the assembly line, eliminating shipping cost and time. The lack of dependence on cheap labour can result in the reduction of outsourcing, resulting in manufacturing returning to the countries from which the designs originate. All these will shorten and simplify the supply chain [10], reducing costs of production and time taken to reach consumers. It also means that businesses may need to rethink their current business models, and that new business models with a focus on innovation and mass customisation may become more prevalent.

1.4.2.2 *Benefits to the consumer*

The consumer can buy products which more closely suit individual needs and wants. Firstly, there is a much greater diversity of offerings to choose from. Secondly, one can buy (and even contribute to the design of) affordable built-to-order products. Furthermore, the consumer will enjoy lower prices, since the manufacturers' savings will ultimately be passed on.

1.5 Standards on AM

The ASTM Committee F42 on Additive Manufacturing Technologies was formed in 2009 to develop standards for AM. The development of international standards is vital if AM is going to be more widely adopted for manufacturing in industry. Standards ensure that parts produced via AM are of a certain quality and have material properties that are sufficient for its intended purpose. This is vital especially for the manufacturing of parts in the aerospace and automobile industry, where safety issues must be addressed. Standards also give researchers and developers of AM technologies a clearer goal to strive towards, potentially increasing the speed at which AM develops; this will consequentially hasten the adoption of AM as a manufacturing tool. Internationally recognised standards ensure that AM files, parts, etc. are cross-compatible internationally and trade in AM parts is made easier as manufacturers all over the world conform to the same standard.

To date, the F42 committee has published several standards covering terminology, AMF (Additive Manufacturing File, a proposed standard for storing digital models to be built with AM) and part production of several metals using AM. In 2011, the ASTM International signed a Partner Standards Development Organisation (PSDO) cooperative agreement with the International Standards Organisation (ISO) [11]. The collaboration between two of the most prominent international standards developers is expected to result in more widespread adoption of developed standards worldwide. As of now, the partnership has resulted in an agreement on additive manufacturing terminology and the ISO/ASTM 52915:2013 Standard specification for AMF format Version 1.1, among other deliverables.

ISO has also proposed a new standard for AM files known as Standard for the Exchange of Product (STEP) data; this will be covered in greater detail in section 6.6.

1.6 Commonly Used Terms

There are many terms used by the engineering communities around the world to describe this technology. Perhaps this is due to the versatility and continuous development of the technology. In the public domain, the most commonly used term is 3D Printing. Though this term originally described only printers that deposit material using a print head or nozzle, it is now more commonly used to refer to additive manufacturing than the official industry standard term [5]. Previously, the most commonly used term was Rapid Prototyping. The term was apt as AM could *rapidly* create a physical model. However, this technology is increasingly being used in other areas and the term is now considered outdated.

Some of the less commonly used terms include *Direct CAD Manufacturing, Desktop Manufacturing* and *Instant Manufacturing.* The rationale behind these terms is also based on AM's speed and ease, though in reality it is hardly direct or instant. *CAD Oriented Manufacturing* is another term and provides an insight into the issue of orientation, often a key factor influencing the output of a prototype made by AM methods.

Another group of terms emphasises the unique characteristic of AM – layer-by-layer addition as opposed to traditional manufacturing methods such as machining which is material removal from a block. This group includes *Layer Manufacturing, Material Deposit Manufacturing, Material Addition Manufacturing* and *Material Incress Manufacturing.*

There is yet another group which chooses to focus on the words "solid" and "freeform" - *Solid Freeform Manufacturing* and *Solid Freeform Fabrication. Solid* is used because while the initial state may be liquid, powder, individual pellets or laminates, the end result is a solid, 3D object, while *freeform* stresses the ability of AM to build complex shapes with little constraint on its form.

References

[1] Wheelwright, S. C., & Clark, K. B. (1992). *Revolutionizing Product Development: Quantum Leaps in Speed, Efficiency, and Quality*. New York: The Free Press.

[2] Ulrich, K. T., & Eppinger, S. D. (2011). *Product Design and Development*. Boston: McGraw Hill, 5th Edition.

[3] ASTM F2792 - 12a, Standard Terminology for Additive Manufacturing Technologies, ASTM International, West Conshohocken, PA, 2012, DOI: 10.1520/F2792-12A, www.astm.org.

[4] Greenberg, A. (May 2013). *$25 Gun Created With Cheap 3D Printer Fires Nine Shots (Video)*. Forbes. Retrieved from http://www.forbes.com/sites/andygreenberg/2013/05/20/25-gun-created-with-cheap-3d-printer-fires-nine-shots-video/

[5] Wohlers, T., & Caffrey, T. (2013). History of additive manufacturing. In *Wohlers Report 2013* (pp. 15-17). Colorado: Wohlers Associates.

[6] Kochan, D. (1992). Solid freeform manufacturing: possibilities and restrictions. *Computers in Industry,* 20, 133-140.

[7] Jacobs, P. F. (1993). *Stereolithography 1993: Epoxy Resins, Improved Accuracy & Investment Casting*. Dearborn, Michigan.

[8] Metelnick, J. (1991). *How today's model/prototype shop helps designers use rapid prototyping to full advantage.* Society of Manufacturing Engineers.

[9] Kochan, D., & Chua, C. K. (1995). State-of-the-art and future trends in advanced rapid prototyping and manufacturing. *International Journal of Information Technology,* 1(2):173-184.

[10] Tuck, C., Hague, R., & Burns, N. (2007). Rapid manufacturing: impact on supply chain methodologies and practice. *International Journal of Services and Operations Management*, 3(1), 1-22.

[11] ASTM International - Standards Worldwide, ASTM and ISO Sign Additive Manufacturing PSDO Agreement. N.p., n.d. July 2013. Retrieved from http://www.astm.org/standardization-news/outreach/astm-and-iso-sign-additive-manufacturing-psdo-agreement-nd11.html.

Problems

1. Describe the historical development of Additive Manufacturing and related technologies.

2. Despite the increase in relative complexity of product's shape and form, project time has been kept relatively shorter. Why?

3. What are the fundamentals of Additive Manufacturing?

4. What is the *Additive Manufacturing (AM) Wheel*? Describe its four primary aspects. Is the *Wheel* a static representation of what AM is today? Why?

5. Describe the advantages of Additive Manufacturing in terms of its beneficiaries such as the product designers, tool designers, manufacturing engineers, marketers and consumers.

6. Name three Additive Manufacturing Systems that are liquid-based.

7. How can a liquid form be converted to a solid form using these liquid-based Additive Manufacturing Systems?

8. What form of materials for Additive Manufacturing Systems can be classified as solid-based? Name three such systems.

9. What is the mechanism behind powder-based Additive Manufacturing Systems?
10. Why are international standards important for AM?
11. What are some of the standards already developed for AM?
12. Many terms have been used to refer to Additive Manufacturing. Discuss three of such terms and explain why they have been used in place of Additive Manufacturing.
13. Give a few examples of the applications of Additive Manufacturing.
14. What are the implications of Additive Manufacturing technology for different industries?

Chapter 2

ADDITIVE MANUFACTURING PROCESS CHAIN

2.1 Fundamental Automated Processes

There are three fundamental fabrication processes [1, 2] as shown in Figure 2.1. They are *Subtractive, Additive and Formative* Fabricators.

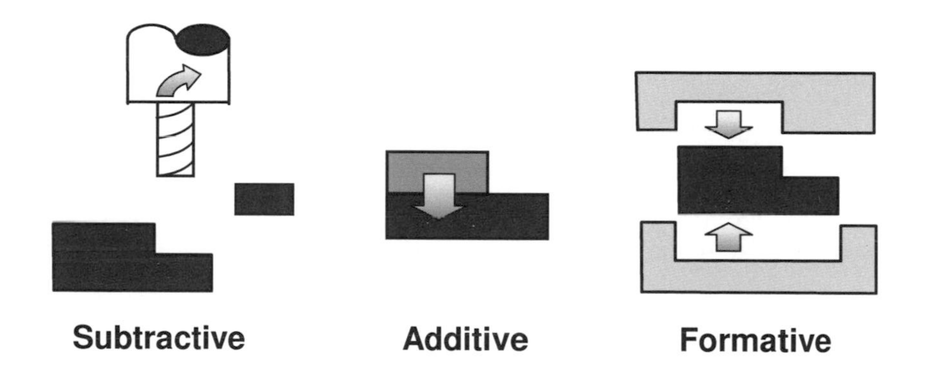

Fig. 2.1. Three types of fundamental fabrication processes.

In a subtractive process, one starts with a single block of solid material larger than the final size of the desired object. Portions of the material are removed until the desired shape is reached.

In contrast, an additive process is the exact reverse in that the end product is much larger than the initial material. Materials are manipulated so that they are successively combined to form the desired object.

Lastly, the formative process is one where mechanical forces or restricting forms are applied on a material so as to form it into the desired shape.

There are many examples of each of these fundamental fabrication processes. Subtractive fabrication processes include most forms of machining processes - CNC or otherwise. These include milling, turning, drilling, planning, sawing, grinding, EDM, laser cutting, water-jet cutting and the like. Most forms of 3D printing and additive manufacturing processes such as Stereolithography (SLA) and Selective Laser Sintering (SLS) fall into the additive fabrication processes category. Examples of formative fabrication processes are: bending, forging, electromagnetic forming and plastic injection moulding. These include both bending of sheet materials and moulding of molten or curable liquids. The examples given are not exhaustive but indicative of the range of processes in each category.

Hybrid machines combining two or more fabrication processes are also possible. For example, in progressive press-working, it is common to see a hybrid of subtractive (as in blanking or punching) and formative (as in bending and forming) processes.

2.2 Process Chain

As described in section 1.3, all AM techniques adopt the same basic approach. As such, all AM systems generally have a similar sort of process chain. Such a generalised process chain is shown in Figure 2.2 [3]. There are a total of five steps in the chain and these are: Step 1: 3D modelling; Step 2: data conversion and transmission, Step 3: checking and preparing, Step 4: building and Step 5: postprocessing. Depending on the quality of the model and part in steps 3 and 5 respectively, the process may be iterated until a satisfactory model or part is achieved.

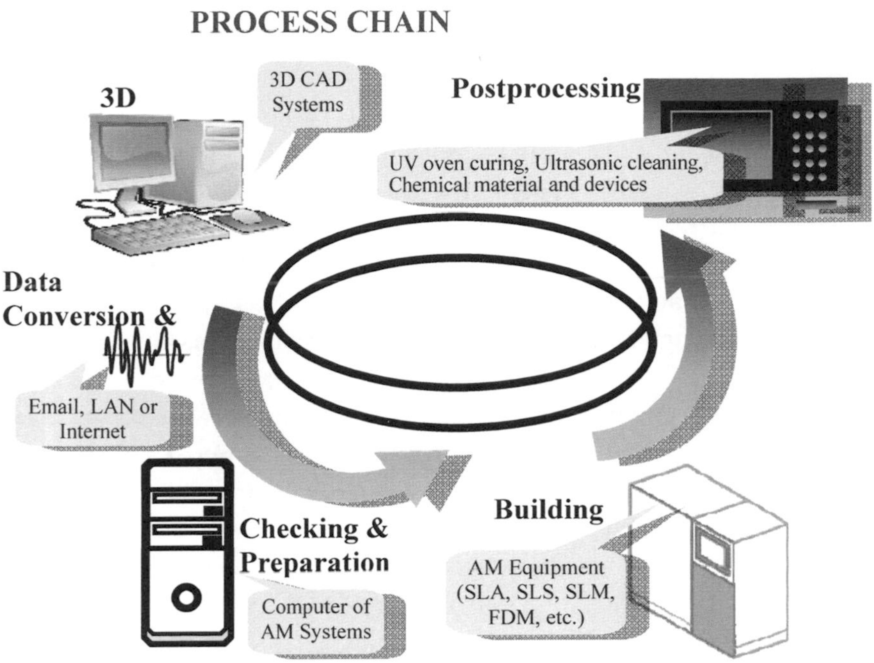

Fig. 2.2. Process chain of Additive Manufacturing systems.

However, like other fabrication processes, process planning is important before AM commences. In process planning, the steps of the AM process chain are listed. The first step is 3D geometric modelling. In this instance, the requirement would be a computer and a CAD modelling system. The factors and parameters which influence the performance of each operation are examined and decided upon. For example, if SLA is used to build the part, the orientation of part is an important factor which would, amongst others, influence the quality of the part and the speed of the process. Thus an operation sheet used in this manner requires proper documentation and sound guidelines. Good documentation, such as a process logbook, allows future examination and evaluation and subsequent improvements can be implemented in process planning. The five steps are discussed in the following sections.

2.3 3D Modelling

Advanced 3D CAD modelling is a general prerequisite in AM processes and is usually the most time-consuming part of the entire process chain. It is most important that such 3D geometric models can be shared by the entire design team for many different purposes, such as interference studies, stress analysis, finite element method (FEM) analysis, detail design and drafting, planning for manufacturing, including numerical control (NC) programming, etc. Many CAD/CAM systems now have a 3D geometrical modeller facility with these special purpose modules.

There are two common misconceptions amongst new users of AM. Firstly, unlike NC programming, AM requires a closed volume of the model, whether the basic elements are surfaces or solids. This confusion arises because new users are usually acquainted with the use of NC programming where a single surface or even a line can be an NC element. Secondly, new users also usually assume *what you see is what you get*. These two misconceptions often lead to the user under-specifying process parameters to AM systems, resulting in poor performance and non-optimal utilisation of the system. Examples of considerations that have to be taken into account include orientation of parts, supports for the part, difficult-to-build part structures such as thin walls, small slots or holes and overhanging elements. Therefore, AM users have to learn and gain experience by working on the AM system. The problem is usually more complex than one can imagine because there are many different AM machines which have different requirements and capabilities.

2.4 Data Conversion and Transmission

Currently, for most AM systems, the solid or surface model to be built is converted into a format dubbed the .STL file format. This format originates from 3D Systems, which pioneered the Stereolithography system. The STL file format approximates the surfaces of the model using tiny triangles. Highly curved surfaces must employ many more triangles, which means that STL files for curved parts can be very large. The STL file format will be discussed in detail in Chapter 6.

Almost, if not all, major CAD-CAM vendors supply the CAD-STL interface. Since 1990, all major CAD-CAM vendors have developed and integrated this interface into their systems.

This conversion step is probably the simplest and shortest of the entire process chain. However, for a highly complex model coupled with an extremely low-performance workstation or PC, the conversion can take several hours. Otherwise, the conversion to STL file should take only several minutes. Where necessary, supports are also converted to a separate STL file. Supports can alternatively be created or modified in the next step by third party software which allows verification and modifications of models and supports.

A new standard, called the ISO/ASTM 52915:2013 and based primarily on the STL format, is the Additive Manufacturing File format (AMF) Version 1.1. This new standard is issued jointly under a collaboration between ASTM and ISO in 2013 which answers the growing need of a standard interchangeable file format that can provide detailed properties of the target product. This is discussed in detail in Chapter 6.

The transmission step is also fairly straightforward. The purpose of this step is to transfer the STL files stored in the CAD computer to the computer of the AM system. It is typical that the CAD computer and the AM system are situated in different locations. The CAD computer, being a design tool, is typically located in a design office. The AM system, on the other hand, is a process or production machine and is usually located on the shop floor. More recently however, there are several AM systems, usually the Concept Modellers (sometimes called 3D printers), which can be located in the design office. Data transmission via agreed data formats such as STL or IGES may be carried out through a thumb drive, email (electronic mail) or LAN (local area network). No validation of the quality of the STL files is carried out at this stage.

2.5 Checking and Preparing

The computer term, *garbage in garbage out*, is also applicable to AM. Many first-time users are frustrated at this step to discover that their STL files are faulty. However, more often than not, it is due to both the errors of CAD models and the non-robustness of the CAD-STL interface. Unfortunately, today's CAD models, whose quality is dependent on CAD systems, human operators and postprocesses, are still afflicted with a wide spectrum of problems, including the generation of unwanted shell-punctures (i.e. holes, gaps, cracks, etc.). These problems, if not rectified, will result in the frequent failure of applications downstream. These problems are discussed in detail in the first few sections of Chapter 6.

Many of the CAD model errors are corrected by human operators assisted by specialised software such as Magics, a software developed by Materialise, Belgium [4]. Magics software enables users to import a wide variety of CAD formats and to export STL files ready for AM, tooling and manufacturing. Its applications include repairing and optimising 3D models; analysing parts; making process-related design changes on the STL files; documentation; and production planning [5-7]. The manual repair process is, however, still very tedious and time-consuming, especially considering the great number of geometric entities (e.g. triangular facets) encountered in a CAD model. The types of errors and their possible solutions are discussed in Chapter 6.

Once the STL files are verified to be error-free, the AM system's computer analyses the STL files that define the model to be fabricated and slices the model into cross-sections. The cross-sections are systematically recreated through the solidification of liquids, binding of powders or fusing of solids, to form a 3D model.

In SLA, for example, each output file is sliced into cross-sections, between 0.12 mm (minimum) to 0.50 mm (maximum) in thickness. Generally, the model is sliced into the thinnest layer (approximately 0.12 mm) as they have to be very accurate. The supports can be created using coarser settings. An internal cross hatch structure is generated between the inner and the outer surface boundaries of the part. This serves to hold up the walls and entrap liquid that is later solidified using UV light.

Preparing building parameters for positioning and stepwise manufacturing in the light of many available possibilities can be difficult if not accompanied by proper documentation. These possibilities include determination of the geometrical objects, building orientation, spatial assortments, arrangement with other parts, necessary support structures and slice parameters. They also include the determination of technological parameters such as cure depth, laser power and other physical parameters as in the case of SLA. It means that user-friendly software for ease of use and handling, user support in terms of user manuals, dialogue mode and on-line graphical aids will be very helpful to users of the AM system.

Many vendors are continually working to improve their systems and their operation software. For example, 3D Systems' Buildstation 5.5 software [8] enables users to simplify the process of setting parameters for the SLA. In early SLA systems, parameters (such as the location in the 250 mm x 250 mm box and the various cure depths) had to be set manually. This was very tedious for up to 12 parameters had to be keyed in. These parameters are shown in Table 2.1.

However, the job is now made simpler with the introduction of default values that can be altered to other specific values. These values can be easily retrieved for use in other models. This software also allows the user to orientate and move the model such that the whole model is in the positive axis' region (the SLA uses only positive numbers for calculations). Thus the original CAD design model can also be in "negative" regions when converting to STL format.

Table 2.1. Parameters used in the SLA process.

1.	X-Y shrink
2.	Z shrink
3.	Number of copies
4.	Multi-part spacing
5.	Range manager (add, delete, etc.)
6.	Recoating
7	Slice output scale
8.	Resolution
9.	Layer thickness
10.	X-Y hatch-spacing or 60/120 hatch-spacing
11	Skin fill spacing (X, Y)
12.	Minimum hatch intersecting angle

2.6 Building

For most AM systems, this step is fully automated. Thus, it is usual for operators to leave the machine on to build a part overnight. The building process may take up to several hours depending on the size and number of parts required. The number of identical parts that can be built is subject to the overall build size and constrained by the build volume of the AM system. Most AM systems come with user alert systems that inform the users remotely via electronic communication, e.g. cellular phone, once the building of the part is complete.

2.7 Postprocessing

The final task in the process chain is postprocessing. At this stage, generally some manual operations are necessary. As a result, the danger of damaging a part is particularly high. Therefore, the operator for this last process step has a high responsibility for the successful realisation of the entire process. The necessary postprocessing tasks for some major AM systems are shown in Table 2.2.

The cleaning task refers to the removal of excess parts which may have remained on the part. Thus, for SLA parts, this refers to excess resin residing in entrapped portions such as the blind hole of a part, as well as the removal of supports. Similarly, for SLS parts, excess powder has to be removed. Likewise for LOM, pieces of excess wood-like blocks of paper which acted as supports have to be removed.

Table 2.2. Essential postprocessing tasks for different AM processes.

	Additive Manufacturing Technologies			
Postprocessing tasks	SLS[1]	SLA[2]	FDM[3]	LOM[4]
1. Cleaning	√	√	X	√
2. Postcuring	X	√	X	X
3. Finishing	√	√	√	√

(√ = required; X = not required)
[1] SLS - Selective Laser Sintering
[2] SLA - Stereolithography Apparatus
[3] FDM - Fused Deposition Modelling
[4] LOM - Layered Object Manufacturing

As shown in Table 2.2, the SLA postprocessing procedures require the highest number of tasks. More importantly, for safety reasons, specific recommendations for postprocessing tasks have to be prepared, especially for the cleaning of SLA parts. It was reported by Peiffer that accuracy is related to post-treatment process [9]. Specifically, Peiffer referred to the swelling of SLA-built parts with the use of cleaning solvents. Parts are typically cleaned with solvents to remove unreacted photosensitive resin. Depending upon the 'build style' and the extent of crosslinking in the resin, the part can be distorted during the cleaning process. This effect was particularly pronounced with the more open 'build styles' and aggressive solvents. With the 'build styles' approaching a solid fill and more solvent-resistant materials, damage with the cleaning solvent has been minimised. With newer cleaning solvents like TPM (Tripropylene Glycol Monomethyl Ether), which was introduced by 3D Systems, part damage due to the cleaning solvent can be reduced or even eliminated [9].

For reasons which will be discussed in Chapter 3, SLA parts are built with pockets of liquid embedded within the part. Therefore, postcuring is required. All other non-liquid AM methods do not need to undergo this task.

Finishing refers to secondary processes such as sanding and painting, which are used primarily to improve the surface finish or aesthetic appearance of the part. It also includes additional machining processes such as drilling, tapping and milling to add necessary features to the parts.

References

[1] Burns, M. (1994). Research Notes. *Rapid Prototyping Report,* **4**(3), 3-6.

[2] Burns, M. (1993). *Automated Fabrication.* New Jersey: PTR Prentice Hall.

[3] Kochan, D., & Chua, C. K. (1994). Solid freeform manufacturing - assessments and improvements at the entire process chain. *Proceedings of the International Dedicated Conference on Rapid Prototyping for the Automotive Industries.* ISATA94, Aachen, Germany.

[4] Materialise, N.V. (2008). Magics 3.01 Software for the RP&M Professional. Retrieved from http://www.materialise.com/materialise/view/en/92074-Magics.html.

[5] Chua, C. K., Gan, J. G. K., & Mei Tong. (1997). Interface between CAD and Rapid Prototyping Systems Part I: A Study of Existing Interfaces. *The International Journal of Advanced Manufacturing Technology,* **13**(8), 566–570.

[6] Chua, C. K., Gan, J. G. K., & Mei Tong. (1997). Interface between CAD and Rapid Prototyping Systems Part II: LMI – An Improved Interface. *The International Journal of Advanced Manufacturing Technology,* **13**(8), 571–576.

[7] Gan, J. G. K., Chua, C. K., & Mei Tong. (1999). Development of New Rapid Prototyping Interface. *Computers in Industry,* **39**(1), 61–70.

[8] 3D Systems, SLA System Software. (Sep 2008). Retrieved from http://www.3dsystems.com/products/software/3d_lightyear/index.asp.

[9] Peiffer, R. W. (1993). The laser stereolithography process - photosensitive materials and accuracy. *Proceedings of the First International User Congress on Solid Freeform Manufacturing*. Germany.

Problems

1. What are the three types of automated fabricators? Describe them and give two examples of each.

2. Each one of the following manufacturing processes/methods in Table 2.3 belongs to one of 3 basic types of fabricators. Tick [√] under the column if you think it belongs to that category. If you think that it is a hybrid machine, you may tick more than 1 category.

S/No	Manufacturing Process	Subtractive	Additive	Formative
1.	Press-working			
2.	*SLS			
3.	Plastic Injection Moulding			
4.	CNC Nibbling			
5.	*CNC CMM			
6.	*LOM			

*For a list of abbreviations used, please refer to the front part of the book.

3. Describe the five steps involved in a general AM process chain. Which steps do you think are likely to be iterated?

4. After 3D geometric modelling, a user can either make a part through NC programming or through additive manufacturing. What are the basic differences between NC programming and AM in terms of the CAD model?

5. STL files are problematic. Is it fair to make this statement? Discuss.

6. Preparing for building appears to be fairly sophisticated. In the case of SLA, what are some of the considerations and parameters involved?

7. Distinguish between cleaning, postcuring and finishing which are the various tasks of postprocessing. Name two AM processes that do not require postcuring and one that does not require cleaning.

8. Which step in the entire process chain is, in your opinion, the shortest? Most tedious? Most automated? Support your choice.

Chapter 3

LIQUID-BASED ADDITIVE MANUFACTURING SYSTEMS

Most liquid-based additive manufacturing systems build parts in a vat of photocurable liquid resin, an organic resin that cures or solidifies under the effect of exposure to light, usually in the UV range. The light cures the resin near the surface, forming a thin hardened layer. Once the complete layer of the part is formed, it is lowered by an elevation control system to allow the next layer of resin to be coated and similarly formed over it. This continues until the entire part is completed. The vat can then be drained and the part removed for further processing, if necessary. There are variations to this technique by the various vendors and they are dependent on the type of light or laser, method of scanning or exposure, type of liquid resin, and type of elevation and optical system used. Another method is by jetting drops of liquid photopolymer onto a build tray via a print head, akin to inkjet printing, and curing them with UV light. Again there are variations to the technique depending on the types of resin, exposure, elevation and so on.

3.1 3D Systems' Stereolithography Apparatus (SLA)

3.1.1 *Company*

3D Systems was founded in 1986 by inventor Charles W. Hull and entrepreneur Raymond S. Freed. Among all the commercial AM systems, the Stereolithography Apparatus, or SLA® as it is commonly called, is the pioneer with its first commercial system marketed in 1988. The company has grown significantly through increased sales and acquisitions, most notably of EOS GmbH's stereolithography business in

1997 and DTM Corp., the maker of the Selective Laser Sintering (SLS) System in 2001. By 2007, 3D Systems had grown into a global company that delivered advanced rapid prototyping solutions to every major market around the world. It has a global portfolio of nearly 400 U.S. and foreign patents, with additional patents filed or pending in the US and several other major industrialised countries. 3D Systems Inc. is headquartered in 333 Three D Systems Circle Rock Hill, SC 29730 USA.

3.1.2 *Products*

3D Systems produces a wide range of AM machines to cater to various part sizes and throughput. There are several models available, including those in the series of ProJet® 6000 SD, HD, and MP, ProJet® 7000 SD, HD, and MP, iPro™ 8000, iPro™ 9000 (see Fig. 3.1) and ProX™ 950. The ProJet® 6000 and ProJet® 7000 series are low-cost machines that can produce SLA parts [1, 2]. The ProJet printers have two different sizes, three high definition print configurations and a wide range of VisiJet® SL print materials such as Tough, Flexible, Black, Clear, HiTemp, Impact and Jewel.

For larger build envelopes, the iPro™ 8000, iPro™ 9000 and ProX™ 950 are available. These three machines are used to create casting patterns, moulds, end-use parts and functional prototypes. These products' surface smoothness, feature resolution, edge definition and tolerances are enhanced. The ProX™ 950 has single-part durability allowing for product build up to 1.5m wide in one piece without any assembly required. Furthermore, it also uses material efficiently, whereby all unused material will remain in the system resulting in minimal waste. Specifications of these machines are summarised in Table 3.1(a) and Table 3.1(b).

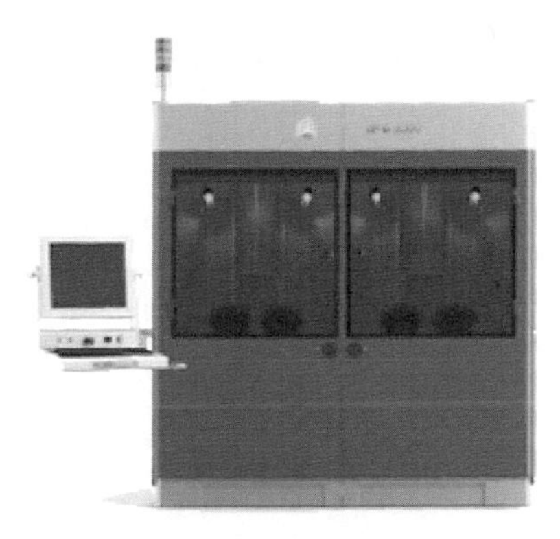

Fig. 3.1. iPro™ 9000 printer. (Courtesy of 3D Systems)

Table 3.1(a). Summary specifications of the Viper ProJet® 6000 SD, HD, and MP, ProJet® 7000 SD, HD, and MP machines. (Courtesy of 3D Systems)

Models	ProJet® 6000 SD	ProJet® 6000 HD	ProJet® 6000 MP	ProJet® 7000 SD	ProJet® 7000 HD	ProJet® 7000 MP
Net Build Volume (X x Y x Z)	Tall 10 x 10 x 10 in (250 x 250 x 250 mm) Medium 10 x 10 x 5 in (250 x 250 x 125 mm) Short 10 x 10 x 2 in (250 x 250 x 50 mm)			Tall 15 x 15 x 10 in (380 x 380 x 250 mm) Medium n/a Short 15 x 15 x 2 in (380 x 380 x 50 mm)		
Resolution	HD - 0.125 mm layers UHD - 0.100 mm layers	HD - 0.125 mm layers UHD - 0.100 mm layers XHD - 0.050 mm layers	HD - 0.125 mm layers UHD - 0.100 mm layers XHD - 0.050 mm layers	HD - 0.125 mm layers UHD - 0.100 mm layers	HD - 0.125 mm layers UHD - 0.100 mm layers XHD - 0.050 mm layers	HD - 0.125 mm layers UHD - 0.100 mm layers XHD - 0.050 mm layers
Accuracy	0.001 - 0.002 inch (0.025 - 0.05 mm) per inch of part dimension Accuracy may vary depending on build parameters, part geometry and size, part orientation and postprocessing methods					

Table 3.1(a). (*Continued*) Summary specifications of the Viper ProJet® 6000 SD, HD, and MP, ProJet® 7000 SD, HD, and MP machines. (Courtesy of 3D Systems)

Models	ProJet® 6000 SD	ProJet® 6000 HD	ProJet® 6000 MP	ProJet® 7000 SD	ProJet® 7000 HD	ProJet® 7000 MP
Materials	VisiJet® SL Flex VisiJet® SL Tough VisiJet® SL Clear VisiJet® SL Black VisiJet® SL Impact VisiJet® SL HiTemp	VisiJet® SL Flex VisiJet® SL Tough VisiJet® SL Clear VisiJet® SL Black VisiJet® SL Impact VisiJet® SL HiTemp VisiJet® SL Jewel	VisiJet® SL Flex VisiJet®SL Tough VisiJet® SL Clear VisiJet® SL Black VisiJet® SL Impact VisiJet® SL HiTemp VisiJet® SL e-Stone™ VisiJet® SL Jewel	VisiJet® SL Flex VisiJet® SL Tough VisiJet® SL Clear VisiJet® SL Black VisiJet® SL Impact VisiJet® SL HiTemp	VisiJet® SL Flex VisiJet® SL Tough VisiJet® SL Clear VisiJet® SL Black VisiJet® SL Impact VisiJet® SL HiTemp VisiJet® SL Jewel	VisiJet® SL Flex VisiJet®SL Tough VisiJet® SL Clear VisiJet® SL Black VisiJet® SL Impact VisiJet® SL HiTemp VisiJet® SL e-Stone™ VisiJet® SL Jewel
Material Packaging	Material in clean no drip 2.0 litre cartridges System auto fills print tray between builds					
Electrical	100 - 240 VAC, 50/60Hz, single-phase, 750W					
Dimensions (WxDxH)	3D Printer Crated 66 x 35 x 79 in (1676 x 889 x 2006 mm) 3D Printer Uncrated 31 x 29 x 72 in (787 x 737 x 1829 mm)			3D Printer Crated 73.5 x 38.5 x 81.5 in (1860 x 982 x 2070 mm) 3D Printer Uncrated 39.0 x 34.0 x 72 in (984 x 854 x 1829 mm)		
Weight	3D Printer Crated 600 lb (272 kg) 3D Printer Uncrated 400 lb (181 kg)			3D Printer Crated 800 lb (363 kg) 3D Printer Uncrated 600 lb (272 kg)		
3D Manage Software	Easy build job setup, submission and job queue management Automatic part placement and build optimisation tools Part stacking and nesting capability Extensive part editing tools Automatic support generation Job statistics reporting					
MP Auto Software	Automation utility for rapid manufacturing applications. Included only with the ProJet 6000 MP			Automation utility for rapid manufacturing applications. Included only with the ProJet 7000 MP		
Network Compatibility	Network ready with 10/100 Ethernet interface 4MB					

Table 3.1(a). (*Continued*) Summary specifications of the Viper ProJet® 6000 SD, HD, and MP, ProJet® 7000 SD, HD, and MP machines. (Courtesy of 3D Systems)

Models	ProJet® 6000 SD	ProJet® 6000 HD	ProJet® 6000 MP	ProJet® 7000 SD	ProJet® 7000 HD	ProJet® 7000 MP
3D Manage Hardware Recommendation	Core 2 Duo 1.8 GHz with 4GB RAM (OpenGL support 128 Mb video RAM)					
3D Manage Operating System	Windows XP Professional, Windows Vista, Windows 7					
Input Data File Formats Supported	STL and SLC					
Operating Temperature Range	64 - 82 °F (18 - 28 °C)					
Noise	< 65 dBa estimated					
Operational Accessories	UV Curing Units, Parts Washer and Right Height Table, ProJet® Cart Station			UV Curing Units, ProJet® Cart Station		

Table 3.1(b). Summary specifications of the iPro™ 8000, and ProX™ 950 machines. (Courtesy of 3D Systems)

Models	iPro™ 8000	ProX™ 950
SteadyPower™ Imager		
Type	Solid state frequency tripled Nd:YVO4	
Wavelength	354.7 nm	
Power (nominal) - at head	1450 mW (1000 mW at resin surface under nominal optical path condition)	
Laser Warranty	10,000 hours or 18 months (whichever comes first), replacement at 800 mW	
Zephyr™ Recoating System		
Process	Removable blade	
Adjustment	Self-levelling; self-correcting	
Layer thickness	0.05 mm (0.002 in) to 0.15 mm (0.006 in)	
ProScan™ Scanning System		
Border spot (diameter at $1/e^2$)	Standard mode nominal 0.13 mm (0.005 in)	
Large hatch spot	Nominal 0.76 mm (0.030 in)	
Maximum Part Drawing Speed		
Border spot	3.5 m/sec (150 ips)	
Large hatch spot	25 m/sec (1000 ips)	
Build Envelope Capacity Interchangeable quick change RDMs with integrated elevator and recoater blade		

Table 3.1(b). (*Continued*) Summary specifications of the iPro™ 8000, and ProX™ 950 machines. (Courtesy of 3D Systems)

Models	iPro™ 8000	ProX™ 950
RDM 650M	650 x 350 x 300 mm (25.6 x 13.7 x 11.8 in); 148 l (39.1 U.S. gal)	n/a
RDM 750SH	650 x 750 x 50 mm (25.6 x 29.5 x 1.97 in); 95 l (25.09 U.S. gal)	n/a
RDM 750H	650 x 750 x 275 mm (25.6 x 29.5 x 10.8 in); 272 l (39.1 U.S. gal)	n/a
RDM 750 F	650 x 750 x 550 (25.6 x 29.5 x 21.65 in); 414 l (109.3 U.S. gal)	n/a
RDM 950 (ProX™ 950)	n/a	1500 x 750 x 550 mm (59 x 30 x22 in)
Maximum part weight	75 kg (165 lbs)	150 kg (330 lbs)
	Resin Delivery Modules (RDMs), Size Options show maximum build envelope capacity (WxDxH); then fill volume	
Electrical Requirements		
With single RDM	200 - 240 VAC 50/60 Hz, single-phase, 30 amps	n/a
With dual RDM	200 - 240 VAC 50/60 Hz, single-phase, 50 amps	
Operating Environment		
Temperature range	20 - 26 °C (68 - 79 °F)	
Maximum change rate	1 °C/hour (1.8 °F/hour)	
Relative humidity	20 - 50% non-condensing	
Space Requirements		
Size (WxDxH)	126 x 220 x 228 cm (50 x 86 x 89 in)	220 x 160 x 226 cm (86.6 x 63 x 89 in)
Weight, crated no RDM module	1590 kg (3500 lbs)	2404 (5300 lbs)
Accessories		
Platform change carts	Manual offload cart optional	
Processing and finishing	ProCure™ 750 UV Finisher	ProCure™ 1500 UV Finisher
System Warranty	One-year warranty, under 3D Systems' Purchase Terms and Conditions	
Control System & Software		
Controller and Part Preparation Software	3DPrint™ and 3DManage™	Print3D Pro and 3DManage™

Table 3.1(b). (*Continued*) Summary specifications of the iPro™ 8000, and ProX™ 950 machines. (Courtesy of 3D Systems)

Models	iPro™ 8000	ProX™ 950
Operating Systems	Windows® XP Professional (SP2)	Windows® 7 or Windows® 8
Input data file format	.stl, .slc	
Network type and protocol	Ethernet, IEEE 802.3 using TCP/IP and NFS	

The Zephyr™ system was introduced in 1996 as a product enhancement in all the SLA systems [3]. The Zephyr™ system eliminates the need for the traditional "deep dip" in which a part is dunked into the resin vat after each layer and then raised to within one layer's depth of the top of the vat. With the deep dip, a wiper blade sweeps across the surface of the vat to remove excess resin before scanning the next layer. The Zephyr™ system has a vacuum blade that picks up resin from the side of the vat and applies a thin layer of resin as it sweeps across the part. This speeds up the build process by reducing the time required between layers and greatly reduces the problems involved when building parts with trapped volumes.

These machines use one-component, photocurable liquid resins as the material for building. There are several grades of resins available and usage is dependent on the machine's laser and the mechanical requirements of the part. Specific details on the correct type of resins to be used are available from the manufacturer. The other main consumable used by these machines is the cleaning solvent which is required to clean the part of any residual resin after the building of the part is completed on the machine.

3.1.3 *Process*

3D Systems' stereolithography process creates three-dimensional plastic objects directly from CAD data. The process begins with the vat filled with the photocurable liquid resin and the elevator table set just below the surface of the liquid resin (see Fig. 3.2). The 3DManage™loads a three-dimensional CAD solid model file into the system. 3DPrint™ software takes the output files from the 3DManage™ software, and then provides control of the build via the attached controller PC on the SLA system. Supports are designed to stabilise the part during building. The translator converts the CAD data into a STL file. The control unit slices the model and support into a series of cross-sections from 0.025 to 0.5 mm (0.001 to 0.020 in.) thick. The computer-controlled optical scanning system then directs and focuses the laser beam so that it solidifies a two-dimensional cross-section corresponding to the slice on the surface of the photocurable liquid resin to a depth greater than one layer thickness. The elevator table then drops enough to cover the solid polymer with another layer of the liquid resin. A levelling wiper or vacuum blade (for Zephyr™ recoating system) moves across the surfaces to recoat the next layer of resin on the surface. The laser then draws the next layer. This process continues building the part from bottom up, until the system completes the part. The part is then raised out of the vat and cleaned of excess polymer.

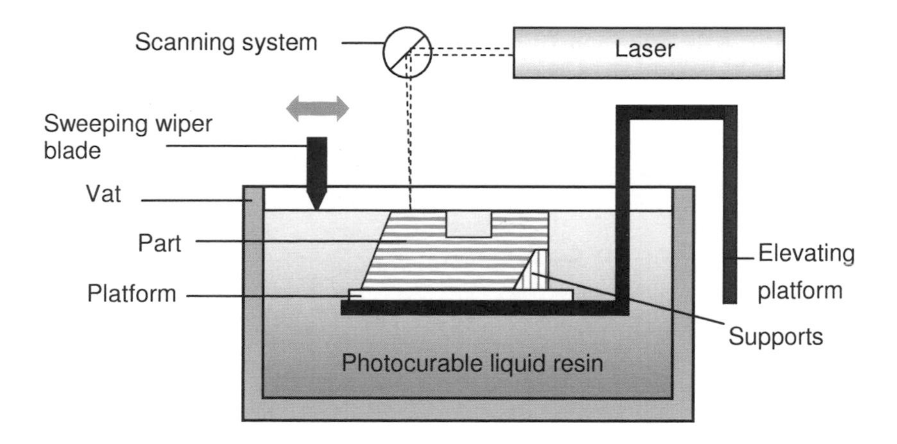

Fig. 3.2. Schematic of SLA process.

3.1.4 *Principle*

The SLA process is based fundamentally on the following principles [4]:

(1) Parts are built from a photocurable liquid resin that cures when exposed to a laser beam (basically, undergoing the

photopolymerisation process) which scans across the surface of the resin.

(2) The building is done layer by layer, each layer being scanned by the optical scanning system and controlled by an elevation mechanism which lowers at the completion of each layer.

These two principles will be discussed briefly in this section to lay the foundation for the understanding of AM processes. They are mostly applicable to the liquid-based AM systems described in this chapter. This first principle deals mostly with photocurable liquid resins, which are essentially photopolymers, and the photopolymerisation process. The second principle deals mainly with CAD data, the laser, and the control of the optical scanning system as well as the elevation mechanism.

3.1.4.1 *Photopolymers*

There are many types of liquid photopolymers that can be solidified by exposure to electro-magnetic radiation, including wavelengths in the gamma rays, X-rays, UV and visible range, or electron-beam (EB) [5, 6]. The vast majority of photopolymers used in commercial AM systems, including 3D Systems' SLA machines are curable in the UV range. UV-curable photopolymers are resins which are formulated from photoinitiators and reactive liquid monomers. There are a large variety of them and some may contain fillers and other chemical modifiers to meet specified chemical and mechanical requirements [7]. The process through which photopolymers are cured is referred to as the photopolymerisation process.

3.1.4.2 *Photopolymerisation*

Loosely defined, polymerisation is the process of linking small molecules (known as monomers) into chain-like larger molecules (known as polymers). When the chain-like polymers are linked further to one another, a cross-linked polymer is said to be formed. Photopolymerisation is polymerisation initiated by a photochemical process whereby the starting point is usually the induction of energy from an appropriate radiation source [8].

Polymerisation of photopolymers is normally an energetically favourable or exothermic reaction. However, in most cases, the formulation of photopolymer can be stabilised to remain unreacted at ambient temperature. A catalyst is required for polymerisation to take place at a reasonable rate. This catalyst is usually a free radical which may be generated either thermally or photochemically. The source of a photochemically generated radical is a photoinitiator, which reacts with an actinic photon to produce the radicals that catalyse the polymerisation process.

The free radical photopolymerisation process is schematically presented in Figure 3.3 [9]. Photoinitiator molecules, P_i, which are mixed with the monomers, M, are exposed to a UV source of actinic photons, with energy of $h\nu$, where h is the Planck constant and ν is the frequency of the radiation. The photoinitiators absorb some of the photons and are in an excited state. Some of these are converted into reactive initiator molecules, P•, after undergoing several complex chemical energy transformation steps. These molecules then react with a monomer molecule to form a polymerisation initiating molecule, PM•. This is the chain initiation step. Once activated, additional monomer molecules go on to react in the chain propagation step, forming longer molecules, PMMM• until a chain inhibition process terminates the polymerisation reaction. The longer the reaction is sustained, the higher will be the molecular weight of the resulting polymer. Also, if the monomer molecules have three or more reactive chemical groups, the resulting polymer will be cross-linked, and this will generate an insoluble continuous network of molecules.

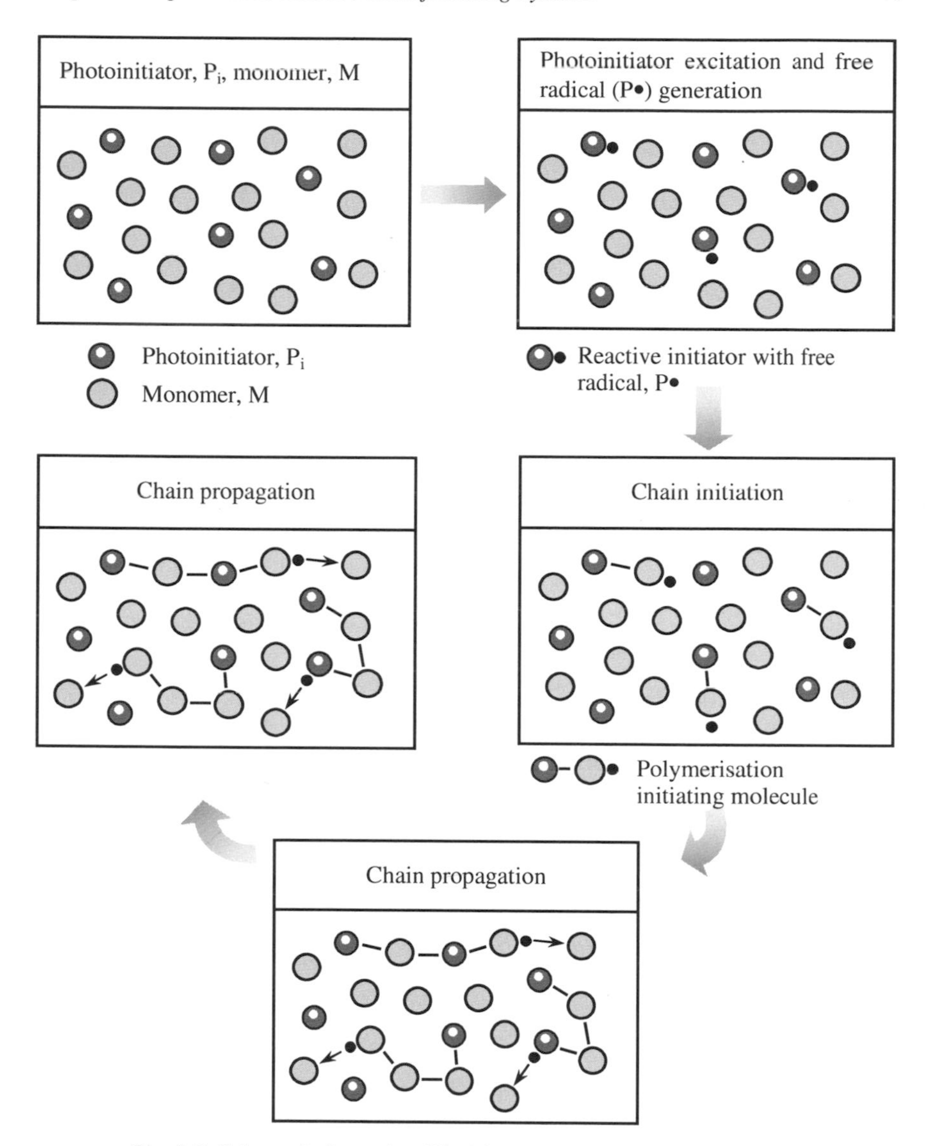

Fig. 3.3. Schematic for a simplified free radical photopolymerisation.

During polymerisation, it is important that the polymers are sufficiently cross-linked so that the polymerised molecules do not re-dissolve back into the liquid monomers. The photopolymerised molecules must also possess sufficient strength to remain structurally sound while the cured resin is subjected to various forces during recoating.

While free radical photopolymerisation is well-established and yields polymers that are acrylate-based, there is another newer 'chemistry' known as cationic photopolymerisation [10]. It relies on cationic initiators, usually iodinium or sulfonium salts, to start polymerisation. Commercially available cationic monomers include epoxies, the most versatile of cationally polymerisable monomers, and vinylethers. Cationic resins are attractive as prototype materials as they have better physical and mechanical properties. However the process may require higher exposure time or a higher power laser.

3.1.4.3 *Layering technology, laser and laser scanning*

Almost all AM systems use layering technology in the creation of prototype parts. The basic principle is the availability of computer software to slice a CAD model into layers and reproduce it in an "output" device like a laser scanning system. The layer thickness is controlled by a precision elevation mechanism. It will correspond directly to the slice thickness of the computer model and the cured thickness of the resin. The limiting aspect of the AM system tends to be the curing thickness rather than the resolution of the elevation mechanism.

The important component of the building process is the laser and its optical scanning system. The key to the strength of the SLA is its ability to rapidly direct focused radiation of appropriate power and wavelength onto the surface of the liquid photopolymer resin, forming patterns of solidified photopolymer according to the cross-sectional data generated by the computer [11]. In the SLA, a laser beam with a specified power and wavelength is sent through a beam expanding telescope to fill the optical aperture of a pair of cross axis, galvanometer driven, and beam scanning mirrors. These form the optical scanning system of the SLA. The beam comes to a focus on the surface of a liquid photopolymer, curing a predetermined depth of the resin after a controlled time of exposure (inversely proportional to the laser scanning speed).

The solidification of the liquid resin depends on the energy per unit area (or "exposure") deposited during the motion of the focused spot on the surface of the photopolymer. There is a threshold exposure that must be exceeded for the photopolymer to solidify.

To maintain accuracy and consistency during part building using the SLA, the cure depth and the cured line width must be controlled. As such, accurate exposure and focused spot size become essential.

Parameters which influence performance and functionality of the parts are physical and chemical properties of resin, speed and resolution of the optical scanning system, the power, wavelength and type of the laser used, the spot size of the laser, the recoating system, and the post-curing process.

3.1.5 *Strengths and Weaknesses*

The main strengths of the SLA are:

(1) *Round the Clock Operation.* The SLA can be used continuously and unattended round the clock.
(2) *Build Volumes.* The different SLA machines have build volumes ranging from small (250 x 250 x 250 mm) to large (1500 x 750 x 550 mm) to suit the needs of different users.
(3) *Good Accuracy.* The SLA has good accuracy and can thus be used for many application areas.
(4) *Surface Finish.* The SLA can obtain one of the best surface finishes among AM technologies.
(5) *Wide Range of Materials.* There is a wide range of materials, from general-purpose materials to specialty materials for specific applications.

The main weaknesses of the SLA are:

(1) *Requires Support Structures.* Structures that have overhangs and undercuts must have supports that are designed and fabricated together with the main structure.
(2) *Requires Postprocessing.* Postprocessing includes removal of supports and other unwanted materials, which is tedious, time-consuming and can damage the model.
(3) *Requires Post-Curing.* Post-curing may be needed to cure the object completely and ensure the integrity of the structure.

3.1.6 *Applications*

The SLA technology provides manufacturers cost justifiable methods for reducing time to market, lowering product development costs, gaining greater control of their design process and improving product design. The range of applications includes:

(1) Models for conceptualisation, packaging and presentation.
(2) Prototypes for design, analysis, verification and functional testing.
(3) Parts for prototype tooling and low volume production tooling.
(4) Patterns for investment casting, sand casting and moulding.
(5) Tools for fixture and tooling design, and production tooling.

Software developed to support these applications includes QuickCast™, a software tool which is used in the investment casting industry. QuickCast™ enables highly accurate resin patterns that are specifically used as an expendable pattern to form a ceramic mould to be created. The expendable pattern is subsequently burnt out. The standard process uses an expendable wax pattern which must be cast in a tool. QuickCast™ eliminates the need for the tooling use to make the expendable patterns. QuickCast™ produces parts which have a hard thin outer shell and contain a honeycomb-like structure inside, allowing the pattern to collapse when heated instead of expanding, which would crack the shell.

3.1.7 *Examples*

3.1.7.1 *3D Systems helps Lotus F1 Team with mass production of race-ready parts*

In 1998, Lotus F1 started using 3D Systems SLA® 5000 to perform additive manufacturing tasks. In a sport where aerodynamic surfacing limits internal race car components, the ability to produce complex and complicated components is key to winning the race. The first SLA® System parts were installed in a race car in 2001 and since then Lotus has continued to explore the boundaries of the technology. 3D Systems' technology became an integral part of Lotus manufacturing process as it allows them to reduce both cycle time and cost, adding invaluable benefits to the team. The team also manufactures gearbox and suspension components, all of which requires accurate casting patterns. The SLA® process follows the exact blueprint of the CAD designs removing any restrictions when it comes to designing complex parts.

3.1.7.2 *DePuy speeds down the information highway*

The demand for rapid prototyping models for medical application is growing rapidly and fabricating models eases the working procedure for surgeons to identify the problems which eventually reduces the risk of the operation. DePuy, one of the world's largest suppliers of orthopaedic appliances and surgical tooling, has long learnt about the advantages of building models by stereolithography. Time is always the crucial parameter when it comes to medical services, not only the total fabrication time but also the time taken for the transferring of the models' data. Together with stereolithography and the internet, things can be made possible. DePuy Inc. has a total of 3 SLA systems: one in Leeds and the other two in Warsaw, Indiana. Transferring SLA files from Leeds to Warsaw can be very efficient and cost saving as the process of transmitting data in Engineering Modelling System, (EMS) or STL is fast and the downloading of a file can be completed in a short time, depending on its size.

With the reduction in development time, new appliances and tooling can be created and this benefits the surgeons and patients. Figure 3.4 shows an example. Over the years, DePuy has always updated itself with the latest technologies to provide the best services to its existing and future customers.

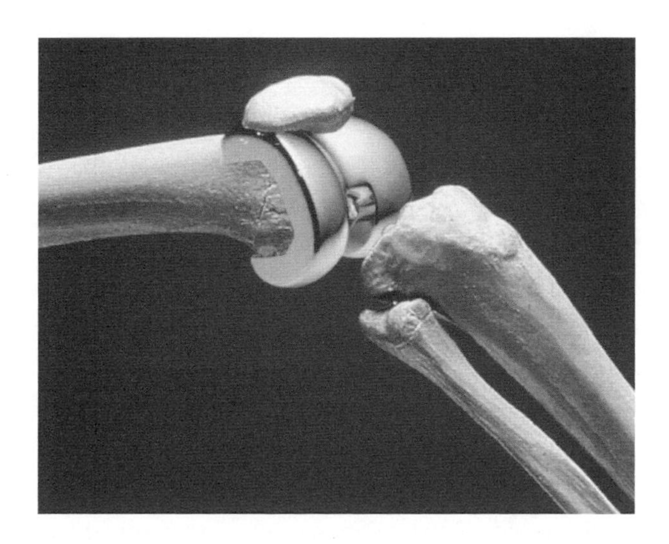

Fig. 3.4. Bone Joint made using SLA system and QuickCast^{TM}. (Courtesy of 3D Systems)

3.1.7.3 *Stereolithography makes a strategic difference at Xerox by helping designers muffle copier noise*

Mufflers are what copiers have in common with automobile vehicles. To filter the copier noise, the air turbulence must be tightly controlled. Stereolithography was the key driver to generating a new copier's muffler assembly design at Xerox's AM laboratories. To conduct a simulation, the central muffler components were built into two pieces using stereolithography and then assembled together (see Fig. 3.5). The components that were built using stereolithography were then used for silicone moulding. Through this process, Xerox was able to optimise the budget for design as well as meet the limitation of the tight schedule.

Stereolithography has helped Xerox to hold firm in their position in a number of areas:

(1) Confirmation in assemblies with the manufactured parts.
(2) Rapid, easy, low cost method for testing new design ideas.
(3) Money saving before embarking on tooling investing.
(4) Usage of silicone moulding.

In the fabrication process, the design drawings were first produced by a Xerox project group which used Intergraph wireframes to represent the assembly. The file was then converted to solids using Intergraph's EMS. At times, when a large portion of a component was to be constructed in one SLA system, the other portion was constructed on another SLA system to shorten the time taken. With SLA systems, two machines can be integrated together to provide greater versatility.

Fig. 3.5. Muffle copier noise using SLA. (Courtesy of 3D Systems)

3.1.8 *Research and Development*

To stand as a leading company in the competitive market and to improve customers' needs, 3D Systems' research has made great improvement not only in developing new materials, and applications, but also in software applications.

At the time of writing, 3D Systems announced they are collaborating with the White House, UI Labs, the Department of Defence and other industries and organisations to deliver their latest Geomagic® perceptual design, manufacturing tools and inspection products to allow unparalleled access to the most advanced design to manufacturing digital thread for the manufacturing industry [12].

The Digital Lab for Manufacturing is an applied research institute geared to transform the manufacturing chain through development, demonstration and deployment of digital manufacturing technologies.

3D Systems has also recently bought Xerox's colour printing technology and their development labs where Xerox's engineering and development teams for product design and materials science will be integrated into 3D Systems' portfolio [13]. This acquisition would bolster their technology and product development to allow 3D Systems to bring affordable and high performance new products to market.

3.2 Stratasys' PolyJet

3.2.1 *Company*

Stratasys Inc. was founded in 1989 in Delaware and developed the company's AM systems based on Fused Deposition Modelling (FDM) technology. In December 2012, Stratasys Inc. merged with Objet [1]. Objet was founded in 1998 and has established itself as the leading platform for high-resolution 3-dimensional printing. Using its patented and market-proven PolyJet™ inkjet-head technology, it is able to print out the most complex 3D models with exceptionally high quality. As a company, Stratasys Inc. is located at 7665 Commerce Way, Eden Prairie, MN 55344-2080, USA.

3.2.2 *Products*

Stratasys manufactures 3D printing equipment and materials that create physical objects based on digital data. Its systems range from affordable desktop 3D printers to large, advanced 3D production systems. All Stratasys 3D printers build parts layer-by-layer.

3.2.2.1 *Design Series — Precision 3D Printers*

Stratasys' Design Series 3D printers aim to build a close link between design and engineering. They speed products to market and reduce costly mistakes. Under Design Series, there are two categories of 3D printers — Precision 3D Printers and Performance 3D Printers. Precision 3D Printers set the industry standard for making final products. Based on PolyJet 3D Printing technology, these systems produce good surface quality and fine details together with the widest range of material properties on the market and the ability to print multi-material models. On the other hand, Performance 3D Printers are further discussed in Chapter 4.1.

Design Series — Precision 3D Printers mainly consists of the Eden™ family and Connex family. The Eden™ family is a group of three machines that can deliver high-resolution prototypes within an office environment [14]. The Eden™ family consists of the Eden 500V™, Eden 350™/350V™ and Eden 260V™, giving options to the users in terms of build size, productivity and budget requirements. For economical and effective small models, Eden 260V™ is able to fit in a small office. Eden 260V™ (see Fig. 3.6) consists of 8 units of Single Head Replacement (SHR) to jet identical amounts of resin resulting in better and more even surface finish. Eden 350™/350V™ are the medium build professional machines in the Eden series which feature printing modes (HQ and HS) and higher material capacity. The Eden 500V™ is the largest build system with a build volume of 490 x 390 x 200mm. It has the best features including dual printing modes, 8 units of Single Head Replacement and an automatic function to switch between cartridges. Specifications of the Eden™ family of machines and the Design Series — Precision 3D Printers are summarised in Table 3.2 and Table 3.3 respectively.

Fig. 3.6. Eden 260V™. (Courtesy of Stratasys Inc.)

Table 3.2. Summary Specifications of Eden™ family printers. (Courtesy of Stratasys Inc.)

Models	Eden 500V™	Eden 350/350V™	Eden 260V™
Tray Size ($X \times Y \times Z$), mm	500 x 400 x 200	350 x 350 x 200	260 x 260 x 200
Net Build Size ($X \times Y \times Z$), mm	490 x 390 x 200	340 x 340 x 200	255 x 252 x 200
Print resolution- *X-axis:* *Y-axis:* *Z-axis:*	 600 dpi: 42 μm 600 dpi: 42 μm 1600 dpi: 16 μm	 600 dpi: 42 μm 600 dpi: 42 μm 1600 dpi: 16 μm	 600 dpi: 42 μm 600 dpi: 84 μm 1600 dpi: 16 μm
Accuracy, μm	20-85 (for features below 50 mm) up to 200 (for full model size)		
Materials cartridges	Sealed 4 x 3.6kg	Sealed 2 x 3.6kg/ 4 x 3.6kg	Sealed 4 x 3.6kg cartridges

Table 3.2. (*Continued*) Summary Specifications of Eden™ family printers. (Courtesy of Stratasys Inc.)

Models	Eden 500V™	Eden 350/350V™	Eden 260V™
Model materials	Transparent rigid (VeroClear) Rubber-like (Tango family) Transparent general-purpose (FullCure RGD720) Rigid Opaque (Vero family) Polypropylene-like (DurusWhite) High Temperature (RGD525) FullCure®705 Support		
Support materials	FullCure 705 non-toxic gel-like photopolymer support		
Input format	STL and SLC File	STL and SLC File	STL and SLC File
Software	Objet Studio™		
Jetting heads	SHR (Single Head Replacement), 8 units	SHR (Single Head Replacement), 8 units	SHR (Single Head Replacement), 8 units
Machine Dimensions (W × D × H), mm	1320 x 990 x 1200	1320 x 990 x 1200	870 x 735 x 1200
Weight, kg	410		
Operational environment	Temperature 18°C to 25°C Relative Humidity 30-70%		

Table 3.3. Summary Specifications of other Precision Series Printers. (Courtesy of Stratasys Inc.)

Models	Objet260 Connex	Objet350 Connex	Objet500 Connex	Objet1000
Model materials	Transparent rigid (VeroClear) Rubber-like (Tango family) Transparent general-purpose (FullCure RGD720) Rigid Opaque (Vero family) Polypropylene-like (DurusWhite)			
Support materials	FullCure 705 non-toxic gel-like photopolymer support			
Net build size, mm (in.)	255 x 252 x 200 (10.0 x 9.9 x 7.9)	342 x 342 x 200 (13.4 x 13.4 x 7.9)	490 x 390 x 200 (19.3 x 15.4 x 7.9)	1000 x 800 x 500 (39.3 x 31.4 x 19.6)
Layer thickness, mm (in.)	Horizontal build layers as fine as 16 microns (0.0006)			
Printing modes	Digital Material (DM): 30-micron (0.001 in.) High Quality (HQ): 16-micron (0.0006 in.) High Speed (HS): 30-micron (0.001 in.)			
Input format	STL, OBJDF and SLC File			

Table 3.3. (*Continued*) Summary Specifications of other Precision Series Printers. (Courtesy of Stratasys Inc.)

Models	Objet260 Connex	Objet350 Connex	Objet500 Connex	Objet1000
Size, mm (in.)	870 x 735 x 1200 (34.3 x 28.9 x 47.2)	1420 x 1120 x 1130 (55.9 x 44.1 x 44.5)		2800 x 1800 x 1800 (110.3 x 70.9 x 70.9)
Weight, kg (lbs)	264 (582)	500 (1102)		1950 (4300)
Print heads	8 units			
Software	Objet Studio			
Power requirement	110–240 VAC 50/60 Hz; 1.5 KW single-phase			240 VAC 50/60 Hz; 32 A single-phase
Operating environment	Temperature 18 C-25 C (64 F-77 F); relative humidity 30-70% (non-condensing)			Temperature 18 C-22 C (64.5 F-71.5 F); relative humidity 30-70%

3.2.2.2 *Materials*

Stratasys Inc. offers a range of additive manufacturing materials, including clear, rubber-like and biocompatible photopolymers, and tough high performance thermoplastics. This variety enables customers to maximise their benefits of 3D printing throughout the product development cycle. From fast, affordable concept modelling to detailed, super-realistic functional prototyping, through certification testing and into agile, low-risk production, Stratasys materials help designers and engineers succeed at every stage. PolyJet photopolymers offer a high level of creating all the details and final products. With chameleon-like ability to simulate clear, flexible and rigid materials and engineering plastics, combined with multiple material properties into one model, prototypes can be made as final products quickly.

3.2.3 *Process*

PolyJet 3D printing is similar to inkjet document printing. But instead of jetting drops of ink onto paper, PolyJet 3D printers jet layers of liquid photopolymer onto a build tray and cure them with UV light. The layers build up one at a time to create a 3D model or prototype. Fully cured models can be handled and used immediately, without additional post-curing. Along with the selected model materials, the 3D printer also jets a gel-like support material specially designed to uphold overhangs and complicated geometries. It is easily removed by hand and with water.

Stratasys' PolyJet process creates 3D objects with the use of Objet Studio Software. The designer loads the three-dimensional CAD solid model file into the system which is compatible with Windows XP and Windows 2000. The Objet Studio will convert the CAD data into an STL or SLC file. The designer will also have to set the orientation arrangement of the designed part on the build tray.

Before the actual building process commences, the designer has to ensure that the build tray and the two types of material cartridges are inserted into the machine. The two types of material cartridges consist of the part material and the supporting material. When the procedure is done, the jetting heads, based on Stratasys' patented PolyJet inkjet technology, will move along the x-axis and lay the first layer of material onto the tray. Depending on the size of the part, the jetting head will move on the y-axis and move on to the x-axis to lay the next layer after the first layer is completed. During the printing process, the jetting head will release the actual amounts of part material and support material. The materials will be immediately cured by the UV light from the jetting head. Whenever the materials are about to be used up, the material cartridges can be easily replaced without interrupting the fabrication process. Once the jetting head cures the first two-dimensional cross-section, the build tray will drop by one layer thickness of 16μm. The jetting head will repeat the process continuously until the system completes the part. The part will be raised up and can be taken out for postprocessing. The support material can then be removed easily by the water jet and the part is complete.

3.2.4 *Principle*

Stratasys' PolyJet™ technology creates high quality models directly from the computerised three-dimensional files. Complex parts are produced with the combination of Stratasys' Studio software and the jetting head.

The process is based on the following principles:

(1) Jetting heads release the required amount of material which shares the same method as the normal inkjet printing method. At the same time when the material is printed on the tray, the material is cured by the UV light which is integrated with the jetting head. Parts are built layer by layer, from a liquid photopolymer where a similar polymerisation process as described in Section 3.1.4 takes place.
(2) Jetting heads are moved only along the XY-axes and each slice of the building process is the cross-section of the parts arranged in the software.
(3) With the completion of a cross-section layer, the build tray will be lowered for the next laycr to bc laid. Thc Z-height of the elevator is levelled accurately so that the corresponding cross-section data can be calculated for that layer.
(4) Both the part material and support material will be fully cured when they are exposed to the UV light and most importantly the non-toxic support material can be removed easily by the water jet.

3.2.5 *Strengths and Weaknesses*

Stratasys' PolyJet™ system has the following strengths [15]:

(1) *High quality*. The PolyJet™ can build layers as thin as 16μm in thickness with accurate details depending on the geometry, part orientation and print size.

(2) *High accuracy.* Precise jetting and build material properties enable fine details and thin walls (600μm or less depending on the geometry and materials).

(3) *Fast process speed.* Certain AM systems require draining, resin stripping, polishing and others whereas Eden™ systems only require an easy wash of the support material which is a key strength.

(4) *Smooth surface finish.* The models built have smooth surface and fine details without any postprocessing.

(5) *Wide range of materials.* Stratasys has a range of materials suited for different specifications, ranging from tough acrylic-based polymer, to polypropylene-like plastics (Duruswhite) to the rubber-like Tango materials.

(6) *Easy usage.* The Eden™ family utilises a cartridge system for easy replacement of build and support materials. Material cartridges provide an easy method for insertion without having any risk of contact with thc materials.

(7) *SHR technology.* The Eden™ machines' nozzles consist of heads and nozzles. With SHR (Single Head Replacemcnt) thesc individual nozzles can be replaced instead of replacing the whole unit whenever the need arises.

(8) *Safe and clean process.* Users are not exposed to thc liquid resin throughout the modelling process and the photopolymer support is non-toxic. PolyJet™ systems can be installed in the office environment without increasing the noise level.

(9) *Combination of the materials.* Based on PolyJet technology, Connex 3D printers from Stratasys are the only additive manufacturing systems that can combine different 3D printing materials within the same 3D printed model, in the same print job.

The PolyJet™ system has the following weaknesses:

(1) *Postprocessing.* A water jet is required to wash away the support material used in PolyJet™, meaning that water supply must be nearby. This is somewhat a let-down to the claim that the machine is suitable for an office environment. In cases where the parts built are

small, thin or delicate, the water jet can damage these parts, so care in postprocessing must be exercised.

(2) *Wastage*. The support material which is washed away with water cannot be reused, meaning additional costs are added to the support material.

3.2.6 *Applications*

The applications of Stratasys' systems can be divided into different areas:

(1) *General applications*. Models created by Stratasys' systems can be used for conceptual design presentation, design proofing, engineering testing, integration and fitting, functional analysis, exhibitions and pre-production sales, market research, and inter-professional communication.
(2) *Tooling and casting applications*. Parts can also be created for investment casting, direct tooling, and rapid, tool-free manufacturing of plastic parts. Also they can be used to create silicon moulding, aluminium epoxy moulds, VLT Moulding (alternative rubber mould) and vacuum forming.
(3) *Medical imaging*. Diagnostic, surgical [16], operation and reconstruction planning and custom prosthesis design. Parts built by PolyJet™ have outstanding detail and fine features which can make the medical problems more visible for analysis and surgery simulation. Due to its fast building time, prototype models are always built for trauma or tumours. Most importantly, it reduces the surgical risks and provides a communication bridge for the patients.
(4) *Jewellery industry*. Presentation of concept design, actual display, design proof and fitting. Pre-market survey and market research can be conducted using these models.
(5) *Packing*. Vacuum Forming is an easy method to produce inexpensive parts and it requires a very short time for the part to be formed.

3.2.7 *Examples*

3.2.7.1 *adidas Group uses Stratasys' PolyJet™ to produce prototypes*

adidas Group is always among the top athletic footwear manufacturers worldwide. To keep up with the highly competitive market, product enhancements and timely product presentations to the market are very crucial. Shortening development, production and commercialisation periods are also the key challenges that adidas Group is facing. The search for systems to produce high quality models ended with the presence of Stratasys' PolyJet™ inkjet technology. Physical models are directly produced from STL files in a short period of time. Stratasys AM systems' assistance has brought adidas Group closer to their company's vision of streamlined digital process for sharing among all business units. Physical models have proven to adidas that they would greatly benefit from using the Stratasys' machines where parts can be used for design verification, development review and production tooling in a short development time.

Stratasys' PolyJet™ has given adidas Group the flexibility to collaborate with each factory in Asia on specific product enhancements while reducing the time necessary. Stratasys' PolyJet™ successfully helped adidas Group to save time in printing and building stenotype. It also enabled adidas Group to identify error at the beginning of designing process. adidas Group is now working with Stratasys to search for ways to apply this technology into other aspects of their products (Fig. 3.7).

3.2.8 *Research and Development*

Stratasys is doing research in developing faster processing, higher performance, greater resolution graphics, smoother and more accurate details. One way of improving the accuracy of the PolyJet procedure can be done by optimising the scaling factor [17].

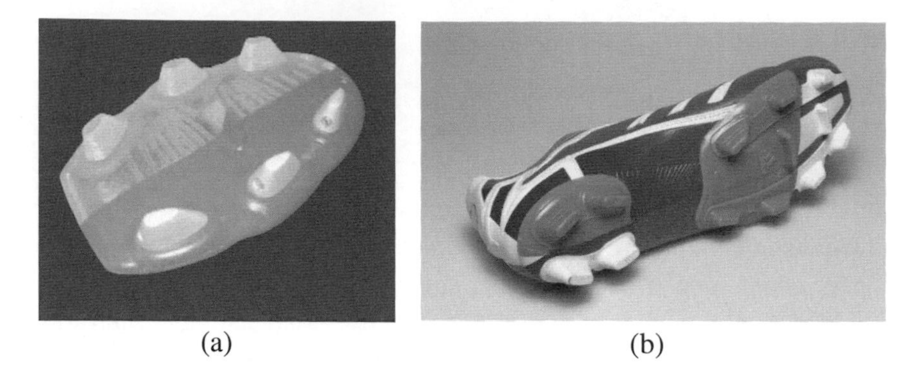

(a) (b)

Fig. 3.7. High quality finishing part that (a) appeared close to a real part and (b) fitted onto a real soccer boot. (Courtesy of Stratasys Inc.)

3.3 3D Systems' MultiJet Printing System (MJP)

3.3.1 *Company*

3D Systems is one of the earliest and largest additive manufacturing companies in the world. Among its many products, 3D Systems also manufactures 3D-Printing systems meant mainly for the office environment. The MultiJet Printing system (MJP) was formerly known as MultiJet Modelling (MJM), first launched in 1996 as a concept modeller for the office to complement the more sophisticated SLA machines (see Section 3.1) and SLS machines (see Section 5.1). The company's details are found in Section 3.1.1.

3.3.2 *Products*

The ProJet™ Series machines are based on MJP technology with thermal material application and UV curing. The newer ProJet™ Series machines were launched in 2008. The ProJet™ 3500 Printer (see Fig. 3.8) is aimed at offering higher productivity, including the high speed printing mode, and larger high definition prints, for the production of functional plastic parts for product design and manufacturing applications. The ProJet™ 5000 printer (see Fig. 3.9) is designed for larger build envelopes to achieve maximum productivity with a net build volume of 553 x 381 x 300 mm. In addition, ProJet™ 5500 produces the highest quality, most accurate and toughest multi-material composites based on MJP technology among all the printers produced by 3D System. It simultaneously prints and fuses flexible and rigid material composites layer by layer together. It uses newly developed VisiJet® composite family of materials to print all the products. [2] The ProJet™ Series machines and their specifications are shown in Table 3.4.

Fig. 3.8. ProJet™ 3500 Printer. (Courtesy of 3D Systems)

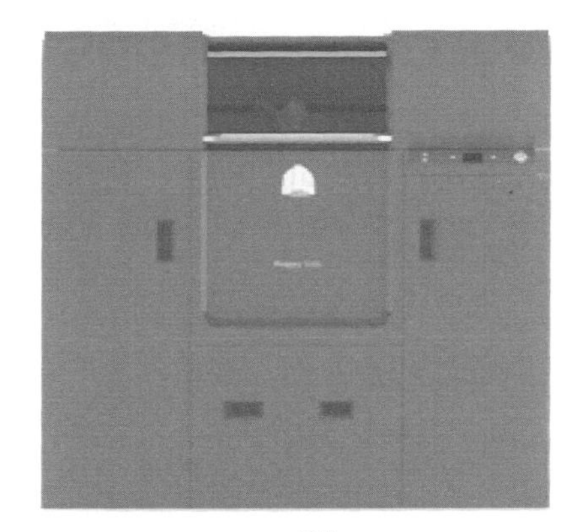

Fig. 3.9. ProJet™ 5000 Printer. (Courtesy of 3D Systems)

Table 3.4. Specifications of ProJet™ 3500 SD & HD series and ProJet™ 5000 & 5500X.

Models	ProJet™ 3510 SD	ProJet™ 3510 HD	ProJet™ 3510 HD*Plus*	ProJet™ 3500 HD*Max*	ProJet™ 5000	ProJet™ 5500X
Printing Modes	HD - High Definition - - -	HD - High Definition - UHD - Ultra-High Definition -	HD - High Definition - UHD- Ultra-High Definition XHD - Xtreme High Definition	HD - High Definition HS - High Speed UHD - Ultra-High Definition XHD - Xtreme High Definition	n/a	n/a

Table 3.4. (*Continued*) Specifications of ProJet™ 3500 SD & HD series and ProJet™ 5000 & 5500X.

Models	ProJet™ 3510 SD	ProJet™ 3510 HD	ProJet™ 3510 HD*Plus*	ProJet™ 3500 HD*Max*	ProJet™ 5000	ProJet™ 5500X
Net Built Volume (xyz) HD Mode HS Mode UHD Mode XHD Mode	11.75 x 7.3 x 8 in (298 x 185 x 203 mm) - - -	11.75 x 7.3 x 8 in (298 x 185 x 203 mm) - 5 x 7 x 6 in (127 x 178 x 152 mm) -	11.75 x 7.3 x 8 in (298 x 185 x 203 mm) - 5 x 7 x 6 in (127 x 178 x 152 mm) 5 x 7 x 6 in (127 x 178 x 152 mm)	11.75 x 7.3 x 8 in (298 x 185 x 203 mm) 11.75 x 7.3 x 8 in (298 x 185 x 203 mm) 11.75 x 7.3 x 8 in (298 x 185 x 203 mm) 11.75 x 7.3 x 8 in (298 x 185 x 203 mm)	21 x 15 x 11.8 in (533 x 381 x 300 mm) - - -	21 x 15 x 11.8 in (533 x 381 x 300 mm) - 21 x 15 x 11.8 in (533 x 381 x 300 mm) -
Resolution HD Mode HS Mode UHD Mode XHD Mode	375 x 375 x 790 DPI (xyz); 32µ layers - - -	375 x 375 x 790 DPI (xyz); 32µ layers - 750 x 750 x 890 DPI (xyz); 29µ layers -	375 x 375 x 790 DPI (xyz); 32µ layers - 750 x 750 x 890 DPI (xyz); 29µ layers 750 x 750 x 1600 DPI (xyz); 16µ layers	375 x 375 x 790 DPI (xyz); 32µ layers 375 x 375 x 790 DPI (xyz); 32µ layers 750 x 750 x 890 DPI (xyz); 29µ layers 750 x 750 x 1600 DPI (xyz); 16µ layers	375 x 375 x 790 DPI (xyz); 32µ layers 375 x 375 x 395 DPI; 64µ layers 750 x 750 x 890 DPI (xyz); 32µ layers -	375 x 375 x 790 DPI (xyz); 32µ layers - 750 x 750 x 890 DPI (xyz); 29µ layers -
Accuracy (typical)	0.001 - 0.002 inch (0.025 - 0.05 mm) per inch of part dimension. Accuracy may vary depending on build parameters, part geometry and size, part orientation, and postprocessing.					

Table 3.4. (Continued) Specifications of ProJet™ 3500 SD & HD series and ProJet™ 5000 & 5500X.

Models	ProJet™ 3510 SD	ProJet™ 3510 HD	ProJet™ 3510 HD*Plus*	ProJet™ 3500 HD*Max*	ProJet™ 5000	ProJet™ 5500X
E-Mail Notice Capability	Yes	Yes	Yes	Yes	Yes	
Tablet/ Smartphone connectivity	Yes	Yes	Yes	Yes	Yes	
5 Year Printhead Warranty	Optional	Standard	Standard	Standard	Yes	
Build Materials	VisiJet M3 X VisiJet M3 Black VisiJet M3 Crystal VisiJet M3 Proplast VisiJet M3 Navy VisiJet M3 Techplast -	VisiJet M3 X VisiJet M3 Black VisiJet M3 Crystal VisiJet M3 Proplast VisiJet M3 Navy VisiJet M3 Techplast VisiJet M3 Procast	VisiJet M3 X VisiJet M3 Black VisiJet M3 Crystal VisiJet M3 Proplast VisiJet M3 Navy VisiJet M3 Techplast VisiJet M3 Procast	VisiJet M3 X VisiJet M3 Black VisiJet M3 Crystal VisiJet M3 Proplast VisiJet M3 Navy VisiJet M3 Techplast VisiJet M3 Procast	VisiJet M5-X VisiJet M5-Black VisiJet M5-MX	VisiJet CR-CL VisiJet CR-WT VisiJet CF-BK
Support Material	VisiJet S300	VisiJet S300	VisiJet S300	VisiJet S300	VisiJet S300	VisiJet S500
Material Packaging Build and Support Materials	In clean 4.41 lbs (2 kg) bottles (machine holds up to 2 with auto-switching)				In clean 2.0 kg cartridges. Printer can hold up to eight cartridges with additional material bays (optional)	Build materials in clean 2.0 kg cartridges and support material in clean 1.75kg cartridges (printer holds 4 build and 4 support cartridges with auto-switching)

Table 3.4. (*Continued*) Specifications of ProJet™ 3500 SD & HD series and ProJet™ 5000 & 5500X.

Models	ProJet™ 3510 SD	ProJet™ 3510 HD	ProJet™ 3510 HD*Plus*	ProJet™ 3500 HD*Max*	ProJet™ 5000	ProJet™ 5500X
Electrical	100 - 127 VAC, 50/60 Hz, single-phase, 15A; 200 - 240 VAC, 50 Hz, single-phase, 10A				115 - 240 VACm, 50/60 Hz, single-phase, 1200 W	100 VAC, 50/60 Hz, single-phase, 15 Amps 115 VAC, 50/60 Hz, single-phase, 15 Amps 240 VAC, 50/60 Hz, single-phase, 8 Amps
Dimensions (WxDxH) 3D Printer Crated 3D Printer Uncrated	32.5 x 56.25 x 68.5 in (826 x 1429 x 1740 mm) 29.5 x 47 x 59.5 in (749 x 1194 x 1511 mm)	32.5 x 56.25 x 68.5 in (826 x 1429 x 1740 mm) 29.5 x 47 x 59.5 in (749 x 1194 x 1511 mm)	32.5 x 56.25 x 68.5 in (826 x 1429 x 1740 mm) 29.5 x 47 x 59.5 in (749 x 1194 x 1511 mm)	32.5 x 56.25 x 68.5 in (826 x 1429 x 1740 mm) 29.5 x 47 x 59.5 in (749 x 1194 x 1511 mm)	72 x 45.5 x 78 in (1828 x 1155 x 1981 mm) 60.3 x 35.7 x 57.1 in (1531 x 908 x 1450 mm)	80 x 48 x 78 in (2032 x 1219 x 1981 mm) 67 x 35.4 x 65 in (1700 x 900 x 1650 mm)
Weight 3D Printer Crated 3D Printer Uncrated	955 lbs, 434 kg 711 lbs, 323 kg	955 lbs, 434 kg 711 lbs, 323 kg	955 lbs, 434 kg 711 lbs, 323 kg	955 lbs, 434 kg 711 lbs, 323 kg	1555 lbs, 708 kg 1180 lbs, 538 kg	2550 lbs, 1157 kg 2060 lbs, 934 kg
ProJet™ Accelerator Software	Easy build job setup, submission and job queue management; Automatic part placement and build optimisation tools; Part stacking and nesting capability; Extensive part editing tools; Automatic support generation; Job statistics reporting tools					
Print3D App	Remote monitoring and control from tablet, computers and smartphones				-	-

Table 3.4. (*Continued*) Specifications of ProJet™ 3500 SD & HD series and ProJet™ 5000 & 5500X.

Models	ProJet™ 3510 SD	ProJet™ 3510 HD	ProJet™ 3510 HD*Plus*	ProJet™ 3500 HD*Max*	ProJet™ 5000	ProJet™ 5500X
Network Compatibility	Network ready with 10/100 Ethernet interface					
Client Hardware Recommen-dation	1.8 GHz with 1GB RAM (OpenGL support 64 mb video RAM) or higher					1.7 GHz or better with 4GB RAM OpenGL 1.1 Compatible 1280 x 1024 resolution or better
Client Operating System	Windows XP Professional, Windows Vista, Windows 7					Windows 7, Windows 8 or Windows 8.1
Input Data File Formats Supported	STL and SLC	STL and SLC	STL and SLC	STL and SLC	STL and SLC	STL and SLC
Operating Temperature Range	64 - 82 °F (18 - 28 °C)	64 - 82 °F (18 - 28 °C)	64 - 82 °F (18 - 28 °C)	64 - 82 °F (18 - 28 °C)	64 - 82 °F (18 -28 °C)	64 - 82 °F (18 - 28 °C)
Noise	< 65dBa estimated (at medium fan setting)					

3.3.3 *Process*

The ProJet™ Series uses the ProJet™ Accelerator Software to set up the 'sliced' STL files for the system from the CAD software. The ProJet™ Accelerator Software is a powerful software which allows users to verify the preloaded STL files and auto-fix any errors where necessary. The software also helps users to position the parts with its automatic part placement features so as to optimise building space and time. The software also has automatic support generation capabilities to ease the operation of the print of the 3D model. After all details have been finalised, the data is placed in a queue, ready for the ProJet™ machine to build the model. The InVision™ Series uses InVision™ Print Client Software instead, and it has similar functions to the ProJet™ Accelerator Software for the ProJet™ Series.

During the build process, the head is positioned above the platform. The head begins building the first layer by depositing materials as it moves in X-direction. As the machine's print head contains hundreds of jets, it is able to deposit material fast and efficiently. After a single layer pass is completed, the platform is lowered for the head to work on the next layer while the UV lamp floods the work space to cure the layer. The process is repeated until the part is finished. The schematic of the process is shown in Figure 3.10.

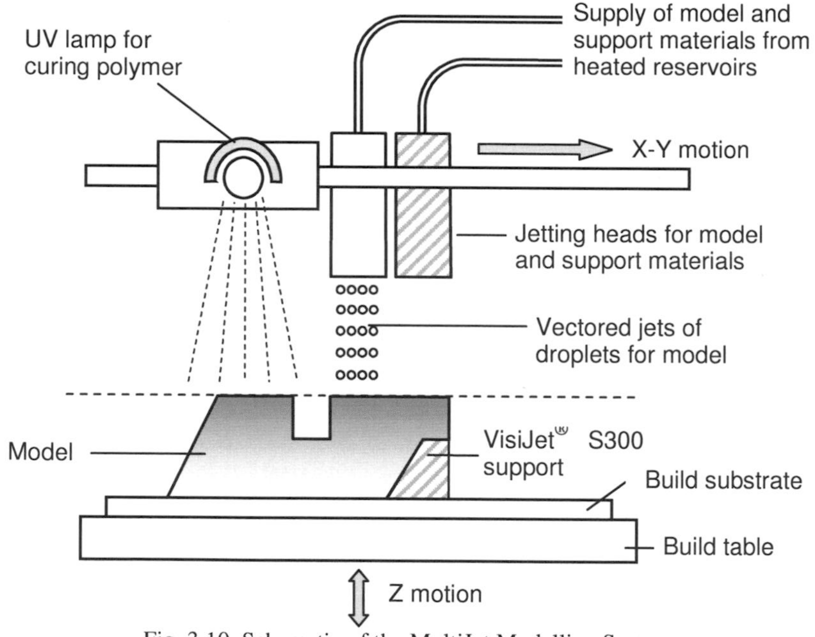

Fig. 3.10. Schematic of the MultiJet Modelling System.

3.3.4 *Principle*

The principle underlying MJP is the layering principle, used in most other AM systems. MJP builds models using a technique akin to inkjet or phase-change printing, applied in three dimensions. The print head jets are oriented in a linear array builds models in successive layers, each jet applying a special thermo-polymer material only where required. The MJP head shuttles back and forth like an inkjet printer (X-axis), building a single layer of what will soon be a 3-dimensional concept model. If the part is wider that the print head, the platform will then reposition (Y-axis) itself to continue building the layer. The UV lamp flashes with each pass to cure the thermo-polymer deposited. When the layer is completed, the platform is distanced from the head (Z-axis) and the head begins building the next layer. This process is repeated until the entire concept model is completed. The main factors that influence the performance and functionalities of the MJP are the thermo-polymer materials, the UV flood lamp curing, the MJP head (number of jets and their arrangements), the X-Y controls, and the Z-controls.

3.3.5 *Strengths and Weaknesses*

The strengths of the MJP technology are as follows:

(1) *Efficient and ease of use.* MJP technology is an efficient and economical way to create concept models. The large number of jets allows fast and continuous material deposition for maximum efficiency. MJP builds models directly from any STL file created with any 3-dimensional solid modelling CAD programs and no file preparation is required.
(2) *High precision.* The new MJP 3D Printer provides best-in-class part quality and accuracy with the choice of both high definition and ultra-high definition build modes in a single system. Its resolution of 750 x 750 x 1600 DPI in xyz orientation is one of the highest in the AM industry.
(3) *Cost-effective.* MJP uses inexpensive thermo-polymer material that provides for cost-effective modelling.
(4) *Fast build time.* As a natural consequence of MJP's raster-based design, geometry of the model being built has little effect on building time. Model work volume (envelope) is the singular determining factor for part build time.

(5) *Office friendly process*. As the system is clean, simple and efficient, it does not require special facilities, thereby enabling it to be used directly in an office environment. Due to its networking capabilities, several design workstations can be connected to the machine just like any other computer output peripherals.

The weaknesses of the MJP technology are as follows:

(1) *Small build volume*. The machine has a relatively small build volume as compared to most other high-end additive manufacturing systems (e.g. ProJet® 6000 SD), thus only small prototypes can be fabricated. The ProJet™ 3500 HD*Max*, being aimed at jewellery and small components, has the smallest build volume.
(2) *Limited materials*. Materials selections are restricted to 3D systems' VisiJet® thermopolymers. This limited range of materials means that many functionally-based concepts that are dependent on material characteristics cannot be effectively tested with the prototypes.

3.3.6 *Applications*

The wide range of uses of the ProJet™ machines, include applications for concept development, design validation, form and fit analysis, production of moulding and casting patterns, direct investment casting of jewellery and other fine feature applications. Specifically for the ProJet™ machines, they can be used in the dental laboratory.

3.3.7 *Examples*

3.3.7.1 *Daimler Trucks drives the ProJet 5000 for design verification*

Daimler Trucks North America LLC, a Daimler company, is the largest heavy-duty truck manufacturer in North America and a leading producer of medium-duty trucks and specialised commercial vehicles. Their key operation is to produce well-known brands such as Freightliner Trucks, Thomas Built Buses, and Detroit Diesel Engines.

Recently, Daimler used ProJet 5000 to print components for their products on a weekly basis for use during their mock up phase to verify critical clearances in the assembled vehicles. After testing, the tough VisiJet MX part material is routinely drilled and tapped and withstand the rough handling conditions characteristic of vehicle manufacturing. For example, it is difficult to visualise and design vehicle pipe routing in CAD to secure the precise clearances. Hence, Daimler used ProJet 5000 to print out the pipes and assemble them onto an engine mock up. This helps them to check every route and associated brackets. Any changes required would be done prior to manufacturing. Furthermore, Daimler also used ProJet 5000 to print long-lead-time components to allow production assembly training in pre-productions to prepare for final production until the components are readily available.

With larger part size capability and higher precision of parts, ProJet 5000 helps Daimler in the production of high quality prints components using tough and stable part material.

3.3.7.2 *3D Systems' ProJet 3500 HD makes its mark at Shachihata*

Shachihata Inc., a leading company in stamps industry, puts product design and innovation as its focus over the years to meet customers' needs creatively at a reduced cost.

In order to improve its ability to satisfy consumers' demand, Shachihata adopted 3D printing into its production line in 2007. More recently, it chose 3D Systems' ProJet 3500 HD 3D Printer to further enhance the working processes and quality of final products.

The ProJet 3500 HD 3D Printer helps Shachihata to efficiently save time. Hence with the shorter production period, Shachihata found it easier to expand to overseas markets and explore diversified products. Furthermore, ProJet 3500 HD 3D Printers not only halved workflow durations but also enhanced the quality of the final products. This further strengthened Shachihata's innovation and creativity.

Aside from improving Shachihata's ability in producing stamps and stationery products, ProJet 3500 HD 3D Printers also helped Shachihata to explore new areas such as products for individuals. After incorporating 3D Printing into workflow, the traditional method of hand drawings' design is replaced by digital 2D/3D CAD method, which accelerates company's expansion and strengthens existing capabilities.

Furthermore, the wax support material by ProJet 3500 HD 3D Printers can be easily and completely removed in an oven. Hence, this support removal process helps Shachihata to save effort and cost in postprocessing. AM also enables Shachihata to add more details into product design and this allows the company to create new concepts faster and more easily.

3.3.8 *Research and Development*

Recently, 3D Systems acquired a portion of Xerox's R&D group which deals with product design, engineering and chemistry. The acquisition brought in an additional 100 engineers and contractors from Xerox into 3D Systems' R&D department. Xerox has long served as a key partner in 3D Systems' most popular line of inkjet based 3D printers, the ProJet series. With this acquisition, the MJP technology will evolve and it serves as a good indication that the ProJet series will get bigger, better and cheaper.

3.4 EnvisionTEC's Perfactory®

3.4.1 *Company*

The company was founded as Envision Technologies GmbH in August 1999 and was re-organised as EnvisionTEC GmbH in 2002 by Mr. Ali El Siblani. EnvisionTEC provides 3D printing systems and solutions to serve customers in a wide variety of applications. EnvisionTEC also deals with software and materials development to increase productivity and cost effectiveness. Currently, it has over 100 patents and patents application pending worldwide. The company has dual headquarters in Gladbeck, Germany and Dearborn, Michigan, US.

3.4.2 *Product*

Perfactory® [18] (see Fig. 3.11) is the AM system built by EnvisionTEC and it is a versatile system suitable for an office environment. Perfactory® requires a file in an STL file format and the STL data is transferred to the system to build the model using the included software package. Resins are cured by photopolymerisation, but Perfactory® uses a different approach from traditional stereolithography in curing the resins. The photopolymerisation process is created by an image projection technology called Digital Light Processing Technology (DLP™) from Texas Instruments and it requires mask projection to cure the resin. The standard system alone can achieve resolutions between 25μm and 150μm, material dependent. Additional components or devices such as the Mini Multi Lens system and the Enhanced

Fig. 3.11. Perfactory® 3 Mini Multi. (Courtesy of EnvisionTEC)

Resolution Module (ERM) enable designers and manufacturers to build smaller figures which require high surface quality. The Perfactory® 3 Mini Multi Lens, fitted with ERM and an 85mm lens is able to create high quality parts with a voxel size as low as 16μm. The specifications of Perfactory® systems are shown in Table 3.5(a) – (c).

Table 3.5(a). Perfactory® 3 Mini Multi Lens specifications. (Courtesy of EnvisionTEC GmbH)

System	Perfactory® 3 Mini Multi Lens		
Lens system, focal length, mm (in.)	60 mm (2.36")	75 mm (2.95")	85 mm (3.35")
Build envelope XYZ, mm, (in.)	84 x 63 x 230 mm 3.31" x 2.48" x 9.06"	59 x 44 x 230 mm 2.32" x 1.73" x 9.06"	44 x 33 x 230 mm (1.73" x 1.3" x 9.06"
Dynamic Voxel thickness Z, μm (in.)	15-150 μm (0.0006 – 0.006)		
Voxel size XY W/ERM, μm (in.)	30 μm (0.0012")	21 μm (0.0008")	16 μm (0.0006")

+Double Perfactory® Resolution

Table 3.5(b). Perfactory® 4 Standard Series System specifications. (Courtesy of EnvisionTEC GmbH)

System	Perfactory® 4 Series	
	Perfactory® 4 Standard	Perfactory® 4 Standard XL
Build envelope XYZ (mm)	160 x 100 x 230 mm 6.3" x 3.9" x 6.3"	192 x 120 x 230 7.6" x 4.7" x 6.3"
Native Voxel Size XY, μm	83 μm 0.0033"	100 μm 0.0039"
ERM Voxel size XY, μm	42 μm 0.0017"	50 μm 0.0020"
Dynamic Voxel thickness Z, μm	25 μm to 150 μm 0.0010" to 0.0059"	25 μm to 150 μm 0.0010" to 0.0059"
Projector Resolution	1920 x 1200 pixels	1920 x 1200 pixels

Table 3.5(c). Perfactory® SXGA+ Standard and Standard UV systems specifications. (Courtesy of EnvisionTEC GmbH)

System	Perfactory® 4 Mini with 60 mm lens	Perfactory® 4 Mini with 75 mm lens	Perfactory® 4 Mini XL with 60 mm lens	Perfactory® 4 Mini XL with 75 mm lens
Build envelope XYZ (mm)	64 x 40 x 230 mm 2.5" x 1.6" x 9.1"	38 x 24 x 230 mm 1.5" x.9" x 91"	115 x 72 x 230 mm 4.5" x 2.8" x 9.1"	84 x 52.5 x 230 mm 3.3" x 2.1" x 9.1"
Native Voxel size XY, μm	33 μm 0.0013"	19 μm 0.0007"	60 μm 0.0023"	44 μm 0.0017"
ERM Voxel size XY, μm	17 μm 0.0007"	10 μm 0.0004"	30 μm 0.0012"	22 μm 0.0009"
Dynamic Voxel thickness Z, μm	15 μm to 150μm 0.0006" to 0.0059"	15 μm to 150 μm 0.0006" to 0.0059"	15 μm to 150 μm 0.0006" to 0.0059"	15 μm to 150 μm 0.0006" to 0.0059"
Projector Resolution	1920 x 1200 pixels	1920 x 1200 pixels	1920 x 1200 pixels	1920 x 1200 pixels

3.4.3 *Process*

To build the part in a systematic manner, Perfactory® undergoes a simple process (see Fig. 3.12) where the 3D model of a solid model is first created with a commercial CAD system. For medical applications, data acquired with MRT or CT systems can be processed directly. The 3D data model in the STL format acquired is voxelised within the Perfactory® software and each voxel data set is converted into a bitmap file with which the mask image is generated. The bitmap image consists of black and white where white represents the material and black represents the void as well as grey which represents partial curing of the voxel. When the image is projected onto the resin with DLP™, the illuminated white portion will cure the resin while the black areas will not and grey area on inner and outer contours will be partially cured in depth thereby eliminating the layering effect.

With the embedded operating system in Perfactory®, it can operate independently and is monitored by the device driver software installed on the PC. The software provides two types of modes, Auto and Expert. The Auto mode allows direct conversion of 3D-CAD data to STL format and other required setups are programmed automatically. The Expert mode is

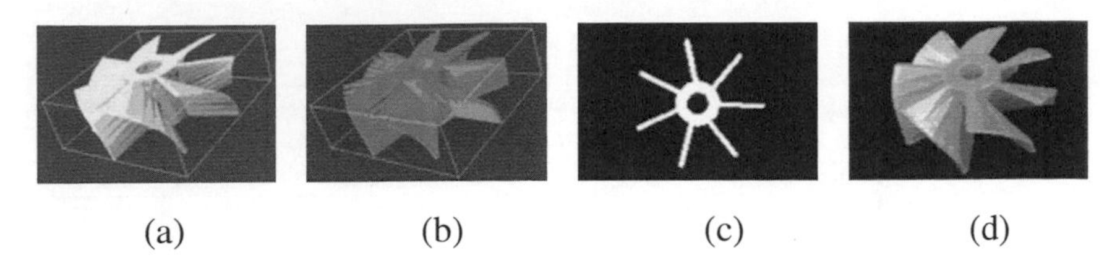

(a) (b) (c) (d)

Fig. 3.12. Perfactory® Process. (a) 3D Data Model (STL), (b) Voxel Data Model, (c) Mask Projection, (d) 3D Solid Model. (Courtesy of EnvisionTEC GmbH)

specially programmed for advanced users to offer them with choice of manual setup according to their experience, needs and preferences.

Unlike almost all other AM systems that build the model from bottom up, Perfactory® builds the model top down (see Fig. 3.13). The build plate or carrier is first immersed into a shallow trough (basement) of acrylate-based photopolymer resin sitting on a transparent contact window. The mask is projected from below the build area onto the resin to cure it. Once the resin is cured, the build platform is raised a single white voxel depth, the voxel thickness being dependent on the material. While the platform is raised, it peels the model away from the transparent contact window, thus allowing fresh resin to flow in through capillary action. The next exposure is then applied and the part is then built in a similar manner. The whole cycle can take as little as 15 seconds without the need for planarisation or levelling for each voxel set cured.

The two key differences of the Perfactory® with other AM systems are the use of mask projection for photopolymerisation and the part is moved upwards with each completed cured layer instead of downwards in other systems. Once the model is built, the user simply has to peel the model off from the platform as the model is stuck to the carrier platform during the entire process.

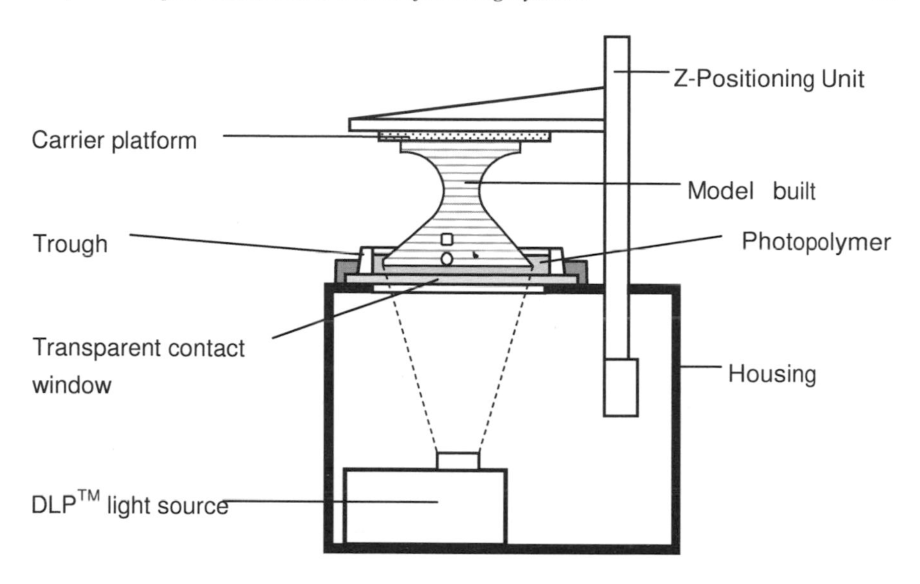

Fig. 3.13. Schematic of the Perfactory® build process with DLP™ Technology. (Courtesy of EnvisionTEC GmbH)

3.4.4 *Principle*

Perfactory® uses the basic principles of stereolithography by undergoing the following stages:

(1) Parts are built from acrylate photopolymer and the user is able to select different material properties with different material colours. Resins are cured when exposed to a mask projection image using Digital Light Processing (DLP™) Technology from Texas Instruments.

(2) Every completion of cured layer is moved away from the build trough containing the resin vertically upwards by a precision linear drive. This is due to the projection system being integrated at the bottom of the AM system. Also, the fabricated part requires minimal support and removing the model from the transparent platform is easy.

3.4.4.1 *Digital Light Processing (DLP™) technology*

Digital Light Process (DLP™) is a projection technology invented by Dr. Larry Hornbeck and Dr. William E. "Ed" Nelson of Texas Instruments in 1987. In DLP™ technology, the key device is the Digital Micromirror Device (DMD), the producer of image. Digital Micromirror Device is an optical semi-conductor and each DMD chip has hundreds of thousands of mirrors arranged in a rectangular array on its surface to steer the photons with great accuracy. This means that each mirror is represented as one pixel in a projected image and therefore the resolution of an image depends on the number of mirrors.

The mirrors in the DMD are made of aluminium and are 16μm in size. Each individual mirror is connected to two support posts where it can be rotated ±10-12° in an ON or OFF state. In the ON state, the light source is reflected from the mirrors into the lens making pixels on the screen. For the OFF state, the reflected light is redirected to the other direction allowing the pixels to appear in a dark tone. During each mask projection, the cross-section of each layer is projected by mirrors in the ON state and resins are cured by the visible light projected from below the transparent contact window.

Every single mirror is mounted on a yoke by compliant torsion hinges with its axle fixed on both ends and able to twist in the middle. Due to its extremely small scale, damages hardly occur and the DMD structure is able to absorb shock and vibration thus providing high stability.

Position control of each mirror is done by two pairs of electrodes as positioning significantly affects the overall image of the cross-section layer. One pair is connected to the yoke while the other is connected to the mirror and every pair has an electrode on each side of the hinge. Most of the time, equal bias charges are applied to both sides to hold the mirror firmly in its current position. In order to move the mirrors, the required state has to load into a static random access memory (SRAM) cell connecting to the electrodes and mirror. Once all the SRAM cells have been loaded, the bias voltage is removed, allowing every individual movement of the mirrors through released charges from the SRAM cell. When the bias is restored, all of the mirrors will be held in their current position to wait for the next loading into the cell. Bias voltage enables instant removal from the DMD chip so that every mirror can be moved together providing more accurate timing while requiring a lower amount of voltage for the addressing of the pixel.

3.4.5 *Strengths and Weaknesses*

The main strengths of Perfactory® systems are:

(1) *High building speed.* The use of a mask image directly exposed to the resin enables a part to be built at speeds as quick as approximately 15mm height per hour at 100μm pixel height. This speed is independent of part size and geometry and is one of the fastest systems in the AM market.
(2) *Office friendly process.* Perfactory® allows operation in an office environment as its foot print is under 0.3 square meters. The curing of the photopolymer does not use UV light and there is no need for special facility. It operates with low noise and zero dust emissions.
(3) *Small quantity of resin during build.* The shallow trough of resin means that the amount of material in use at any one time is small (about 200ml). This means that should a number of different resins be required, they can be swapped out easily with minimal wastage.
(4) *No wiper or leveller.* When the carrier platform is raised with the model, there is no need for planarisation or levelling. This eliminates the possibility of causing problems to the stability of the parts during the build, e.g. a wiper damaging a small detail on the part during the wiping action.
(5) *Less shrinkage.* Due to immediate curing of a controlled layer (based on the voxel thickness) there is less shrinkage during the process.
(6) *Additional components.* Perfactory® is able to build even higher quality tough parts with the use of additional components which can be integrated into the system.

The main weaknesses of Perfactory® systems are:

(1) *Limited building volumes.* Structures are built from the bottom of the build chamber and stuck to the carrier platform; this limits the size of the build.
(2) *Peeling of completed part.* The user has to peel the completed model from the build platform as the model is built on a movable carrier platform which moves vertically upwards. Care has to be taken so as not to damage the model during the peeling process.
(3) *Requires postprocessing.* After the model is completed, cleaning and postprocessing, sometimes including post-curing are required.

3.4.6 *Applications*

The application areas of the Perfactory® systems include the following:

(1) Concept design models for design verification, visualisation, marketing, and commercial presentation purposes.
(2) Working models for assembly purpose, simple functional tests and for conducting experiments.
(3) Master models and patterns for simple moulding and investment casting purposes.
(4) Building and limited production of completely finished parts.
(5) Medical [19] and dental applications. Creating exact physical models of patient's anatomy from CT and MRI scans.
(6) Jewellery applications. Creating designs for direct casting or moulding.

3.4.7 *Research and Development*

EnvisionTEC has been researching on building larger build envelope machines and new casting materials. Recently the company announced two new casting materials called Epic and Epic M. Epic is a wax-based material and offers the quality and detail of the plastic casting materials but with the properties and adaptability of its wax products [20].

3.5 CMET'S Solid Object Ultraviolet Laser Printer

3.5.1 *Company*

CMET (Computer Modelling and Engineering Technology) Inc. was established in November 1990 by the Mitsubishi Corporation as the major shareholder and two other substantial shareholders, NTT Data Communication Systems and Asahi Denak Kogyo K.K. However, in 2001, CMET Inc. merged with Opto-Image Company, Teijin Seiki Co., Ltd. The company's address is CMET Inc., Sumitomo Fudosan Shin-Yokohama Building, 2-5-5 Shin-yokohama Kohoku-ku Yokohama, Kanagawa 222-0033 Japan.

3.5.2 *Products*

3.5.2.1 *Models and specifications*

CMET has four models of machines: EQ-1, Rapid Meister ATOMm-4000, Rapid Meister 6000 II and Rapid Meister 3000. Each of them serves individual designers and manufacturers with different parameters. EQ-1 (Fig. 3.14) is the successor model to the Rapid Meister 6000 II and it claims to be the world's fastest scanner developed in-house. EQ-1 has

Fig. 3.14. EQ-1 machine. (Courtesy of CMET Inc.)

a higher speed scanner system and a different recoater. It comes equipped with the state-of-the-art laser system which enhances the transparency of the model. The digital scanner system uses ultra-high speed scan to create parts in a short time with scanning speeds up to 40m/s. The central working section provides a high position precision suited to very small models like connectors, and solid structure models like components. A summary table of the machines available is given in Table 3.6.

Table 3.6. Rapid Meister series' specifications. (Courtesy of CMET Inc.)

Models	EQ-1	ATOMm-4000	Rapid Meister 3000	Rapid Meister 6000 II
Laser type	Solid state laser 2.0W 120MHz	Solid state laser 400mW 40KHz	LD Laser 400mW 25kHz	LD Laser 800mW 60kHz
Scan mode	Digital scanner mirror (TSS4)	Digital scanner mirror (TSS4)	Digital scanner System	Digital scanner System
Max. Scanning Speed	40m/s	30m/s	12m/s	22m/s
Beam Diameter	Variable system			
Maximum Modelling size X x Y x Z (mm)	610 x 610 x 500 mm	400 x 400 x 300 mm	300 x 300 x 250 mm	610 x 610 x 500 mm
Minimum Build Layer	0.05mm	0.025mm	50 μm	
Vat	Interchangeable			
Recoater	Vent recoater	Blade recoater	Braid recoater	Multi-purpose High speed(MH) recoater
Power	AC100V Single-Phase (Control unit 30A/Heater 10A)	AC100V Single-phase 15A	AC100V single-phase 30A	AC100V single-phase x Two circuits (Main part 30A Heater part 10A)
Dimension X x Y x Z (mm)	1160 x 1790 x 2100	1565 x 1050 x 1860	1430 x 1045 x 1575	1020 x 2045 x 2050
Weight	1200 kg (without resin)	550 kg (without resin)	400kg (without resin)	1400 kg (without resin)

3.5.2.2 *Advantages and disadvantages*

The main advantages of the Rapid Meister systems are:

(1) *New recoating system.* The new recoating system provides a more accurate Z layer and shorter production time.
(2) *Software system.* The software system allows for real time processing.
(3) *High scanning speed.* It has scanning speeds of up to 40 m/sec.
(4) *Variety of resins.* Rapid Meister has a variety of materials with different properties for users to select from (ABS performance resin, high precision resin, rubber-like resin and etc.)

The main disadvantages of the Rapid Meister systems are:

(1) *Requires support structures.* Structures that have overhangs and undercuts must have supports that are designed and fabricated together with the main structure.
(2) *Requires postprocessing.* Postprocessing includes removal of supports and other unwanted materials which is tedious, time-consuming and can damage the model.
(3) *Requires post-curing.* Post-curing may be needed to cure the object completely and ensure the integrity of the structure.

3.5.3 *Process*

The Rapid Meister process contains the following three main steps:

(1) *Creating a 3D model with CAD system*: The three-dimensional model, usually a solid model, of the part is created with a commercial CAD system. Three-dimensional data of the part are then generated.
(2) *Processing data with the software*: Often data from CAD system are not faultless enough to be used by the AM system directly. Rapid Meister's software can edit CAD data, repair its defects like gaps

and overlaps etc. (see Chapter 6), slice the model into cross-sections and finally generate the corresponding Rapid Meister machine data.
(3) *Making model with the Rapid Meister units*: The laser scans the resin, solidifying it according to the cross-section data from the software. The elevator lowers and the liquid covers the top layer of the part which is recoated and prepared for the next layer. This is repeated until the whole part is created.

Rapid Meister's hardware contains: a communication controller, a laser controller, a shutter/filter controller, a scanner controller (galvanometer mirror unit) and an elevator controller. The Rapid Meister software is a real-time working, multi-user and multi-machine control software. It has functions such as simulation, a convenient editor for editing and error repair, 3D offset, loop scan for filling area between the outlines, delectable structure and automatic support generation.

3.5.4 *Principle*

The Rapid Meister system is based on the laser lithography technology which is similar to that described in Section 3.1.4. The one major difference is in the optical scanning system. The main trade-off is in scanning speed and consequently, the building speed. Parameters which influence performance and functionality are galvanometer mirror precision for the galvanometer mirror machine, laser spot diameter, laser power and speed, slicing thickness and resin properties.

3.5.5 *Applications*

The application areas of the Rapid Meister systems include the following:

(1) Concept models for design verification, visualisation and commercial presentation purposes.
(2) Working models for form fitting and simple functional tests.
(3) Master models and patterns for silicon moulding, lostwax, investment casting, and sand casting.

(4) As completely finished parts.
(5) Medical purposes. Creating close to exact physical models of patient's anatomy from CT and MRI scans.
(6) 3-dimensional stereolithography copy of existing product.

3.5.6 *Research and Development*

CMET focuses its research and development on a new recoating system. The new recoating system can make Z layer with high accuracy even though a small Z slice pitch (0.05 mm) is used. As the down-up process of Z table is not productive, the 'no-scan' time can be shortened by more than 50%. More functions will be added into CMET's software. CMET is also conducting the following experiments for improving accuracy:

(1) Investigation of the relationship between process parameters and curl distortion: This experiment is done by changing laser power, scan speed while fixing other parameters like the Z layer pitch and beam diameter.
(2) Investigation of the effects of scan pattern: This experiment aims to compare building results between using and not using scanning patterns.
(3) Comparative testing of resins' properties on distortion.

3.6 EnvisionTEC'S Bioplotter

3.6.1 *Company*

EnvisionTEC GmbH is the same company that developed the Perfactory® System discussed in Section 3.4. Details of the company can be found in Section 3.4.1.

3.6.2 *Product*

Medical applications have been gaining a foothold in additive manufacturing and Bioplotter [21, 22] has become one of the solutions for computer-aided tissue engineering. Bioplotter Technology was first invented in 1999 at Freiburg Materials Research Centre for producing scaffold structures from various biochemical materials and even living cells. The EnvisionTEC Bioplotter (see Fig. 3.15) has been sold in Europe since 2002. Recently, the 3D-Bioplotter was updated to its 4th generation.

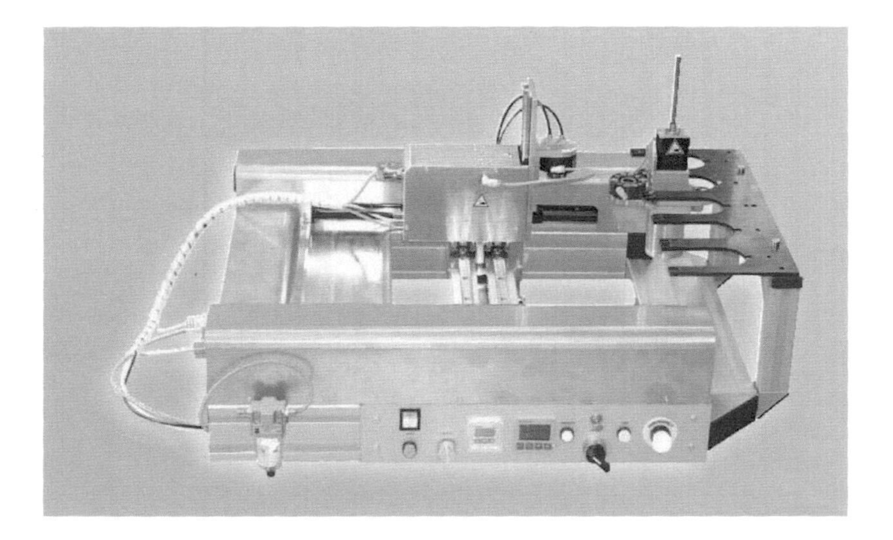

Fig. 3.15. EnvisionTEC Bioplotter. (Courtesy of EnvisionTEC GmbH)

Bioplotter undergoes a simple process of CAD data handling, dispensing material and finally solidification of material to create the structures. The system is based on a 3D dispenser which allows only a certain amount of materials to be dispensed into a plotting medium. Specific 2½D CAD-CAM software is used to handle CAD data and machine/process control. The specifications of the Bioplotter are summarised in Table 3.7.

Table 3.7. Bioplotter system specifications. (Courtesy of EnvisionTEC GmbH)

Model	Bioplotter
Machine size LxWxH, mm	976 x 623 x 733 (38.4 x 24.5 x 30.4 in)
Working area XxYxZ, mm	150 x 150 x 140 (5.91 x 5.91 x 5.51 in)
Resolution in X/Y/Z, µm	1 (0.00004 in)
Speed	0.1 -150 mm/s (0.004 - 5.91 in/s)
Camera Resolution (XY)	0.009 mm (0.00035 in) per Pixel
Weight	130 kg (286.6 lb)
Needle Sensor Resolution (Z)	0.001 mm (0.00004 in)
Minimum Strand Diameter	0.100 mm (0.004 in) - Material Dependent
Electrical Requirements	100 - 240 VAC, F 50/60 Hz
Machine control	Via PC incl. CAD/CAM software with integrated device driver

3.6.3 *Process*

The Bioplotter process contains the following three main steps:

(1) *3D data handling*. Digital data models of the design structure need to be generated which is the first condition for building 3D scaffolds. The 3D scaffolds can be designed on any commercial 3D CAD software or data can be generated from data acquired through MRI, CT or X-ray. Additional software like VoXim from IVS-solutions can also be used for volume data for 3D image processing and then exported in data formats like DXF and CLI.
2½D CAD-CAM software PrimCam® is the main program used to import data formats specially programmed for the Bioplotter system. Data formats are transferred to Primus data to modify single layers of the imported data or to design another structure design for plotting process.

(2) *Control process*: PrimCam® software is designed specially for the Bioplotter system allowing users to calibrate and design machine and process control. Users can input machine setup and process parameters either manually or automatically. Two key functions of

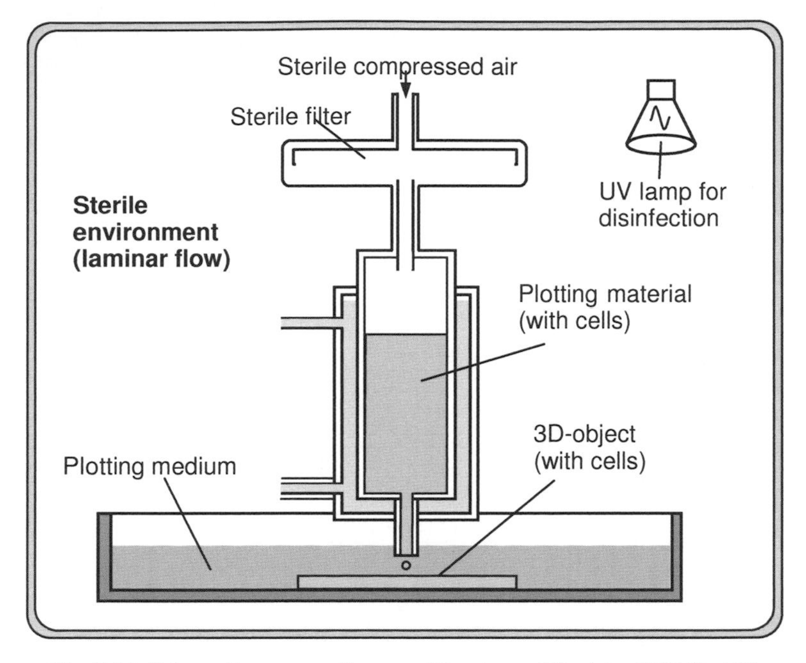

Fig. 3.16. Schematic process diagram. (Courtesy of EnvisionTEC GmbH)

the software enable the selection of the dispensing path in the object layers and the simulation of the dispensing process, which can then be checked before the commencement of the process.

(3) *Dispensing process*: Liquids or solutions are dispensed to build the model in layers and all the plotting material is first stored in a cartridge (see Fig. 3.16). Plotting material is then later forced to extrude through a small dispensing needle of a diameter close to 80 microns into the plotting medium. The purpose of this second medium is to cause the solidification of the first solution through certain reactions when both materials are in contact. The solidification of material depends on the material, medium and the temperature control creating the precipitation reaction, phase transition or chemical reaction [23] (see Fig. 3.17). As mentioned before about the solidification by heating, storage cartridges can be heated up to 230°C while the build platform can be heated up to 100°C.

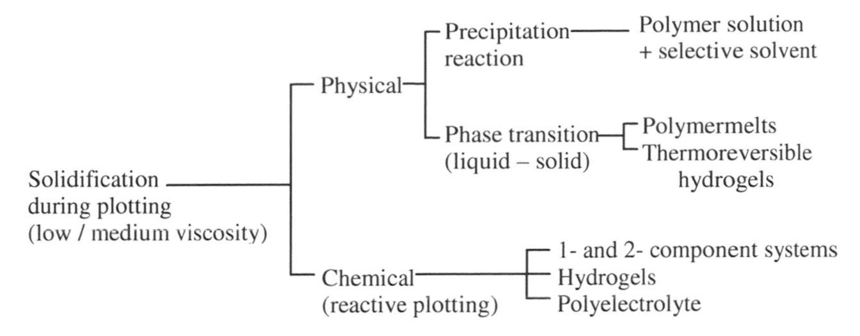

Fig. 3.17. Material reaction flow diagram.

(4) Secondly, the plotting medium acts as a supporting force to create buoyancy to prevent structure from collapsing. Normally when materials are dispensed without a supporting force, gravity will cause the complex scaffold design to collapse.

3.6.4 *Principle*

The design of the Bioplotter system is created for the purpose of tissue engineering applications and its approach in terms of the process will be different even though the object is created layer by layer. The process of Bioplotter system is based on the following principles:

(1) When dispensing materials, complex structures will collapse due to gravity without the plotting medium (see Fig. 3.18). With the Bioplotter, the compensation method is to dispense the plotting materials into a matched plotting medium. This can be done based on the buoyancy force where the densities of both plotting materials and medium are similar and create the required gravity force to prevent the collapse of the structure. Additionally with the use of the medium, support structures are not required in the process which means no waste or unwanted support will be built.

(2) The production or dispensing process can also be carried out in a sterile environment when living cells are to be incorporated into the plotting material.

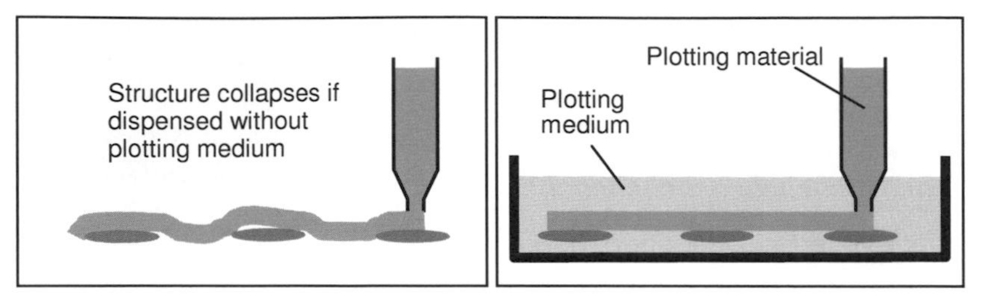

Fig. 3.18. Diagram for compensation due to buoyancy force. (Courtesy of EnvisionTEC GmbH)

3.6.5 *Strengths and Weaknesses*

The main strengths of Bioplotter system are:

(1) *Broad range of biomaterials.* Materials that can be processed on the Bioplotter include natural and thermoplastic polymers, reactive resins and living cells.
(2) *Layer modification.* User is able to modify process parameters within a specific layer.
(3) *No support required.* Bioplotter consists of plotting medium which provides the up thrust and thus acts as a support during the process.

The main weaknesses of Bioplotter system are:

(1) *Sterile environment.* The system requires a sterile environment as materials include biomaterial, biochemical and living cells.
(2) *Temperature control.* Control of temperature is necessary to ensure solidification between plotting material and medium by specific reaction.

3.6.6 *Applications*

The application areas of the Bioplotter system include the following:

(1) Scaffolds models for design verification, visualisation, experimental and fabrication purposes.

(2) Testing and experimental purposes.
(3) Medical purposes like creating implantations for human and generation of living tissues.
(4) Fabrication of biodegradable materials for tissue engineering and drug delivery purposes.

3D Bioplotter is designed with unique capabilities differentiating it from other commercial systems based on its ability to plot biological cells. Due to this capability, it is mainly meant for medical applications, surgical and biomedical research.

3.6.7 *Research and Development*

University of Freiburg and EnvisionTEC are partners in developing and marketing Bioplotter. EnvisionTEC technology supports users by providing the required information and materials for the 3D plotting process. 3D plotting technology is still under further development to bring biomedical research to a higher level in the future.

3.7 RegenHU's 3D Bioprinting

3.7.1 *Company*

RegenHU Ltd was founded in November 2011. The company specialises in making bioprinters and producing 3D organomimetic models for tissue engineering. RegenHU Ltd is headquartered in Z.i. du Vivier 22, CH-1690 Villaz-St-Pierre, Switzerland.

3.7.2 *Products*

RegenHU Ltd produces a wide range of 3D printers to produce bio tissues and bio materials. There are two main models which are the 3DDiscovery® and BioFactory®. 3DDiscovery® printer (see Fig. 3.19) is designed to further study the potential of 3D tissue engineering through the bioprinting approach. Its innovation lies in the spatial controlling of cells and morphogens in a three-dimensional scaffold. The system helps in constructing organotypic in-vitro models of soft and hard tissues. Meanwhile, BioFactory® (see Fig. 3.20) is designed for scientists and specialists to pattern cells, biomolecules and a range of soft and rigid materials in desirable 3D composite structures. Both 3DDiscovery® and BioFactory® use BioInk™ as the building materials because BioInk™ is a semi-synthetic hydrogel supporting cell growth of different cell types. Specifications of 3DDiscovery® printer are summarised in Table 3.8.

Fig. 3.19.

Fig. 3.20.

Table 3.8. Specifications of 3DDiscovery® printer.

Model	3DDiscovery® printer
External dimensions	580 x 540 x 570 mm
Precision	±10μm
Working Range	130 x 90 x 60 mm
4 printing heads for viscosity range	Up to 10,000 mPaS
Software	BioCAD
Temperature	Up to 80 °C
Material	BioInk™

3.7.3 *Process*

RegenHU bioprinting applies the following steps to creating the tissue or cell:

(1) The CAD file of building process is formed in G-Code format. STL importation of 3D objects are generated through medical imaging (computer tomography, MRI) and/or CAD/CAM systems.
(2) RegenHU's bioprinters produce the tissue and cells layer upon layer.
(3) Natural tissue is assembled by the 3D printed cell and extracellular matrix.
(4) The final product is removed from the working plate.

3.7.4 *Principle*

The design of the RegenHU's 3D bioprinters is created for the purpose of tissue engineering applications. Tissues are built up layer upon layer. Cells, signal molecules and biomaterials create a highly dynamic network of proteins and signal transduction pathways. The printed cell closely mimics the natural tissue. Its well-defined structure enables the control and study of biological and mechanical cell/molecule interaction processes.

3.7.5 *Materials*

Generally, the materials used are dependent on the cell or tissues being printed. Cells, proteins, bioactives, thermopolymers, natural hydrogels such as collagen, hyaluronic acid, chitosan and gelatine can be used in the RegenHU bioprinters. RegenHU provides two special materials dedicated for their bioprinters:

3.7.5.1 *BioInk®*

BioInk® (see Fig. 3.21) is a semi-synthetic hydrogel that promotes cell growth of different cells by providing cell adhesion sites and mimicking the natural extracellular matrix. BioInk® is available as a ready-to-use solution to print 3D tissues models.

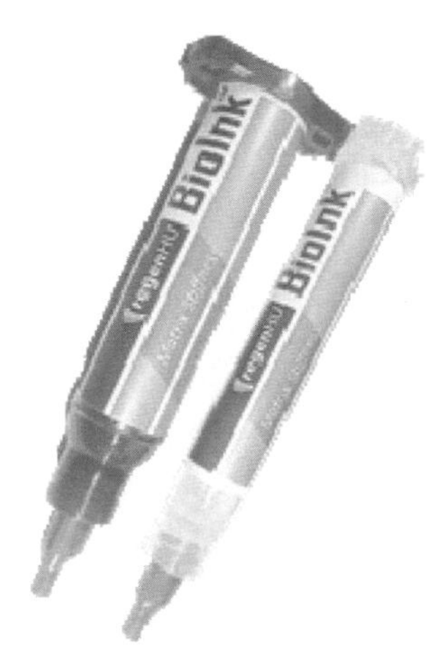

Fig. 3.21. The BioInk®. (Courtesy of RegenHU Ltd)

3.7.5.2 *OsteoInk®*

OsteoInk® is a ready-to-use calcium phosphate paste for structural engineering dedicated to BioFactory® and 3DDiscovery® tissue printers. OsteoInk's chemical composition such as Hydroxyapatite or tricalcium phosphate is optimal for hard tissue engineering such than bone and cartilage. Three-dimensional printing enables the freeform fabrication with controlled pore structure. BioFactory and 3DDiscovery enable composite tissue printing by loading or filling the pores with bioactives such as cells, proteins, blood derivate or hydrogels.

3.7.6 *Strengths and Weaknesses*

RegenHU's bioprinters have the following strengths:

(1) *Unique technology*: Researchers and scientists can design and produce pattern cells, biomolecules and a range of soft and rigid materials in desirable composite structures.

(2) *Cost-effective*: The cost of using such bioprinters is relatively low, which allows researchers to do a few attempts.

It has the following weaknesses:

(1) *Small building volume*: The bio printer's working range is only 130 x 90 x 60 mm, limiting the size of the build.
(2) *Sterile environment:* Keeping the bioprinting process under a sterile environment is crucial as materials include biomaterials, biochemical and living cells.

3.7.7 *Applications*

The general applications areas are given as follows:
(1) *Skin model (Soft tissue engineering)*: RegenHU's BioFactory® can produce full thickness skin models consisting of dermal and epidermal layers. This helps researchers to further study on them.
(2) *Drug discovery*: RegenHU's BioFactory® helps to provide an alternative for researchers to study or analyse the induced effects of active substances on highly dynamic networks of proteins and signal transduction pathways in tissues, cell-cell and cell-extracellular matrix interactions. Traditionally, this kind of research can only be done in a 2D manner.
(3) *System Engineering*: With RegenHU's bioprinters, the risks associated with the reproducibility, traceability and quality control of drugs developments are mitigated.

3.7.8 *Research and Development*

RegenHU is focusing on the development of the hardware of its bioprinters. The materials used by the machines are mainly developed by their key consumers and R&D institutes worldwide.

3.8 Rapid Freeze Prototyping

3.8.1 *Introduction*

Most of the existing additive manufacturing processes are relatively expensive and many of them generate substances such as smoke, dust, hazardous chemicals etc., which are harmful to human health and the environment. Continuing innovation is essential in order to create new rapid prototyping processes that are fast, clean, and of low cost.

Dr. Ming Leu of the University of Missouri-Rolla, has been developing a novel, environmentally benign additive manufacturing process that uses cheap and clean materials and can achieve good layer binding strength, fine build resolution, and fast build speed. They have invented such a process called Rapid Freeze Prototyping (RFP) [24]; it makes 3D ice parts layer by layer by freezing water droplets.

3.8.2 *Objectives*

The fundamental study of this process has three objectives:

(1) Developing a good understanding of the physics of this process, including the heat transfer and flow behaviour of the deposited material in forming an ice part.
(2) Developing a part building strategy to minimise part build time, while maintaining the quality and stability of the build process.
(3) Investigating the possibility of investment casting application using the ice parts generated by this process.

3.8.3 *Process*

The experimental AM system [25] in Fig. 3.22 consists of the following mechanical mechanism: (a) a three-dimensional positioning subsystem; (b) a material depositing subsystem (see Fig. 3.23); (c) a freezing chamber and (d) electronic control device. Tedious modelling and analysis efforts [26] have been carried out to understand and improve the behaviour of material solidification and rate of deposition during the ice part building process. The process starts with software which receives STL file from CAD software and generates the sliced layers of contour information under a CLI file. Together with a CLI file, process parameters like nozzle transverse speed, environment temperature and fluid viscosity are then further processed to generate the NC code. The NC code is sent to the experimental system to control the fabrication of the ice parts.

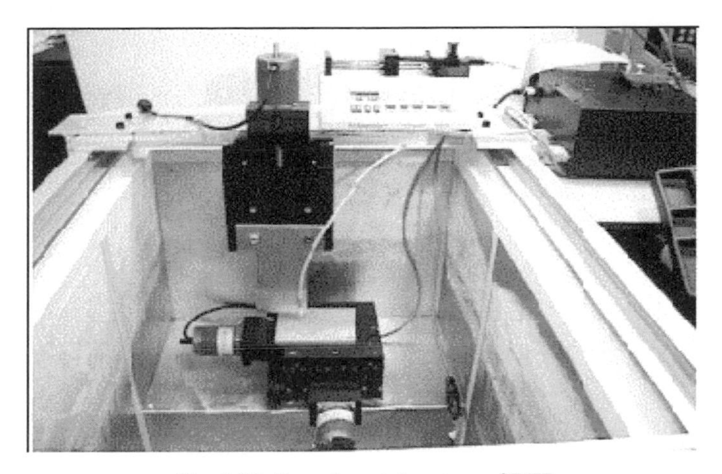

Fig. 3.22. Experimental system of RFP.

The 3D positioning system consists of an XY table and a Z-elevator and it is seated in the freezer at a low temperature of -20°C or lower. The depositing nozzle is mounted on the Z-elevator whereby water droplets are extracted from the nozzle onto the substrate. Once the first layer solidifies, the nozzle will eventually level up and continue depositing for the next layer.

During the fabrication process, building and support solutions are used. The building solution will be water and the support solution is the eutectic sugar solution ($C_6H_{12}O_6 - H_2O$) of a melting temperature -5.6°C [27]. Due to the difference in each solution's melting temperature, the built part is then placed in an environment between 0 °C and the melting point of the support material to melt the support material (see Fig. 3.24).

Experiments have also been conducted where small parts are fabricated with radius of 0.205 mm and thin wall of 0.48 mm in thickness [28]. Fig. 3.25 shows the control system schematics for the system.

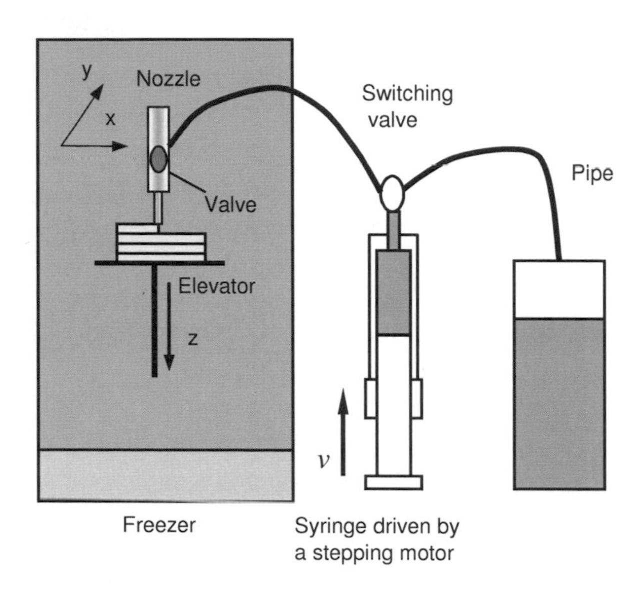

Fig. 3.23. Building environment and water extrusion subsystem.

Fig. 3.24. An ice part before removal of support material and after removal.

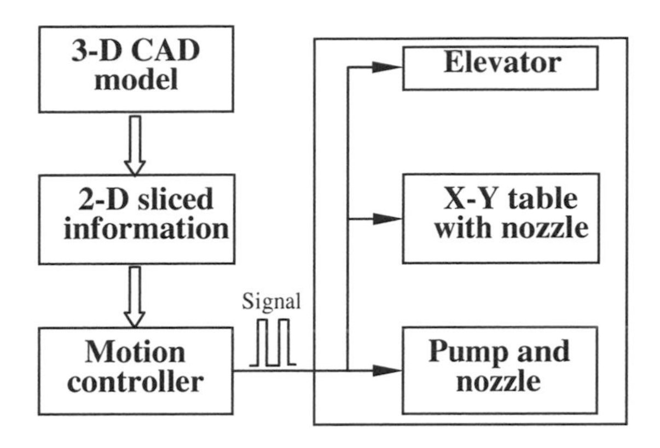

Fig. 3.25. Control system schematics.

3.8.4 *Principles*

The rapid freezing prototyping system is based on the principle of using a water freezing method to build an icy cold prototype. Besides being environmentally friendly, the water solution can also be easily recycled. Before the frozen part can be built, an STL file from a CAD design is required. The software will convert the STL file to a CLI file and finally to NC codes. NC codes together with other process parameters are then transferred to a motion control card to begin fabrication.

The system consists of 3 major hardware equipment: a positioning subsystem, a water ejecting subsystem and an electronic control device. Each of these holds an important role during the building process. Firstly, the XY table is placed in the freezer under a low temperature in order for the extracted water droplets from the nozzle to freeze in a short time. The nozzle is attached at the Z-elevator and it will move in an upwards direction after every completed layer. For the first layer of the ice part, only water droplets are deposited. Every water droplet does not solidify immediately upon deposition as the water droplets are spread and form a continuous water line. The water line will then be frozen by convection through the cold environment and conduction from the previous frozen layer rapidly.

During each experimental process for building an ice part, process parameters set in the water ejection subsystem and electronic control device are important as they affect the overall fabrication accuracy. The key parameters include the ambient build temperature, scan speed, and water feed rate. The ratio of material flow to XY movement speed is kept at the best value to prevent discontinuity in the freezing strands. Each layer thickness and smoothness is determined by the adjustment of nozzle scanning speed and water feed rate rather than the mechanical mechanism [29].

3.8.5 *Strengths and Weaknesses*

RFP has the following strengths:

(1) *Low Running Cost.* The Rapid Freeze Prototyping (RFP) process is cheaper and cleaner than all the other AM processes. The energy utilisation of RFP is low compared to other AM processes such as laser stereolithography or Selective Laser Sintering.
(2) *Good Accuracy.* RFP can build accurate ice parts with excellent surface finish. It is easy to remove the RFP made ice part in a mould making process by simply heating the mould to melt the ice part.
(3) *Good Building Speed.* The build speed of RFP can be significantly faster than other AM processes, because a part can be built by first depositing water droplets to form a waterline and then filling in the enclosed interior with a water stream (see Fig. 3.26). This is possible due to the low viscosity of water. It is easy to build colour and transparent parts with the RFP process.
(4) *Environmentally Friendly Materials.* Both the building material and support material are clean and non-toxic when handled.
(5) *Suitability in Investment Casting.* Eliminating problems created from traditional wax patterns in investment casing like pattern expansion and shelling cracking during pattern removal.

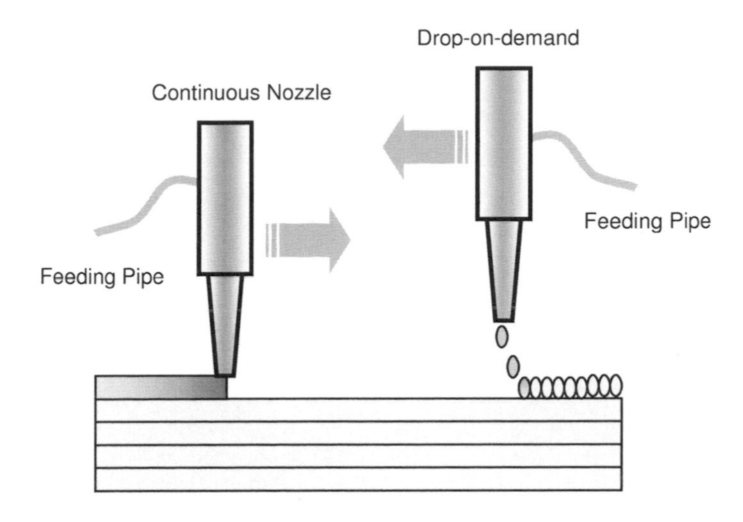

Figure 3.26. Two material extrusion methods.

On the other hand, RFP has also the following weaknesses:

(1) *Requires Cold Environment.* The prototype of RFP is made of ice and hence it cannot maintain its original shape and form at room temperature.
(2) *Needs Additional Processing.* The prototype made with RFP cannot be used directly but has to be subsequently cast into a mould and so on and this increases the production cost and time.
(3) *Repeatability.* Due to the nature of water, the part built in one run may differ from the next one. The composition of water is also hard to control and determine unless tests are carried out.
(4) *Postprocessing.* The final part is required to be placed in a lower temperature environment to remove the support material.

3.8.6 *Potential Applications*

(1) *Part Visualisation.* Parts can be built for the purpose of visualisation. Examples can be seen in Fig. 3.27.

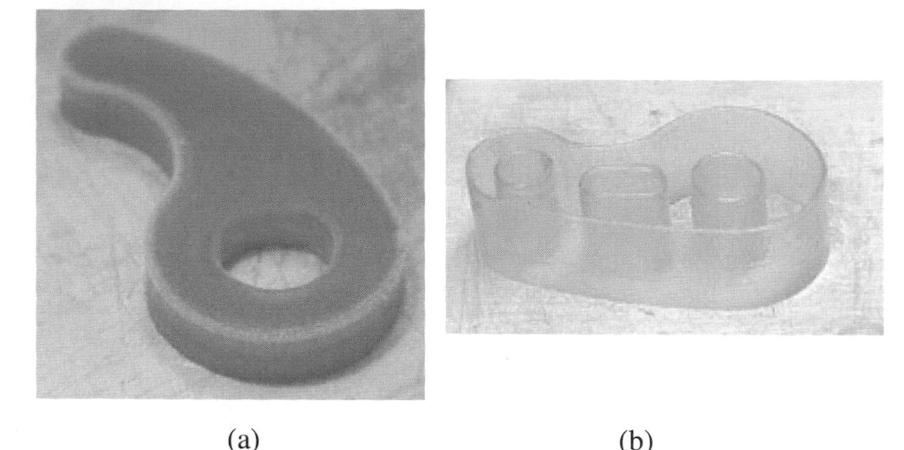

(a) (b)

Fig. 3.27. Examples of visualisation parts: (a) Solid link rod (10 mm height) and (b) contour of a link rod.

(2) *Ice Sculpture Fabrication.* One application of Rapid Freeze Prototyping is making ice sculptures for entertainment purposes. Imagine how much more fun a dinner party can provide if there are colourful ice sculptures that can be prescribed one day before and they vary from table to table. This is not an unlikely scenario because RFP can potentially achieve a very fast build speed by depositing water droplets only for the part boundary and filling in the interior with a water stream in the ice sculpture fabrication process. Fig. 3.28(a) shows the CAD model of an ice sculpture, while Fig. 3.28(b) shows the part made by RFP.

(3) *Silicone Moulding.* The experiments on UV silicone moulding have shown that it is feasible to make silicone moulds with ice patterns (see Fig. 3.29) and further make metal parts from the resulting silicone moulds (see Fig. 3.30). The key advantage of using ice patterns instead of plastic or wax patterns is that ice patterns are easier to remove (without pattern expansion) and no demoulding step is needed before injecting urethane or plastic parts. This property can avoid demoulding accuracy loss and can allow more complex moulds to be made without the time-consuming and experience-dependent design of demoulding lines.

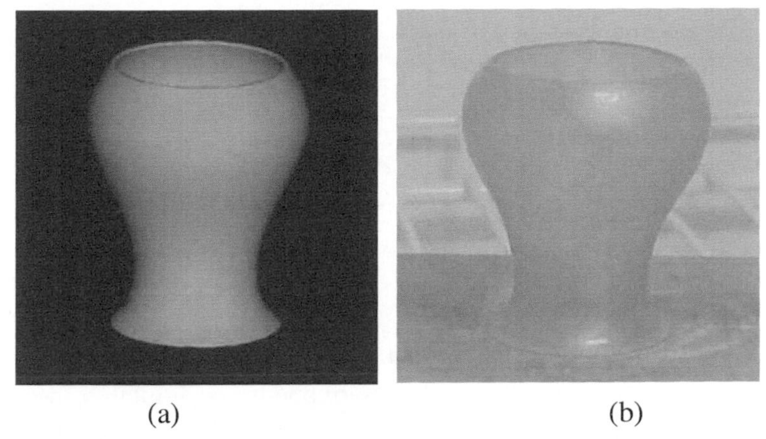

(a) (b)

Fig. 3.28. Ice sculpture: (a) the CAD model and (b) the RFP fabricated part.

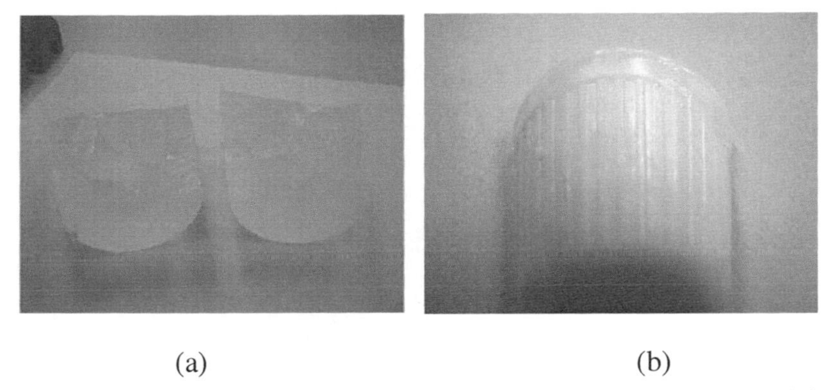

(a) (b)

Fig. 3.29. (a) UV silicone mould made by ice pattern and (b) urethane part made by UV silicone mould.

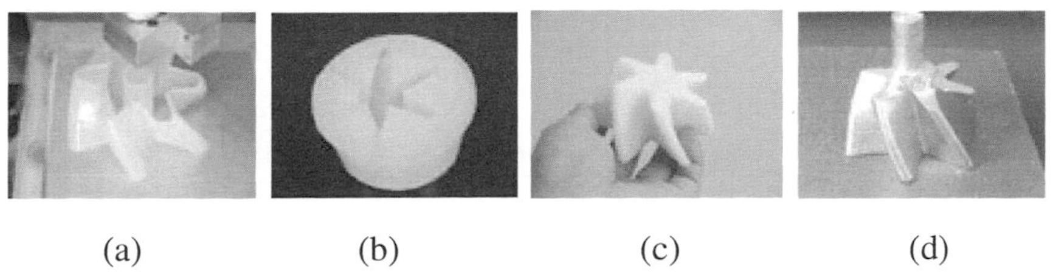

(a) (b) (c) (d)

Fig. 3.30. (a) Making the ice part by RFP, (b) UV silicone mould made by ice part, (c) urethane part made by UV silicone mould and (d) metal part made by UV silicone mould.

(4) *Investment Casting*. A promising industrial application of RFP technology is investment casting. DURAMAX recently developed the Freeze Cast Process (FCP), a technology of investment casting with ice patterns made by moulding. The company has demonstrated several advantages of this process over the competing wax investment and other casting processes, including low cost (35-65% reduction), high quality, fine surface finish, no shell cracking, easy process operation, and faster run cycles. Additionally, there is no smoke and smell in investment casting with ice patterns. Apparently there is a finding that the alcohol base binder can damage the surface of the ice pattern due to the solubility of ice and alcohol binder. Research has been ongoing for the search of an interface agent to solve the problem [30]. Fig. 3.31 shows a metal model made by investment casting using ice patterns from RFP.

Fig. 3.31. Metal model made by investment casting with ice pattern.

3.9 Other Notable Liquid-Based AM Systems

They are several other commercial and non-commercial liquid-based AM methods that are similar in terms of technologies, principles and applications to those presented in the previous sections, but with some interesting variations. Some of these AM systems have not made it commercially in the market while others are no longer available due to the severe competition in the AM market. However their technologies are of significant interest to the AM community.

3.9.1 *Two Laser Beams*

Unlike many of the earlier methods which use a single laser beam, or the DLP method (EnvisionTEC GmbH), another liquid-based AM method is to employ two laser beams. First conceptualised by Wyn Kelly Swainson in 1967, the method involves penetrating a vat of photocurable resin with two lasers [31]. As opposed to SLA and other similar methods, this method no longer works only on the top surface of a vat of resin. Instead, the work volume resides anywhere within the vat.

In the two-laser-beams method, the two lasers are focused to intersect at a desired point in the vat. The principle of this method is a two-step variation of ordinary photopolymerisation: a single photon initiates curing and two photons of different frequencies are required to initiate polymerisation. Therefore, the point of intersection of the two lasers is the desired point of curing. The numerous points of intersection will collectively form the 3-dimensional part.

In this method, it is assumed that one laser excites molecules all along its path in the vat. While most of these molecules will return back to their unexcited state after a short time, those also in the path of the second laser will be further excited to initiate the polymerisation process. On this basis, the resin cures at the point of intersection of the two laser beams, while the rest lying along either of the two paths will remain liquid.

Unfortunately, after more than three decades of work on this method, the research work has not been realised into a commercial system, despite the continuation of Swainson's work by Formigraphic Engine Co. in collaboration with Battell Development Corporation, USA, and another independent research group, French National Centre for Scientific Research (CNRS), France.

Three major problems need to be resolved before the process is deemed to deliver parts of usable size and resolution. The first two problems relate to the control of the laser beams so that they intersect precisely at the desired 3-dimensional point in the vat. Firstly, the focusing problem exists because each laser beam undergoes numerous refractions as its ray moves from the source in air through the liquid resin. Non-uniformities in the refractive index of the resin are difficult to control and monitor because of the following reasons:

(1) Different progress of curing can cause non-uniformities.
(2) Inherent non-uniformities in the resin itself.
(3) Change of temperature in the vat.

Assuming that the resin can be completely homogenous and temperature changes are minimised so that its effects are insignificant, the second problem to contend with is how two laser beams moving at high speed can be focused to a very small and accurate point of intersection. The problem is compounded by the high energies required of the two lasers, which means that the two lasers will have short focal lengths. Thus, the distance is limited and in turn, this limits the size of the part that can be made by this method.

Thirdly, it is not always the case that molecules lying along the paths of either laser beams will return to the liquid state. This means unwanted resins may be formed along the path of either laser beams. Consequently, the cured resins may pose further problems to the rest of the process.

3.9.2 *Cubital's Solid Ground Curing (SGC)*

3.9.2.1 *Product*

The Solid Ground Curing (SGC) System, produced by Cubital Ltd is one of the more sophisticated high-end systems providing wider range and options for various modelling demands. Table 3.9 summarises the specifications of the SGC system.

Cubital's system uses several kinds of resins, including liquid resin and cured resin as materials to create parts, water soluble wax as a support material and ionographic solid toner for creating an erasable image of the cross-section on a glass mask.

Table 3.9. Cubital Ltd's Solider. (Sourced from Cubital Ltd)

Model	Solider
Irradiation medium	High power UV lamp
XY resolution (mm)	Better than 0.1
Surface definition (mm)	0.15
Elevator vertical resolution (mm)	0.1-0.2
Minimum feature size (mm)	0.4 (horizontal, X-Y) 0.15 (vertical, Z) - 0.4 (horizontal, X-Y) 0.15 (vertical, Z)
Work volume, XYZ(mm x mm x mm)	350 x 350 x 350 - 500 x 350 x 500
Production rate (cm^3 /hr)	550 - 1,311
Minimum layer thickness (mm)	0.06
Dimensional accuracy	0.1%
Size of unit, XYZ (m x m x m)	1.8 x 4.2 x 2.9 - 1.8 x 4.2 x 2.9
Data control unit	Data Front End (DFE) workstation

3.9.2.2 *Process*

The Cubital's Solid Ground Curing process comprises three main steps: data preparation, mask generation and model making [32].

(1) *Data Preparation.* In this first step, the CAD model is prepared and the cross-sections are generated digitally and transferred to the mask generator. The software used, Cubital's Solider DFE (Data Front End) software, is a CAD application package that processes data prior to sending them to the Cubital Solider system. DFE can search and correct flaws in the CAD files and render files on-screen for visualisation purposes. Solider DFE accepts CAD files mainly in the STL format.

(2) *Mask Generation.* After data is received, the mask plate is charged through an 'image-wise' ionographic process (see item 1, Fig. 3.32). The charged image is then developed with electrostatic toner.

(3) *Model Making.* In this step, a thin layer of photopolymer resin is spread on the work surface (see item 2, Fig. 3.32). The photo mask from the mask generator is placed in close proximity above the work piece, and aligned under a collimated UV lamp (item 3). The UV light is turned on for a few seconds (item 4). The part of resin layer which is exposed to the UV light through the photo mask is

hardened. Note that the layers laid down for exposure to the lamp are actually thicker than the desired thickness. This is to allow for the final milling process. The unsolidified resin is then collected from the work piece (item 5) by vacuum suction. Melted wax is then spread into the cavities (item 6). Consequently, the wax in the

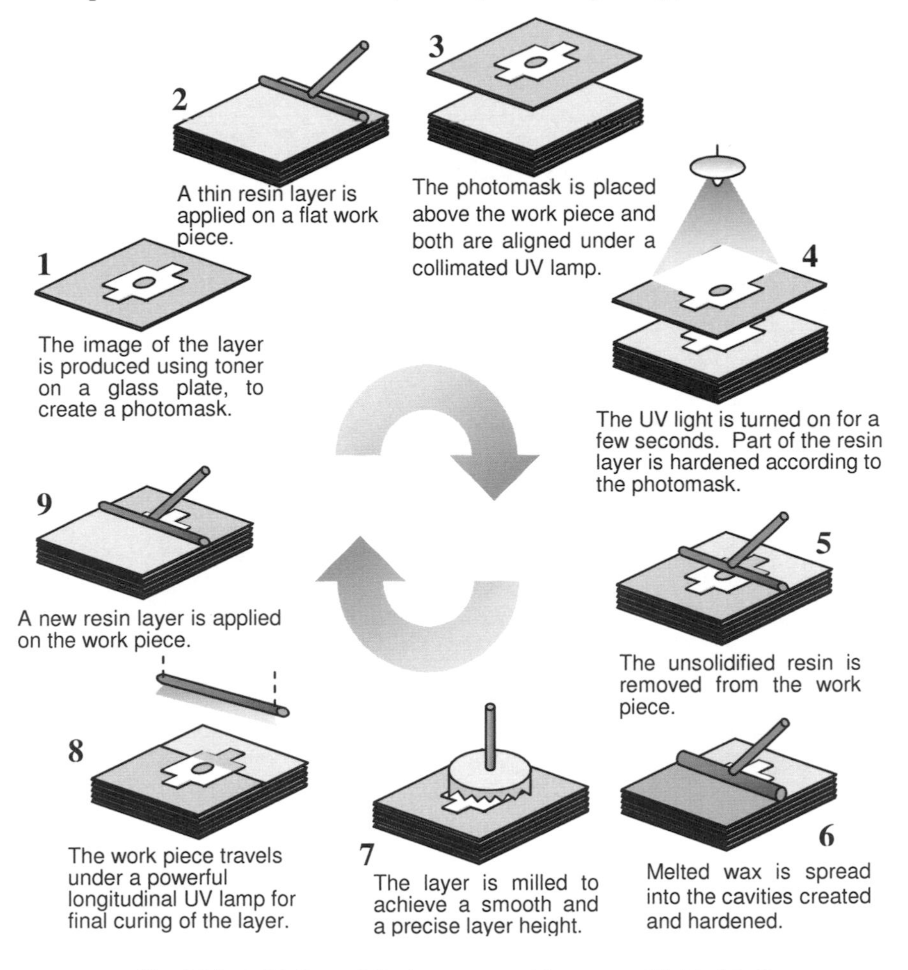

Fig. 3.32. Solid Ground Curing process. (Courtesy of Cubital Ltd)

cavities is cooled to produce a wholly solid layer. The layer is then milled to its exact thickness, producing a flat surface ready to receive the next layer (item 7). In the SGC 5600, an additional step (item 8) is provided for final curing of the layer whereby the work piece travels under a powerful longitudinal UV lamp. The cycle repeats itself until the final layer is completed.

The main components of the Solider system are:
(1) Data Front End (DFE) workstation.
(2) Model Production Machine (MPM). It includes the process engine, operator's console and vacuum generator.
(3) Automatic Dewaxing Machine (optional).

3.9.2.3 *Principle*

Cubital's AM technology creates highly physical models directly from computerised three-dimensional data files. Parts of any geometric complexity can be produced without tools, dies or moulds by Cubital's AM technology. The process is based on the following principles:

(1) Parts are built, layer by layer, from a liquid photopolymer resin that solidifies when exposed to UV light. The photopolymerisation process is similar to that described in Section 3.1.4, except that the irradiation source is a high power collimated UV lamp and the image of the layer is generated by masked illumination instead of optical scanning of a laser beam. The mask is created from the CAD data input and 'printed' on a transparent substrate (the mask plate) by an non-impact ionographic printing process, a process similar to the Xerography process used in photocopiers and laser printers [33]. The image is formed by depositing black powder, a toner which adheres to the substrate electrostatically. This is used to mask the uniform illumination of the UV lamp. After exposure, the electrostatic toner is removed from the substrate for reuse and the pattern for the next layer is similarly 'printed' on the substrate.
(2) Multiple parts may be processed and built in parallel by grouping them into batches (runs) using Cubital's proprietary software.
(3) Each layer of a multiple layer run contains cross-sectional slices of one or many parts. Therefore, all slices in one layer are created simultaneously. Layers are created thicker than desired. This is to allow the layer to be milled precisely to its exact thickness, thus

giving overall control of the vertical accuracy. This step also produces a roughened surface of cured photopolymer, assisting adhesion of the next layer to it. The next layer is then built immediately on the top of the created layer.

(4) The process is self-supporting and does not require the addition of external support structures to emerging parts since continuous structural support for the parts is provided by the use of wax, acting as a solid support material.

3.9.2.4 *Strengths and weaknesses*

The Solider system has the following strengths:

(1) *Parallel Processing*. The process is based on instant, simultaneous curing of the whole cross-sectional layer area (rather than point-by-point curing). It has high speed throughput that is about 8 times faster than its competitors. Its production costs can be 25%-50% lower. It is a time and cost saving process.
(2) *Self-Supporting*. It is user-friendly, fast, and simple to use. It has a solid modelling environment with unlimited geometry. The solid wax supports the part in all dimensions and therefore support structure is not required.
(3) *Fault Tolerance*. It has good fault tolerances. Removable trays allow job changing during a run and layers are erasable.
(4) *Unique Part Properties*. The part that the Solider system produces is reliable, accurate, sturdy, machinable, and can be mechanically finished.
(5) *CAD to AM Software*. Cubital's AM software, Data Front End (DFE), processes solid model CAD files before they are transferred to the machine.
(6) *Minimum Shrinkage Effect*. This is due to the full curing of every layer.
(7) *High Structural Strength and Stability*. This is due to the curing process that minimises the development of internal stresses in the structure. As a result, they are much less brittle.

(8) *No Hazardous Odours Generated.* The resin stays in liquid state for a very short time, and the uncured liquid is wiped off immediately. Thus safety is considerably higher.

The Solider system has the following weaknesses:

(9) *Requires Large Physical Space.* The size of the system is much larger than other systems with similar build volume size.
(10) *Wax Gets Stuck in Corners and Crevices.* It is difficult to remove wax from parts with intricate geometry. Thus, some wax may be left behind.
(11) *Waste Material Produced.* The milling process creates shavings, which have to be cleaned from the machine.
(12) *Noisy.* The Solider system generates a high level of noise as compared to other systems.

3.9.2.5 *Applications*

The applications of Cubital's system can be divided into four areas:

(1) *General Applications.* Conceptual design presentation, design proofing, engineering testing, integration and fitting, functional analysis, exhibitions and pre-production sales, market research, and inter-professional communication.
(2) *Tooling and Casting Applications.* Investment casting, sand casting, and rapid, tool-free manufacturing of plastic parts.
(3) *Mould and Tooling.* Silicon rubber tooling, epoxy tooling, spray metal tooling, acrylic tooling, and plaster mould casting
(4) *Medical Imaging.* Diagnostic, surgical, operation and reconstruction planning and custom prosthesis design.

3.9.2.6 *Examples: New Cubital-based Solicast process offers inexpensive metal prototypes in two weeks*

Schneider Prototyping GmbH created a metal prototype (see Fig. 3.33) for investment cast directly from CAD files in two weeks with SoliCast [34]. Starting with a CAD file, Schneider can deliver a metal prototype in two weeks, whereas similar processes with other AM methods take four to six weeks and conventional methods based on CNC prototyping can take 10 to 16 weeks. In general, Schneider's prototypes can be as much as 50% cheaper than those produced by other AM methods.

Fig. 3.33. Cubital prototypes and metal prototypes created from them. (Courtesy of Cubital America, Inc.)

3.9.3 *Teijin Seiki's Soliform System*

3.9.3.1 *Product*

Teijin Seiki Co., Ltd produces two main series (250 and 500 series) of the Soliform system. The Soliform 250 series machine is shown in Fig. 3.34.

The specifications of the machines are summarised in Table 3.10. The materials used by Soliform are primarily SOMOS photopolymers supplied by Du Pont and TSR resins, its own developed resin.

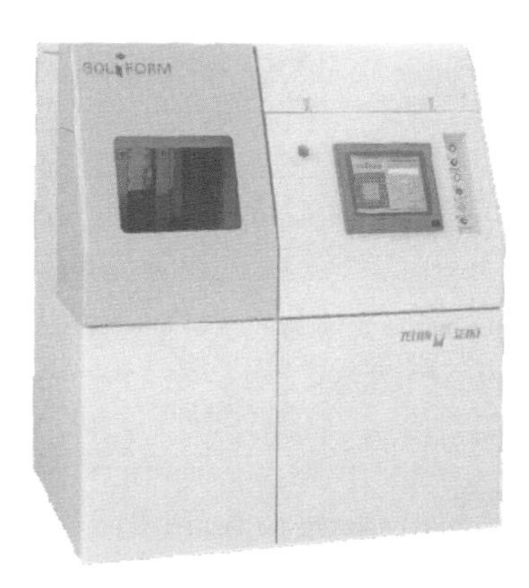

Fig. 3.34. The Soliform 250B solid forming system. (Courtesy of Teijin Seiki Co., Ltd)

Table 3.10. System specifications of Soliform. (Courtesy of Teijin Seiki Co., Ltd)

Model	SOLIFORM 250B	SOLIFORM 250EP	SOLIFORM 500C	SOLIFORM 500EP
Laser type	Solid state		Solid state	
Laser power (mW)	200	1000	400	1000
Scanning system	Digital scanner mirror		Digital scanner mirror	
Maximum scanning speed (m/s)	12 (max)		24 (max)	
Beam Diameter	Fixed: 0.2 mm Variable: 0.1 - 0.8 mm (Option)		Fixed: 0.2 mm Variable: 0.2 - 0.8 mm (Option)	
Max. build envelope, xyz (mm)	250 x 250 x 250		500 x 500 x 500	
Min. build layer (mm)	0.05		0.05	
Vat	Interchangeable		Interchangeable	
Recoating system	Dip coater system		Dip coater system	
Resin level controller	Counter volume system		Counter volume system	
Power supply	100 V_{AC}, single-phase, 30 A		200 V_{AC}, 3 phase, 30 A	
Size of unit, xyz(mm)	1430 x 1045 x 1575		1850 x 1100 x 2280	
Weight (kg)	400		1200	

3.9.3.2 *Process*

The process of the Soliform system comprises of the following steps: concept design, CAD design, data conversion, solid forming and plastic model.

(1) *Concept design*: Product design engineers create the concept design. This may or may not necessarily be done on a computer.
(2) *CAD design*: The 3D CAD model of the concept design is created in the SUN-workstation.
(3) *Data conversion*: The 3D CAD data is transferred to the Soliform software to be tessellated and converted into the STL file.
(4) *Solid forming*: This is the part building step. A solid model is formed by the ultraviolet laser layer by layer in the machine.
(5) *Plastic model*: This is the postprocessing step where the completed plastic part is cured in an oven to be further hardened.

The Soliform system contains the following hardware: a SUN-EWS workstation, an argon ion laser, a controller, a scanner to control the laser trace of scanning and a tank which contains the photopolymer resin.

3.9.3.3 *Principle*

The Soliform system creates models from photocurable resins based essentially on the principles described in Section 3.1.4. The resin developed by Teijin is an acrylic-urethane resin with a viscosity of 40,000 centipoise and a flexural modulus of 52.3MPa compared to 9.6MPa for a grade used to produce conventional prototype models.

Parameters which influence performance and functionality are generally similar to those described in Section 3.1.4, but for Soliform, the properties of the resin and the accuracy of the laser beam are considered more significant.

3.9.3.4 *Strengths and weaknesses*

The Soliform system has several technological strengths:

(1) *Fast and Accurate Scanning*. Its maximum scanning speed of 24m/s is faster than all other AM systems. Its accurate scan system is controlled by digital encoder servomotor technique.

(2) *Good Accuracy*. It has exposure control technology for producing highly accurate parts.
(3) *Photo Resins*. It has a relatively wide range of acrylic-urethane resins for various applications.

The Soliform system has the following weaknesses:

(1) *Requires Support Structures*. Structures that have overhangs and undercuts must have supports that are designed and fabricated together with the main structure.
(2) *Requires Postprocessing*. Postprocessing includes removal of supports and other unwanted materials, which is tedious, time-consuming and can damage the model.
(3) *Requires Post-Curing*. Post-curing may be needed to cure the object completely and ensure the integrity of the structure.

3.9.3.5 *Applications*

The Soliform system has been used in many areas, such as injection modelling (low cost die, see Fig. 3.35), vacuum moulding, casting, and lost wax moulding.

Fig. 3.35. AM model (left), the ABS injection moulding (centre) and the low cost metal die (right). (Courtesy of Teijin Seiki Co., Ltd)

(1) *Injection moulding (low cost die)*. The Soliform can be used to make injection moulding or low cost die based on a process described in Figure 3.36. Comparing this method with the

conventional processes, this process has shorter development and execution time. The created CAD data can be used for mass production die and no machining is necessary. Photopolymers that can be used include SOMOS-2100 and SOMOS-5100. An example of the product moulded in ABS is shown together with the injection mould and the SOLIFORM pattern in Figure 3.35.

(2) *Vacuum moulding*. Vacuum moulding moulds are made by the Soliform system using a process similar to that illustrated in Figure 3.36. The photopolymers that can be used include SOMOS-2100, SOMOS-3100 and SOMOS-5100.

(3) *Casting*. The process of making casting part form is again similar to the process illustrated in Figure 3.36 with the exception of using casting sand instead of the injection mould. The photopolymer recommended is SOMOS-3100.

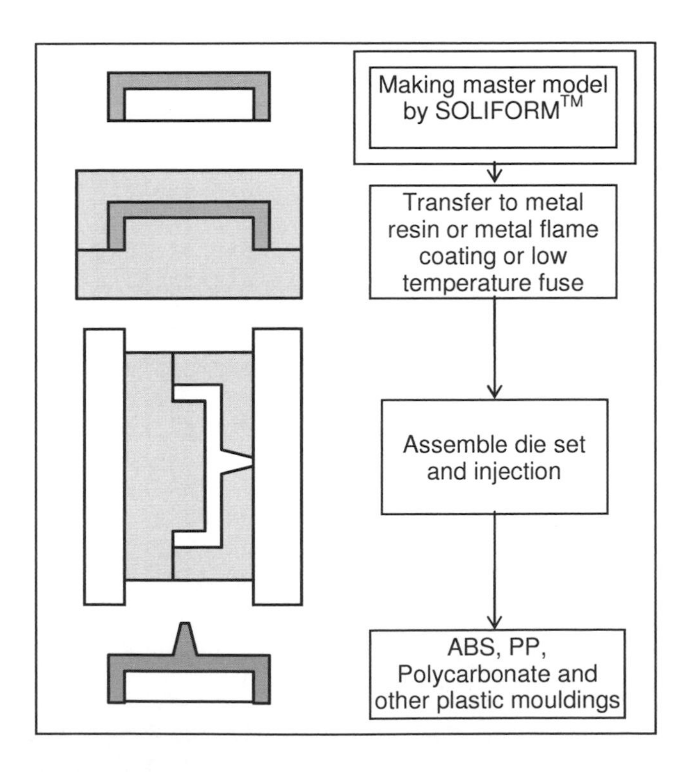

Fig. 3.36. Creating injection moulding parts. (Adapted from Teijin Seiki Co., Ltd)

(4) *Lost wax moulding.* Again this process is similar to the one described in Figure 3.36. The main difference is that the wax pattern is made and later burnt out before the metal part is moulded afterwards. The recommended photopolymer is SOMOS-4100.

(5) *Making injection and vacuum moulding tools directly.* The other application of SOLIFORM is the process of making the tools for injection moulding and vacuum moulding directly. From the CAD data of the part, the CAD model of the mould inserts are created. These new CAD data are then transferred the usual way to SOLIFORM to create the inserts. The inserts are then cleaned and post cured before the gates and ejector pin holes are machined. They are then mounted on the mould sets and subsequently on the injection moulding machine to mould the parts. Fig. 3.37 and Fig. 3.38 show the injection mould cavity insert and the moulding for a handy telephone respectively while Fig. 3.39 describes schematically the process flow.

A process similar to that described in Figure 3.39 can also be used for creating moulds and dies for vacuum moulding. Figure 3.40 shows a mouse cover that is moulded using vacuum moulding from tools created by SOLIFORM.

Fig. 3.37. Core and cavity mould inserts on moulding machine of the handy telephone. (Courtesy of Teijin Seiki Co., Ltd)

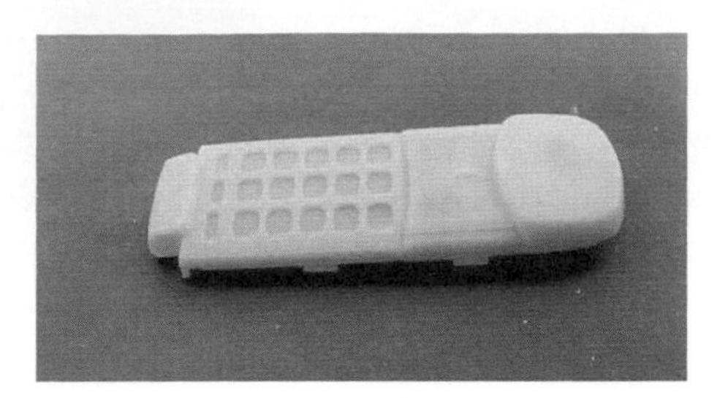

Fig. 3.38. Injection moulded part from the SOLIFORM moulds of the handy telephone. (Courtesy of Teijin Seiki Co., Ltd)

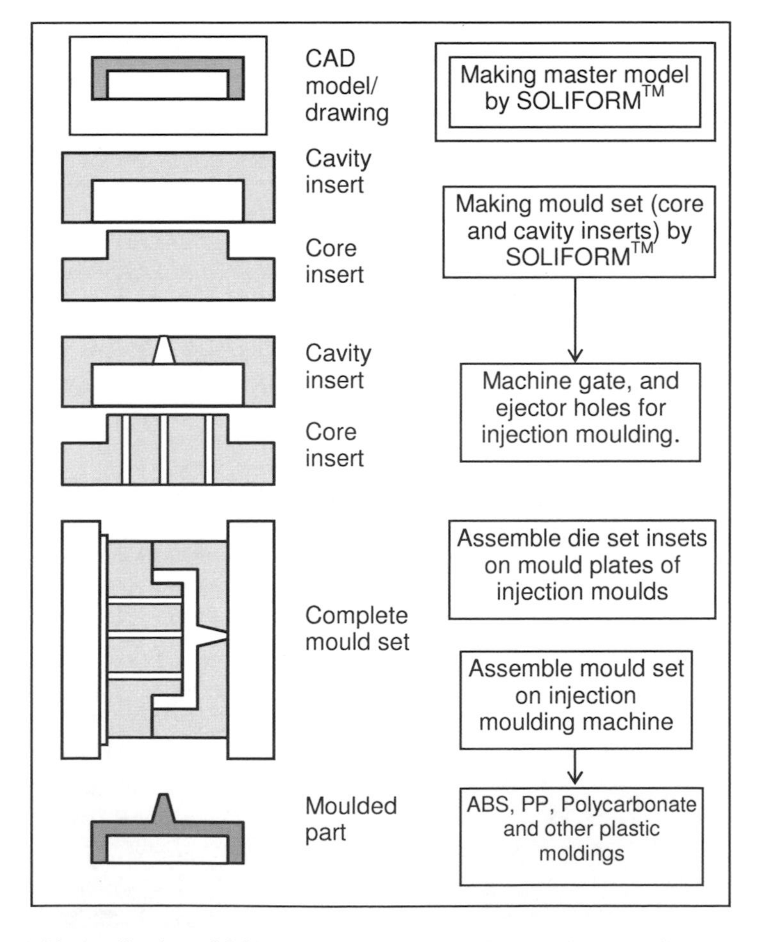

Fig. 3.39. Application of SOLIFORM to directly make injection moulding dies.

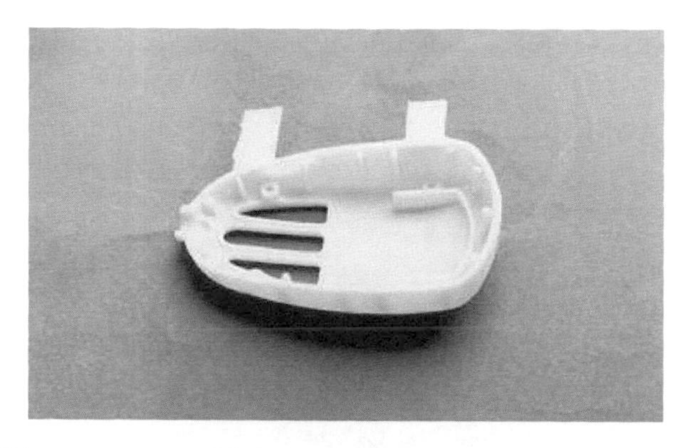

Fig. 3.40. Direct vacuum moulded mouse cover. (Courtesy of Teijin Seiki Co., Ltd)

3.9.4 *Meiko's Additive Manufacturing System for the Jewellery Industry*

3.9.4.1 *Company*

Meiko RPS Co., Ltd., a company making machines for analysing components of gases such as CO_2, NO_X and O_2, factory automation units such as loaders and unloaders for assembly lines, and two arm robots, was founded in June 1962. The system for the jewellery industry was developed together with Yamanashi-ken Industrial Technical Centre in the early 1990s. Meiko RPS Co., Ltd was established in 2000 selling 3D software and additive manufacturing systems. The company's address is Meiko RPS Co., Ltd., 2F, 51-3 Oyamakanaicho, Itabashi-ku, Tokyo 173-0024, Japan.

3.9.4.2 *Product*

Meiko's LC-510 is an optical modelling system that specialises in building prototypes for small models such as jewellery. LCV-700 (see Fig. 3.41) and LCV-810 both can provide a model with a maximum size of 60 x 60 x 60 mm. The two main differences between the systems are that LCV-810 has a scanning speed of 4 times higher than LCV-700 but LCV-700 can provide a better surface finish than LCV-810 due to the difference in spot diameter. The specifications of the LC-510, LCV-700 and LCV-800 are summarised in Table 3.11.

Fig. 3.41. LCV-700 AM system. (Courtesy of Meiko RPS Co., Ltd)

Table 3.11. Specifications of Meiko's systems. (Courtesy of Meiko RPS Co., Ltd)

Model	LC-510	LCV-700	LCV-800
Laser type	HeCd	Semi-Conductor	
Laser power, mW	5	30	60
Spot size, mm	0.08	0.04	0.05
Method of laser operation	X-Y plotter		
Scanning Speed (Max), mm/min	1,000	500	2000
Resolution in X-Y direction, mm	0.01		
Max work volume, XYZ, mm	100 x 100 x 60	60 x 60 x 60	
Slice thickness, mm	0.2 – 0.01		
Operating System	Windows 98/Me/2000/XP/NT		
Power supply	100 V_{AC} 15 A	100 V_{AC} 10 A	
CAD format	JSD, DXF, STL		

Parts of different properties can be fabricated due to the types of photocurable resins offered by Meiko Co., Ltd. The photocurable resins are in translucent yellowish, greenish and bluish-violet colours. The consumable components in the systems are the laser tube and the photocurable resin. The 3D CAD software, which is named JCAD3/Takumi is customised software for 3D printing.

3.9.4.3 *Process*

The process building models by the system MEIKO is illustrated in Fig. 3.42.

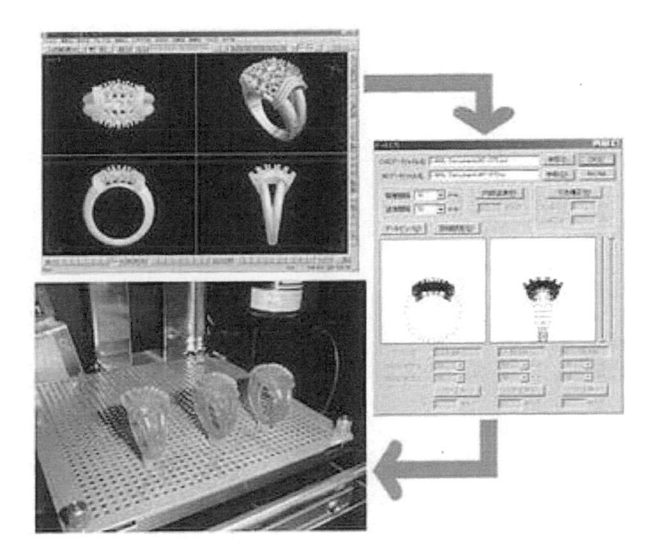

Fig. 3.42. Modelling jewellery on JCAD3/Takumi and MEIKO. (Courtesy of Meiko Co., Ltd)

(1) *Designing a model using the exclusive 3 dimensional CAD software*: A model is created on the computer using JCAD3/Takumi, specifically created for jewellery design.
(2) *Creating NC data from the CAD data directly by the exclusive CAM module*: From the CAD data, NC data is generated using the CAM module. This is based on the standard software and methods used in CNC machines.
(3) *Forwarding the NC data to the LC-510 machine*: The NC data and codes are transferred to the NC controller in the LC-510 machine via RS-232C cable. This transfer is similar to those used in CNC machines.
(4) *Building the model by the LC-510*: From the downloaded NC data, the LC-510 optical modelling system will create the jewellery prototype using the laser and the photocurable resin layer by layer.

3.9.4.4 *Principle*

The fundamental principle behind the method is the laser solidification process of photocurable resins. Similar to other liquid-based systems, its principles are like those described in Section 3.1.4. The main difference is in the controller of the scanning system. The system MEIKO uses an X-Y (plotter) system with NC controller instead of the galvanometer mirror scanning system.

Parameters which influence performance and functionality of the system are the properties of resin, the diameter of the beam spot, and the XY resolution of the machine.

3.9.4.5 *Strengths and weaknesses*

The strengths of the product and process are as follows:

(1) *Good Accuracy.* Highly accurate modelling that is primarily targeted for jewellery.
(2) *Low Cost.* The cost of prototyping with the system is relatively inexpensive.
(3) *Cost Saving.* Lead time is cut and cost is saved by labour saving.
(4) *Exclusive CAM Software.* The CAM software is for jewellery and small parts.
(5) *Ease of Manufacturing Complicated Parts.* Intricate geometries can be easily made using JCAD3/Takumi and LC-510 in a single run.

The weaknesses of the product and process are as follows:

(1) *Requires Support Structures.* Structures that have overhangs and undercuts must have supports that are designed and fabricated together with the main structure.
(2) *Requires Postprocessing.* Postprocessing includes removal of supports and other unwanted materials, which is tedious, time-consuming and can damage the model.

(3) *Require post-curing.* Post-curing may be needed to cure the object completely and ensure the integrity of the structure.

3.9.4.6 *Applications*

The general application areas of the system MEIKO are as follows:

(1) Production of jewellery
(2) Production of other small models (such as hearing-aids and a part of spectacle frames)
(3) Production of machine parts
(4) Production of medical/dental parts

Users of the system MEIKO produce resin models using the exclusive resin UVM-8001 as its material as shown in Fig. 3.43.

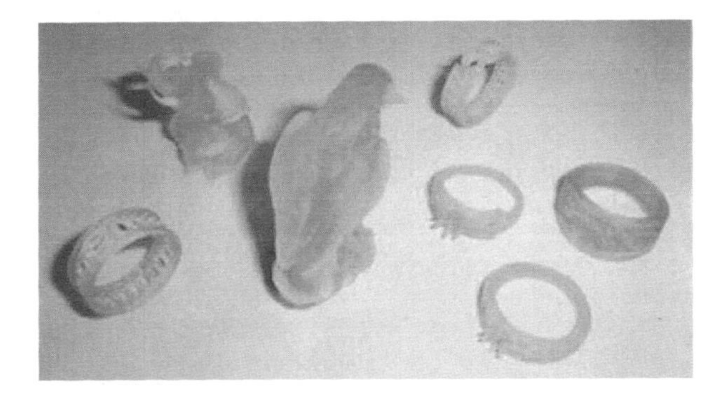

Fig. 3.43. Resin models made by the system MEIKO. (Courtesy of Meiko Co., Ltd)

There are two ways to create actual products from these resin models as follows:

(1) In the traditional way (see Fig. 3.44), rubber models are made, and wax models are fabricated (see Fig. 3.45). Then using the lost wax method, actual products can be manufactured (see Fig. 3.46).
(2) These resin models are used for casting, the same as wax models.

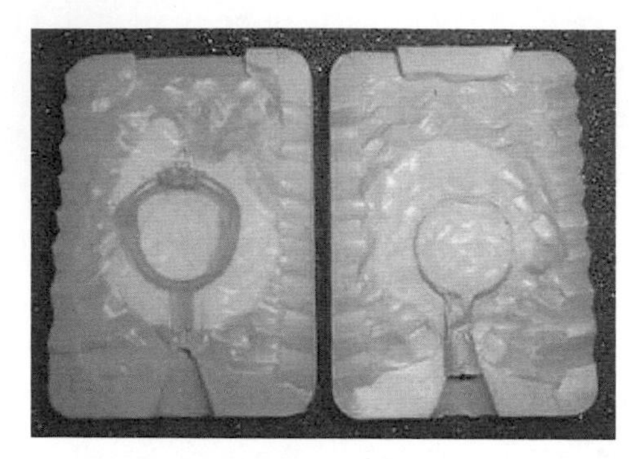

Fig. 3.44. Making rubber mould from resin model rings. (Courtesy of Meiko Co., Ltd)

Fig. 3.45. Resin ring and metal casted ring models. (Courtesy of Meiko Co., Ltd)

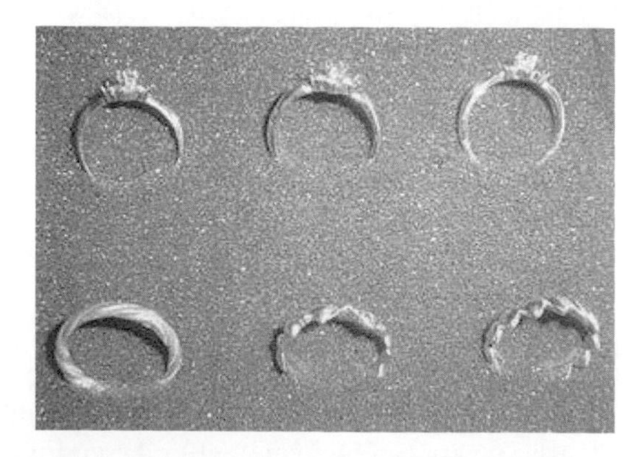

Fig. 3.46. Actual precious metal rings from casted from originals of resin ring models. (Courtesy of Meiko Co., Ltd)

3.9.4.7 *Research and development*

Meiko Co., Ltd. researches on improving the existing machines and a new edition of the JCAD3/Takumi, cooperating with Yamanashi University. Meiko will invest more manpower into producing more machines in order to widen its market.

3.9.5 *Others*

Other similar liquid-based systems are summarised as follows:

(1) *SLP (Solid Laser-diode Plotter).* The SLP series of resin-based AM machines is a joint effort of Denken Engineering and Autostrade Co., Ltd assisted by Nippon Kayaku with resin development and by Shimadzu with laser development. The manufacturer of the SLP is Denken Engineering Co, Ltd., 2-1-40, Sekiden-machi Oita City, Oita Prefecture, 870 Japan.

(2) *COLAMM (Computer-Operated, Laser-Active Modelling Machine).* Opposed to most methods where parts are built on a descending platform, the COLAMM is built on an ascending platform [35]. The first (top) layer of the part is scanned by a laser placed below, through a transparent window plate. The layer is built on the platform and since it is upside down, the part is suspended. Next, the platform is raised by the programmed layer thickness to allow a new layer of resin to flow onto the window for scanning. The manufacturer is Mitsui Zosen Corporation of Japan.

(3) *LMS (Layer Modelling System).* The manufacturer is Fockele und Schwarze Stereolithographietechnik GmbH (F&S) of Borchen-Alfen and its address is Alter Kirchweg 34, W-4799 Borchen-Alfen, Germany. A unique feature of the LMS is its low cost, non-optical resin level control system (LCS). The LCS does not contact the resin and therefore allows vats to be changed without system recalibration.

(4) *Light Sculpting.* Not sold commercially, the Light Sculpting device is offered as a bureau service by Light Sculpting of Milwaukee, Wisconsin, USA [E20]. The Light Sculpting technique uses a

descending platform like SLA and many others, and irradiates the resin surface with a masked lamp as in Cubital's SGC. However, the unique feature of this technique is that the resin is cured in contact with a plate of transparent material on which the mask rests. The resulting close proximity of the mask to the resin surface has good potential of ensuring accurate replication of high-resolution patterns.

References

[1] 3D Systems Product brochure. (2014). SLA Production series.

[2] 3D Systems Product brochure. (2014). ProJet® 6000, 7000, 3500, 3510, 5000 and 5500X.

[3] *Rapid Prototyping Report.* (1996). 3D Systems Introduces Upgraded SLA-250 with Zephyr Recoating, **6**(4), CAD/CAM Publishing Inc.

[4] Jacobs, P. F. (1992). *Rapid Prototyping & Manufacturing, Fundamentals of Stereolithography* (pp. 11-18). Society of Manufacturing Engineers.

[5] Wilson, J. E. (1974). *Radiation Chemistry of Monomers, Polymers, and Plastics.* Marcel Dekker, NY.

[6] Lawson, K. (1994). UV/EB Curing in North America, *Proceedings of the International UV/EB Processing Conference*, Florida, USA, May 1-5, 1.

[7] Reiser, A. (1989). *Photosensitive Polymers.* John Wiley, NY.

[8] Jacobs, P. F. (1992). *Rapid Prototyping & Manufacturing, Fundamentals of Stereolithography* (pp. 25-32). Society of Manufacturing Engineers.

[9] Jacobs, P. F. (1996). *Stereolithography and other RP&M Technologies* (pp. 29-35). Society of Manufacturing Engineers.

[10] Jacobs, P. F. (1992). *Rapid Prototyping & Manufacturing, Fundamentals of Stereolithography* (pp. 53-56). Society of Manufacturing Engineers.

[11] Jacobs, P. F. (1992). *Rapid Prototyping & Manufacturing, Fundamentals of Stereolithography* (pp. 60-78). Society of Manufacturing Engineers.

[12] *3D Systems Corporation Joins White House's Manufacturing Initiative.* (Mar 2014). Retrieved from http://www.thestreet.com/story/12529118/1/3d-systems-corporation-joins-white-houses-manufacturing-initiative.html

[13] Molitch-Hou, M. (Dec 2013). *3D Systems Acquires Portion of Xerox R&D Division.* Retrieved from http://3dprintingindustry.com/2013/12/19/3d-systems-acquires-portion-xerox-rd-division/

[14] Stratasys Product brochures. Objet Eden 260V, 350V, 500V, 2014.

[15] Durham, M. (2003). Rapid prototyping - Stereolithography, selective laser sintering, and polyjet, *Advance Materials & Processes*, **161**: 40-42.

[16] Cheng, Y. L., & Chen, S. J. (2006). Manufacturing of Cardiac Models Through Rapid Prototyping Technology for Surgery Planning, *Materials Science Forum*, **505-507**: 1063-1068.

[17] Brajlih, T., Drstvensek, I., Kovacic, M., & Balic, J. (2006). Optimising scale factors of the PolyJet™ rapid prototyping procedure by generic programming, *Journal of Achievement in Materials and Manufacturing Engineering*, **16**: 101-106.

[18] Hendrik, J. *Perfactory® - A Rapid Prototyping system on the way to the "personal factory" for the end user.* Envision Technologies GmbH.

[19] Stampfl, J., Cano Vives, R., Seidler, S., Liska, R., Schwager, F., Gruber, H., Wöβ, A., & Fratzl, P. (2003). Proceedings of the *1st International Conference on Advanced Research in Virtual and Rapid Prototyping*, 1.-4. Leiria, Portugal, 659-666.

[20] Grunewald, S. J. (Mar 2014). *EnvisionTEC Launches its Latest Micro 3D Printer.* Retrieved from http://3dprintingindustry.com/2014/03/20/3d-printer-micro-envisiontec/

[21] Landers, R., John, H., & Mülhaupt, R. *Scaffolds for tissue engineering applications fabricated by 3D printing.* Freiburger Materialforschungszentrum and Institut für makromolekulare Chemie of the Albert-Lubwigs-Universität, Freiburg, Germany, Envision Technologies, Marl, Germany.

[22] Moroni, L., de Wijn, J., & van Blitterswijk, C. A. (2004). 3D Plotted Scaffolds for Tissue Engineering: Dynamical Mechanical Analysis, *European Cells and Materials,* **7**, Suppl. 1: 68.

[23] Mülhaupt, R., Landers, R., & Thomann, Y. (2003). Biofunctional Processing: Scaffold Design, Fabrication and Surface Modification, *European Cells and Materials,* **6**, Suppl. 1: 12.

[24] Zhang, W., Leu, M. C., Ji, Z., & Yan, Y. (1999). Rapid Freezing Prototyping with Water. *Materials and Design* 20 (June): 139-145.

[25] Leu, M. C., Zhang, W., & Sui, G. (2000). An Experimental and Analytical Study of Ice Part Fabrication with Rapid Freeze Prototyping. *Annals of the CIRP* **49**: 147-150.

[26] Sui, G., & Leu, M. C. (2003). Thermal Analysis of Ice Wall Built by Rapid Freeze Prototyping, *ASME Journal of Manufacturing Science and Engineering* **125** (Nov): 824-834.

[27] Bryant, F. D., & Leu, M. C. (2004). Study on Incorporating Support Material in Rapid Freeze Prototyping, *Proceedings of Solid Freeform Fabrication Symposium* (Aug): 2-4.

[28] Leu, M. C., Liu, Q., & Bryant, F. D. (2003). Study of Part Geometric Features and Support Materials in Rapid Freeze Prototyping, *Annals of the CIRP* **52**: 185-188.

[29] Sui, G., & Leu, M. C. (2003). Investigation of Layer Thickness and Surface Roughness in Rapid Freeze Prototyping, *ASME Journal of Manufacturing Science and Engineering* **125** (Aug): 556-563.

[30] Liu, Q., & Leu, M. C. (2006). Investigation of Interface Agent for Investment Casting with Ice Patterns, *ASME Journal of Manufacturing Science and Engineering* **128** (May): 554-562.

[31] Burns, M. (1993). *Automated Fabrication: Improving Productivity in Manufacturing*. PTR Prentice Hall.

[32] Kobe, G. (1992). Cubital's Unknown Solider, *Automotive Industries*, August 1992:54-55.

[33] Johnson, J. L. (1994). *Principles of Computer Automated Fabrication.* Palatino Press, Chapter 2:44.

[34] Cubital News Release. (1995). *New Cubital-based SoliCast Process Offers Inexpensive Metal Prototypes in Two Weeks.*

[35] Burns, M. (1993). *Automated Fabrication: Improving Productivity in Manufacturing*. PTR Prentice Hall.

Problems

1. Describe the process flow of the 3D System Stereolithography Apparatus.

2. Describe the process flow of the Stratasys' Polyjet systems.

3. Compare and contrast the laser-based stereolithography systems and the Stratasys' Polyjet systems. What are the strengths for each of the systems?

4. Describe the main differences between Perfactory and other commercial AM systems which use UV curing process.

5. Which liquid-based machine has the largest work volume? Which has the smallest?

6. Meiko Co., Ltd. produces the LC and LCV series for jewellery prototyping. By comparing the machine specifications with other vendors, discuss what you think are the important specifications that will determine their suitability for jewellery prototyping.

7. What is DMD, as found in Perfactory?

8. Which are the key materials for Bioplotting? Identify the methods of reaction for solidification.

9. Discuss the principle behind the two-laser-beams method. What are the major problems in this method?

10. As opposed to many of the liquid-based AM systems which use a photosensitive polymer, water is used in Rapid Freeze Prototyping (RFP). What are the pros and cons of using water?

11. How is the support of the ice part removed from the actual part?

Chapter 4

SOLID-BASED ADDITIVE MANUFACTURING SYSTEMS

Solid-based additive manufacturing systems are very different from the liquid-based photo-curing systems described in Chapter 3. They are also different from one another, though some of them do use laser in the additive manufacturing process. The basic common feature among these systems is that they all utilise solids (in one form or another) as the primary medium to create the prototype. A special group of solid-based AM systems that uses powder as the additive manufacturing medium will be covered separately in Chapter 5.

4.1 Stratasys' Fused Deposition Modelling (FDM)

4.1.1 *Company*

Stratasys Inc. was founded in 1989 in Delaware and developed the company's AM systems based on Fused Deposition Modelling (FDM) technology. The technology was first developed by Scott Cramp in 1988 and the patent was awarded in the USA in 1992. FDM uses an extrusion process to build 3D models. Stratasys introduced its first AM machine, the 3D modeller®, in early 1992. Stratasys' headquarters in the US are located at 7665 Commerce Way, Eden Prairie, MN 55344-2080, USA.

4.1.2 *Products*

Stratasys manufactures 3D printing equipment and materials that create physical objects directly from digital data. Its systems range from affordable desktop 3D printers to large, advanced 3D production systems, making 3D printing more accessible than ever. All of Stratasys' 3D printers build parts layer-by-layer.

4.1.2.1 *Design Series — Performance 3D Printers*

Stratasys' Design Series 3D Printers are built for resolving issues between design and engineering by realising engineers' design and prototyping the product's design. The Design Series is created as an affordable, office-friendly AM system meant mainly for concept modelling, creating product replicas, and some functional testing capabilities. Production timeline can be shortened and any costly mistakes can be reduced through rigorous testing. Under the Design Series, there are two categories of 3D printers — Precision 3D Printers and Performance 3D Printers. Precision 3D Printers is based on PolyJet 3D Printing technology, which is discussed in Chapter 3.2. On the other hand, the two machines in the Performance 3D Printers, Dimension 1200es (See Fig. 4.1) and Dimension Elite (See Fig. 4.2), are powered by Fused Deposition Modelling (FDM) Technology. Performance 3D

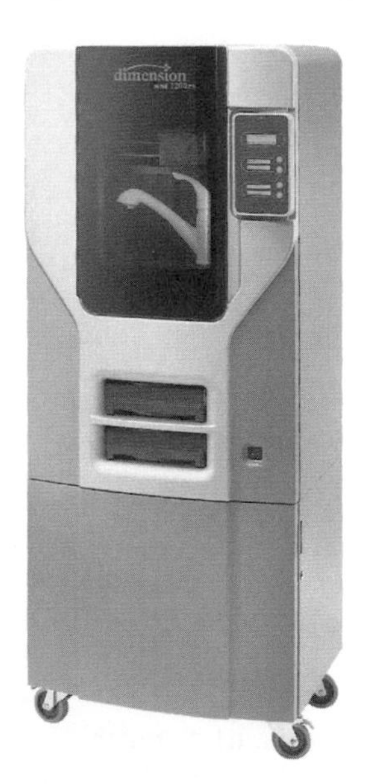

Fig. 4.1. Dimension 1200es 3D Printer. (Courtesy of Stratasys Inc.)

Fig. 4.2. Dimension Elite. (Courtesy of Stratasys Inc.)

Printers deliver 3D modelling in real ABSplus thermoplastic. Parts are durable and dimensionally stable — perfect for tough testing, and materials are affordable, meaning you can work iteratively with frequent 3D modelling. Details of the Design Series — Performance 3D Printers are summarised in Table 4.1.

Table 4.1. Specifications of the Design Series — Performance 3D Printers.

Models	Dimension 1200es	Dimension Elite
Build size, mm (in.)	254 x 254 x 305 (10 x 10 x 12)	203 x 203 x 305 (8 x 8 x 12)
Materials	ABS or ABSplus plastic in standard natural, blue, fluorescent yellow, black, red, olive green, nectarine, or grey colours Custom colours are available (Dimension Elite uses only ABSplus)	
Support structures	BST breakaway support technology or SST soluble support removal process	SST soluble support removal
Material cartridges	One autoload cartridge with 922 cu. cm. (56.3 cu. in.) ABS OR ABSplus material One autoload cartridge with 922 cu. cm. (56.3 cu. in.) Support material	
Layer thickness, mm (in.)	0.254 or 0.33 (0.010 or 0.013)	0.178 or 0.254 (0.007 or 0.010)
Size, mm (in.)	838 x 737 x 1143 (39 x 29 x 45)	686 x 914 x 1041 (27 x 36 x 41)
Weight, kg (lbs)	148 (326)	136 (300)
Software	Catalyst® EX software	
Power requirements	110 - 120 VAC, 60 Hz, minimum 15 A dedicated Circuit or 220 - 240 VAC, 50/60 Hz, Minimum 7 A dedicated circuit	

4.1.2.2 *Idea Series*

Stratasys' Idea Series 3D Printers are compact and affordable desktop 3D printers that enhance users' design capability at the push of a button. With Fused Deposition Modelling (FDM) Technology, they liberate customers' creativity and accelerate the design process. Stratasys' Mojo is shown in Fig. 4.3, and the uPrint SE Plus is shown in Fig. 4.4. These 3D printers bring professional 3D printing to consumers' desktops or small team workspaces. Details of the Idea Series 3D Printers are summarised in Table 4.2.

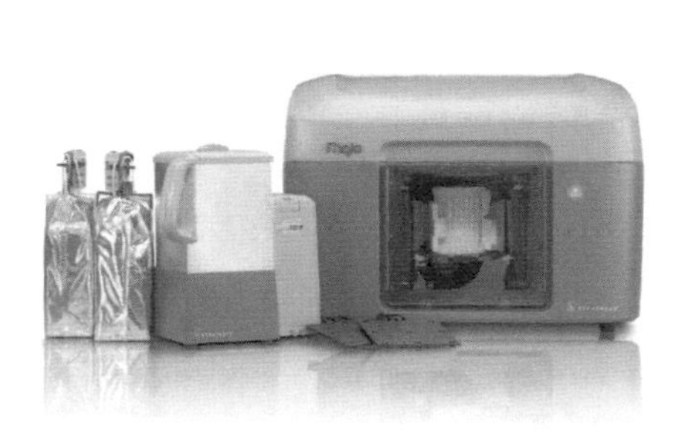

Fig. 4.3. Mojo. (Courtesy of Stratasys Inc.)

Fig. 4.4. uPrint SE Plus 3D Printer. (Courtesy of Stratasys Inc.)

Table 4.2. Specifications of the Idea Series 3D Printers.

Models	Mojo	uPrint SE	uPrint SE Plus
Model materials	ABSplus in ivory, white, blue, fluorescent yellow, black, red, nectarine, olive green or grey	ABSplus in ivory	ABSplus in ivory, white, blue, fluorescent yellow, black, red, nectarine, olive green or grey
Support materials	SR-30 soluble		
Maximum part size, mm (in.)	127 x 127 x 127 (5 x 5 x 5)	203 x 152 x 152 (8 x 6 x 6)	
Layer thickness, mm (in.)	0.178 (0.007)	0.254 (0.010)	0.254 (0.010) or 0.330 (0.013)
Size, mm (in.)	630 x 450 x 530 (25 x 18 x 21)	uPrint SE (Plus) with one material bay 635 x 660 x 787 (25 x 26 x 31) uPrint SE (Plus) with two material bays 635 x 660 x 940 (25 x 26 x 37)	

Table 4.2. *(Continued)* Specifications of the Idea Series 3D Printers.

Models	Mojo	uPrint SE	uPrint SE Plus
Weight, kg (lbs)	27 (60)	uPrint SE with one material bay 76 (168) uPrint SE with two material bays 94 (206)	
Software	Print Wizard	CatalystEX	
Power requirements	100 - 127 VAC, 6 A, 60 Hz or 220 - 240 VAC, 2.5 A, 50 Hz	100 - 127 VAC 50/60 Hz, minimum 15 A dedicated circuit or 220 - 240 VAC 50/60Hz, minimum 7 A dedicated circuit	

4.1.2.3 *Production Series*

Stratasys' Production Series 3D Printers produce products without the oppressive costs and time requirements of tooling. They make changes quickly and affordably — at any stage in the production cycle. Furthermore, they also create low–volume assembly fixtures and jigs directly from CAD data. Thus they can be used for prototyping, tooling and digital manufacturing for designers and engineers. There are four machines in the Production Series, Fortus 250mc, Fortus 360mc, Fortus 400mc and Fortus 900mc (See Fig. 4.5). The Fortus 250mc is the most compact and affordable machine, offering advanced 3D manufacturing capabilities with an office-friendly footprint. The Fortus 900mc 3D Production Systems offers the widest range of Fused Deposition

Fig. 4.5. Stratasys' Production Series 3D Printers, Fortus 250mc (left), Fortus 360mc (centre), Fortus 900mc (right). (Courtesy of Stratasys Inc.)

Modelling materials, including high performance thermoplastics, for durable, accurate parts as large as 914 x 610 x 914 mm (36 x 24 x 36 in.) with predictable mechanical, chemical and thermal properties. Details of the Production Series 3D Printers are summarised in Table 4.3.

Table 4.3. Specifications of the Production Series 3D Printers.

Models	Fortus 250mc	Fortus 360mc	Fortus 400mc	Fortus 900mc
Model materials	ABSplus - P430	ABS-M30 FDM Nylon 12 PC PC-ABS	ABS-ESD7 ABSi ABS-M30 ABS-M30i FDM Nylon 12 PC PC-ABS PC-ISO PPSF ULTEM 9085	
Build envelope (XYZ), mm (in.)	254 x 254 x 305 (10 x 10 x 12)	355 x 254 x 254 (14 x 10 x 10) 406 x 355 x 406 (16 x 14 x 16)		914 x 610 x 914 (36 x 24 x 36)
Layer thicknesses, mm (in.)	0.178 - 0.330 (0.007 - 0.013)	0.127 - 0.330 (0.005 - 0.013)		0.178 - 0.330 (0.007 - 0.013)
Support Structure	Soluble	Soluble or breakaway for PC; soluble for ABS-M30 and PC-ABS	Soluble for most materials; breakaway for PC-ISO, ULTEM and PPSF; soluble or breakaway for PC	
Available colours	Ivory, black, blue, dark grey, fluorescent yellow, nectarine, olive green, red, white, and custom colours	-	-	-
Size, mm (in.)	838 x 737 x 1143 (33 x 29 x 45)	1281 x 896 x 1962 (50.5 x 35.5 x 77.3)		2772 x 1683 x 2027 (109.1 x 66.3 x 79.8)

Table 4.3. *(Continued)* Specifications of the Production Series 3D Printers.

Models	Fortus 250mc	Fortus 360mc	Fortus 400mc	Fortus 900mc
Weight. kg (lbs)	With crate 186 (409) Without crate 148 (326)	With crate 786 (1,511) Without crate 593 (1,309)		With crate 3287 (7247) Without crate 2869 (6325)
Software	All Fortus Systems include Insight and Control Centre job-processing and management software	Insight Software		
Power requirements	110–120 VAC, 60 Hz, minimum 15A dedicated circuit; or 220–240 VAC 50/60 Hz, minimum 7A dedicated circuit	230 VAC, 50/60 Hz, 3 phase, 16A/phase (20 amp dedicated circuit required)		230VAC nominal three-phase service with 5% regulation 230VAC as measured phase-to-phase; 50 Hz or 60Hz; 40 Amp circuit

4.1.2.4 *Materials*

Stratasys Inc. offers a wide range of additive manufacturing materials, including clear, rubberlike and biocompatible photopolymers, and tough high performance thermoplastics. This variety enables customers to maximise the benefits of 3D printing throughout the product-development cycle. The wide range of materials enables designers and engineers to build practically anything, at an accelerated rate; taking them swiftly from fast and affordable concept modelling, to detailed, realistic and functional prototyping, to certification testing, and finally to agile, low-risk production. Furthermore, FDM thermoplastics are used to build tough, durable parts that are accurate, repeatable and stable over time. Materials used are mainly ABS, PC and high performance ULTEM 9085.

4.1.3 *Process*

In Stratasys' patented process [1], a geometric model of a conceptual design is created on CAD software which uses .STL or IGES formatted files. It can then be imported into the workstation where it is processed using InsightTM software, or Catalyst® software for the Dimension series, which automatically generates supports. Within this software, the CAD file is sliced into horizontal layers after the part is oriented for the optimum build position, and any necessary support structures are automatically detected and generated. The slice thickness can be set manually anywhere between 0.178 to 0.356 mm (0.007 to 0.014 in.) depending on the needs of the models. Tool paths of the build process are then generated and downloaded to the FDM machine.

Modelling material is in the form of a filament, very much like a stiffened fishing line, and is stored in a cartridge or in a spool. The filament is fed into an extrusion head and heated to a semi-liquid state. The semi-liquid material is extruded through the head and then deposited in ultra-thin layers from the FDM head, one layer at a time. Since the air surrounding the head is maintained at a temperature below the material's melting point, the exiting material quickly solidifies. Moving on the X-Y plane, the head follows the tool path generated by InsightTM, generating the desired layer. When the layer is completed, the head moves on to create the next layer. Two filament materials are dispensed through a dual tip mechanism in the FDM machine. A primary modeller material is used to produce the model geometry and a secondary material, or release material, is used to produce the support structures. The software handles the location where the support deposition takes place. The release material forms a bond with the primary modeller material with itself and can be washed away upon completion of the 3D models. A schematic diagram of the FDM process is shown in Figure 4.6.

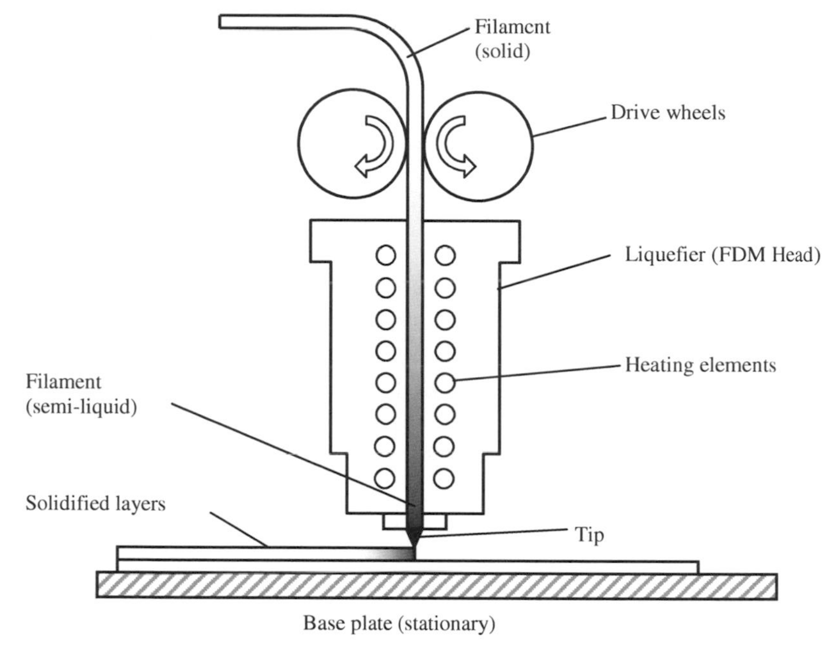

Fig. 4.6. The basic FDM process.

4.1.4 *Principle*

The principle of FDM technology (Fused Deposition Modelling) is based on surface chemistry, thermal energy and layer manufacturing technology. It is known for its reliability and durable parts, and extrudes fine lines of molten thermoplastic that solidify as they are deposited. 3D printers that run on FDM technology build parts layer-by-layer by heating thermoplastic material to a semi-liquid state and extruding it according to computer-controlled paths. FDM uses two materials to execute a print job: modelling material, which constitutes the finished piece, and support material, which acts as scaffolding supports for the finished piece. Material filaments are fed from the 3D printer's material bays to the print head, which moves in X and Y coordinates, depositing material to complete each layer before the base moves down the Z-axis and the next layer begins. Once the 3D printer is done building, the user breaks the support material away or dissolves it in detergent and water, and the part is ready to use.

4.1.5 *Strengths and Weaknesses*

The main strengths of the FDM technology are as follows:

(1) *Fabrication of functional parts*: the FDM process is able to fabricate prototypes with materials that are similar to that of the actual moulded product. With ABS, it is able to fabricate fully functional parts that have 85% of the strength of the actual moulded part and with ABS*plus,* the strength of the parts can rival those that are injection moulded. This is especially useful in developing products that require quick prototypes for functional testing.
(2) *Minimal wastage*: The FDM process builds parts directly by extruding molten semi-liquids onto the model. Thus, only those materials needed to build the part and its support are needed, and material wastages are kept to a minimum. There is also little need to clean up the model after it has been built.
(3) *Ease of support removal*: With the use of Break Away Support System (BASS™) and WaterWorks™ Soluble Support System, or the Breakaway Support Technology (BST) and Soluble Support Technology (SST) on the Dimension series, support structures generated during the FDM building process can be easily snapped off (for BST) or simply washed away in a water-based solution (for SST). This makes it very convenient for users to get to their finished parts very quickly, and there is very little or no postprocessing necessary.
(4) *Ease of material change*: Build materials, supplied in spool or cartridge form, are easy to handle and can be changed readily when the materials in the system are running low. This keeps the operation of the machine simple and maintenance relatively easy.
(5) *Large build volume*: FDM machines, especially the Fortus 900mc, offer larger build volume than most of the other AM systems available.

The weaknesses of the FDM technology are as follows:

(1) *Restricted accuracy*: Parts built with the FDM process usually have restricted accuracy due to the shape of the material used, i.e. the filament form. Typically, the filament used has a diameter of 1.27 mm and this tends to set a limit on how accurate the part can be achieved. The newer FDM machines however have made significant improvements in easing this problem by using better machine control.
(2) *Slow process*: The building process is slow, as the whole cross-sectional area needs to be filled with building materials. Building speed is restricted by the extrusion rate or the flow rate of the build material from the extrusion head. As the build materials used are plastics and their viscosities are relatively high, the build process cannot be easily sped up.
(3) *Unpredictable shrinkage*: As the FDM process extrudes the build material from its extrusion head and cools them rapidly on deposition, stresses induced by such rapid cooling invariably are introduced into the model. As such, shrinkages and distortions caused to the model are common and are usually difficult to predict, though with experience, users may be able to compensate these by adjusting the process parameters of the machine.

4.1.6 *Applications*

FDM models can be used in the following general application areas:

(1) Models and prototypes for conceptualisation and presentation: FDM 3D printers can create models and prototypes for new product design and testing, and build finished goods in low volumes.
(2) Educational use: Educators can use FDM technology to elevate research and learning in science, engineering, design and art.
(3) Customisation of 3D models: Hobbyists and entrepreneurs use FDM 3D Printing to expand manufacturing into the home — creating gifts, novelties, customised devices and inventions.

4.1.7 *Examples*

4.1.7.1 *Xerox uses Dimension for prototype design [2]*

Xerox Limited, with its 350 engineers, develops designs for printers, copiers, multifunction devices and document centres. Before the arrival of the Dimension 3D Printer into the department, Xerox relied on traditional machining from external agents in the prototype design process for moulded components. With the Dimension 3D Printer, the engineers are able to create prototype parts in their own office at a much faster rate, resulting in significant time and cost savings. Within 2 months, with 12 hours of operation every day at 70% utilisation, the Dimension 3D Printer built 254 parts. Part files can be sent from Xerox's partner company in Toronto, Canada, to Xerox in the UK, and are quickly produced by the machine.

Peter Keilty, the facility manager at Xerox headquarters in Welwyn Garden City, England, commented: "Capitalising on dead time gave us a significant advantage, and having the ability to produce the parts overnight when everyone was asleep saved the day. Thinking bigger picture – time to market has been reduced and Dimension has played a significant part in that."

4.1.7.2 *Hyundai uses FDM for design and testing [3]*

Hyundai Mobis is a Korean company that makes original and aftermarket equipment for the automotive industry. To be one of the best in its field, it relies on good prototyping for design verification, airflow evaluation, and functional testing. The company uses an FDM system for components such as instrument panels, air ducts, gear frame bodies, front-end modules, and stabiliser-bar assemblies.

The quality of a vehicle is judged by consumers on many aspects, the most important being component fit and finish. Hyundai Mobis paid a lot of attention to evaluate up to the smallest details, using rapid prototypes to ensure the fit conveys a sense of quality. An example is the design verification of an instrument panel for Kia's Spectra. This instrument panel exceeded the largest build area of the company's Stratasys FDM machines, so it was modelled in four pieces and then assembled to measure 498 x 454 x 1382 mm. To ensure the quality of the model, the design team mounted it on a fixture and scrutinised it with a CMM (coordinate measuring machine) before mounting it to the cockpit assembly. The result was that over a length of 1382 mm, the greatest deviation on the model was just 0.75 mm. Mounted in the cockpit mockup, the 27 design flaws were discovered. Although these errors were minor, collectively they would have added cost and likely delaying the project. The prototype allowed the designers to locate these design issues that were challenging to pick up in 3D CAD. When the FDM part was combined with mating components and subassemblies, the design flaws were quickly detected and repaired. As a result, Kia has garnered accolades from Car and Driver magazine for its Spectra, which wrote "...its interior fit and finish is premium."

Besides meeting Hyundai Mobis' requirements on the tight tolerance of the model, the FDM machine also gives them durable parts they need for assembly and functional testing. The water-soluble support structure is very important due to the time pressures as without it, a complex component like the instrument panel would take many hours, if not days, to post-process.

4.1.8 *Research and Development*

Since its founding in 1989, Stratasys has grown significantly over the last 25 years. Stratasys acquired Solidscape Inc. in May, 2011 and later MakerBot in June 2013. In between, Stratasys merged with Objet Inc. in April, 2012. MakerBot, a subsidiary of Stratasys since 2013, manufactures the company's prosumer desktop 3D printers in Brooklyn, New York. It maintains the Thingiverse design-sharing community and facilitates a wide network of user groups. Stratasys also operates RedEye On Demand, a digital manufacturing service. Through its network of certified resellers, Stratasys delivers responsive, regional support around the globe. The company maintains dual headquarters in Eden Prairie,

Minnesota and Rehovot, Israel. Stratasys holds nearly 500 granted or pending additive manufacturing patents worldwide. It is a public company that trades on NASDAQ under the symbol (SSYS).

4.2 Solidscape's Benchtop System

4.2.1 *Company*

Solidscape, Inc., formerly known as Sanders Prototype Inc., was first incorporated in February 1994 with global headquarters in Merrimack, New Hampshire, USA. It was funded by Sanders Design Inc. (SDI), a research company, which is principally owned and managed by Solidscape's founder, Royden C. Sanders. In 2004, SDI and Sanders transferred all AM technology to Solidscape and ceased any further work in the AM field. In May 2011, Solidscape was acquired by Stratasys. The address of the company is Solidscape, Inc., 316 Daniel Webster Highway, Merrimack, NH 03054-4115, USA.

4.2.2 *Products*

3Z series (3Z MAX, 3Z Pro, 3Z Studio and 3Z Lab) printers share the same design and basic functionality. The 3Z Max and 3Z Pro printers are shown in Figure 4.7. The four 3Z series machines vary mainly by industry-specific software configurations and Z-axis built height. 3Z series printers have highly achievable accuracy with ± 0.0010 inch/inch (±0.0254 mm/mm) across the XYZ dimensions. The footprint size (W x D x H) is 21.4 x 18 x 16 in (558 x 495 x 419 mm). The specifications of the 4 printers are listed in Table 4.4. 3Z series printers use 3Z Model (proprietary material formulated specifically for casting) as the printing material and 3Z Support Organic compound as the support material, which completely dissolves away in a liquid solution, leaving a clean wax part without the need for manual refining. Furthermore, Solidscape® 3D Software (3Z Works, 3Z Analyzer, 3Z Organizer) are designed to process all 3D open source files (.STL, .SLC) and to setup the printer's resolution as well as to control the motion algorithms.

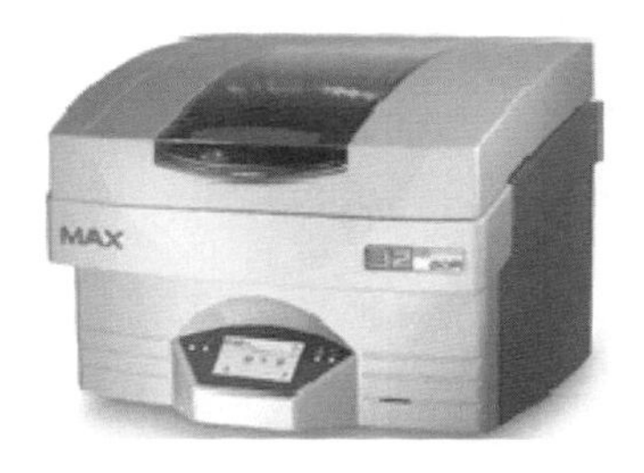

Fig. 4.7. 3Z MAX (left) and 3Z Pro with touch screen (right). (Courtesy of Solidscape)

Table 4.4. Specifications of 3Z series printer (3Z MAX, 3Z Pro, 3Z Studio and 3Z Lab).

Models	3Z Max	3Z Pro	3Z Studio	3Z lab
Accuracy, in. (mm)	± 0.0010 inch/inch (±0.0254 mm/25.4 mm)			
Build envelop (X x Y x Z), mm (in.)	152.4 mm x 152.4 mm x 101.6 mm (6" x 6" x 4")		152.4 mm x 152.4 mm x 50.8 mm (6" x 6" x 2")	
Dimension (W x D x H), mm (in.)	21.4 x 18 x 16 in(558 x 495 x 419 mm)			
Software	Solidscape® 3D Software supporting STL and SLC			
Material	3Z Model			
Support structure	3Z Support Organic compound			
Consumables	VSO, Build Plates, Build & Support material, Paper tape			
Features	Fastest Solidscape 3D wax printer	Produces models of intricate designs	-	Helps lab owners hold the line on rising costs
Cost in US$	49,650	45,650	24,650	-

4.2.3 *Process*

The process can be summarised into 4 major steps. Firstly, like most other commercial systems, a CAD interface such as STL file or SLC file is needed in the computer. Then, the CAD interface is converted by Solidscape®'s 3D Software such as 3Z Works, 3Z Analyzer and 3Z Organizer. Once the printer is ready, the printing process uses two ink-jet type print heads, one depositing the thermoplastic building material (3Z Model) and the other depositing soluble supporting wax (3Z Support Organic) compound. The model is lowered one layer at a time after each layer is built. The liquefied build material cools as it is ejected from the print head and solidifies upon impact with the model. 3Z Support is deposited to provide a flat, stable surface for deposition of build material in the subsequent layers. After each layer is complete, a cutter removes a thin portion off the layer's top surface to provide a smooth, even surface ready for the next layer. Figure 4.8 shows the schematic of the process.

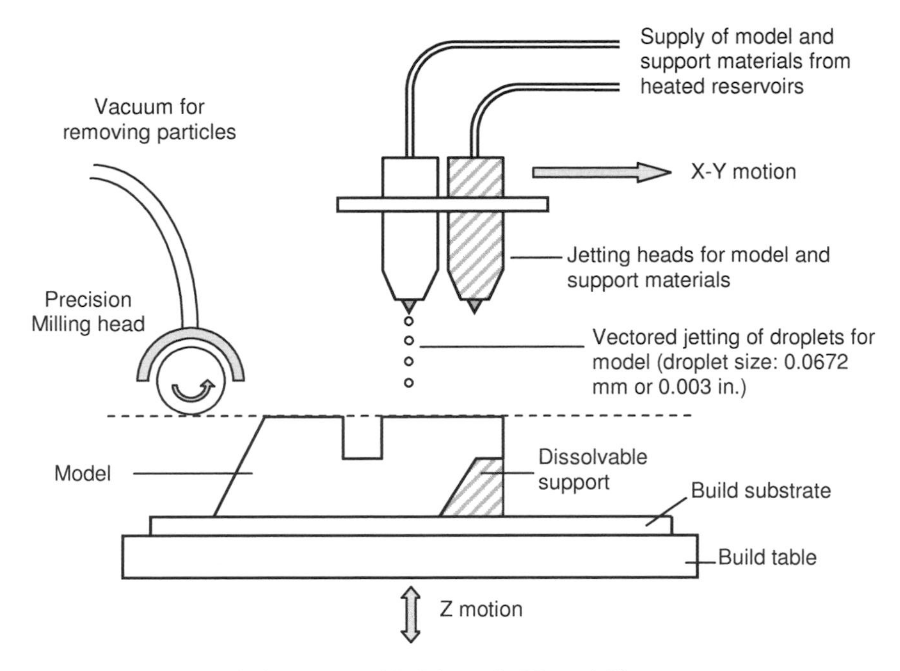

Fig. 4.8. Schematic of Solidscape's 3D modelling system.

After several layers have been deposited, print heads are moved automatically to a monitoring area to check for logging of the jet nozzles. If there is no blockage, the process continues, otherwise the jets are purged and the cutter removes any layers deposited since the last check of jets and restarts the building process from that point. This ensures the accuracy of the overall build.

4.2.4 *Principle*

The principle is that of building parts layer-by-layer, like in most other AM systems. The system builds models via material jetting, an additive manufacturing process in which droplets of build material are selectively deposited. The droplets are dispensed selectively as two inkjet-printing heads move across the build area. Solidscape produces wax patterns for casting small metal and ceramic parts using a propriety thermoplastic ink jetting process combined with high precision milling of each layer. The 3Z models feature 1 button starting interfaced through a touch screen, automatic status monitoring and fault detection, and an auto-calibrated print head. The propriety build materials are specially formulated so that there is no ash, no residue or shrink, no postprocessing work or special equipment needed. The main factors which influence the performance and functionalities are the proprietary jetting technology, which is called Drop on Demand (DoD), the X-Y controls, the Z-controls and the thermoplastic materials.

4.2.5 *Strengths and Weaknesses*

The main strengths are as follows:

(1) *High accuracy*: The systems are able to achieve an accuracy of ± 0.0010 inch/inch (±0.0254 mm/25.4mm) across the X-, Y-, and Z-dimensions. This compares favourably with most AM machines.
(2) *Office-friendly process*: This system can be used in an office environment, without any special facilities as the process is simple, clean and efficient. Moreover, materials used are nontoxic, which makes the process safe for office operation.

The weakness is:

(1) *Limited materials*: Like most concept modellers, materials that can be processed by these systems are restricted to 3Z Model (proprietary material formulated specifically for casting). Thus, it

puts a heavy limitation to the range of material-dependent functional capabilities of the prototypes.

4.2.6 *Applications*

The main applications of Solidscape's system are the production of high-precision master patterns for design verification and production application, especially in advanced research, education, aerospace, automotives, jewellery, orthopaedics, and toys.

4.2.7 *Examples*

4.2.7.1 *Solidscape helps advances cerebral aneurysm research [4]*

The Image Processing Applications Laboratory (IPALab), a lab at Arizona State University, tackles current and important image processing problems in variety of different fields. To remain competitive and to improve humans' standard of living, it carries out a lot of biomedical, industrial and military projects, through the development and use of advanced image processing.

Cerebral aneurysms affect 1 in 50 people and contribute to nearly 20,000 deaths in the USA alone every year. If an aneurysmal sac ruptures in the brain, it becomes a highly lethal condition with a 50% mortality rate per rupture. Hence, IPALab at Arizona State University's research findings are directly applied at participating hospital partners and in the design of improved endovascular medical devices.

The Solidscape machine is right at the heart of the research process. IPALab used Solidscape machines to build the core blood vessel models that can be converted into transparent flow models for their experiments. The final product is a transparent block whereby there is a hollow portion of the model that is an exact replica of a patient's cerebral aneurysm. Over the years, Solidscape machines have helped IPALab to build all of these models to help with their research process.

4.2.7.2 *Miniature masterpiece design powered by Solidscape [5]*

Horbach GmbH, based in Germany, provides one of Europe's premier casting services. The company helps Misko Models create precision car models for discerning collectors. In 1931, Bentley Blower 4.5L LeMans was a head-turner that showcased grace, craftsmanship and precision. After nine decades, it has since been recreated as a 1:5 scale model by Misko Models craftsmen, utilising state-of-the-art CAD design and the precision of a Solidscape 3D printer.

Horbach GmbH undertook casting for critical parts and components of the model based from models created on Solidscape machines. The precision of every single part was essential for the model's many moving parts such as workable brakes, clutch, choke and adjustable ignition. According to Christian Muller, Horbach GmbH's CEO, the quality of the parts created using Solidscape 3D printer is beyond what they could previously achieve with other AM systems, CNC machining and even skilled handwork. He believes that the quality of the parts is among the best in the world.

4.3 MCOR Technologies' Selective Deposition Lamination (SDL)

4.3.1 *Company*

Mcor Technologies Ltd was founded in 2004 and the company's AM systems are developed based on Selective Deposition Lamination (SDL) technology, which is also known as paper 3D printing. This technology was first invented by Dr. Conor MacCormack and Fintan MacCormack in 2003. Mcor Technologies Ltd's printers are the only 3D printers to use ordinary and affordable office papers as the build material. Mcor Technologies Ltd is located at Unit 1, IDA Business Park, Ardee Road, Dunleer, Co Louth, Ireland.

4.3.2 *Products*

Mcor Technologies Ltd manufactures 3D printers to suit customers' demand. It aims to produce eco-friendly and user-friendly 3D printers with lower cost. It consists of two types of 3D printers: Mcor IRIS and Mcor Matrix 300+.

4.3.2.1 *Mcor IRIS*

Mcor IRIS (Figure 4.9) is a colour 3D printer that can produce coloured products with high quality (5760 x 1440 x 508 dpi) with more than one million colours. The full colour printing is ensured by Mcor 3D Ink. Mcor IRIS is also affordable and it can also be used in offices due to its small size and low noise level. Details of Mcor IRIS printer are summarised in Table 4.3.

Fig. 4.9. The Mcor IRIS 3D Printer. (Courtesy of Mcor Technologies Ltd)

4.3.2.2 *Mcor Matrix 300+*

Mcor Matrix 300+ uses letter or A4 paper and a water-based adhesive to produce products at a low cost. Its printing speed has also been improved up to 3 times. Details of the Mcor Matrix 300+ Printer are summarised in Table 4.5.

Table 4.5. Specifications of Mcor IRIS and Mcor Matrix 300+ Printers.

	Mcor IRIS	Mcor Matrix 300+
Build Size	A4 Paper: 256 x 169 x 150 mm Letter Paper: 9.39 x 6.86 x 5.9 inches	
Build Material	A4 80 gsm (160 gsm ply colour only) US Letter Standard Paper 20lbs (43lbs ply colour only)	A4 (80 gsm & 160 gsm) US Letter Standard Paper (20lbs & 43 lbs)
Layer Thickness	0.1 mm (0.004 in) 0.19 mm (0.007 in ply colour only)	0.1 mm (0.004 in) and 0.19 mm (0.007 in)
Resolution	Colour resolution: 5760 x 1440 x 508 dpi Axis Resolution (X x Y x Z): 0.012mm, 0.012mm, 0.1 mm (0.0004 in, 0.0004 in, 0.004 in)	0.1 mm (0.004 in)
Power Requirements	350 W 240 V, 50 Hz or 120 V, 60 Hz	
Network Connectivity	-	TCP/IP 100/10 base T
Regulatory Compliance	CE,UL	
Hardware Requirement	8GB memory and 100Gb hard drive, 2 network cards, one for the printer	
Operating System	64 bit Windows XP, Windows Vista or Windows 7	

4.3.2.3 *ColourIT Software*

ColourIT Software is developed by Mcor Technologies Ltd to enable customers to improve their 3D files with colours by adding them directly to the surfaces of 3D models. It is compatible with most file formats, including STL, WRL, OBJ, 3DS, FBX, DAE and PLY. After enhancing the colours, the model is exported by the software as an ARL file, which can be imported to the Mcor IRIS printers for production.

4.3.3 ***Process***

Selective Deposition Lamination (SDL) is primarily a 'lamination and cut' process which can be described as follows:

(1) A STL or OBJ or VRML (for colour 3D Printing) file is imported into Mcor Technologies Ltd's 3D printers.

(2) With Mcor Technologies Ltd's control software, SliceIT, the CAD file is sliced into printable layers equivalent in thickness to the standard papers. The additional software, ColourIT, is used to add colours onto CAD files.
(3) After the slicing file is ready, the first sheet of paper is attached to the build plate.
(4) The 3D printer deposits drops of adhesive onto the first sheet of paper according to the sliced cross-section.
(5) A new sheet of paper slides in and pressure is exerted by the printer to bond these two sheets together.
(6) An adjustable tungsten carbide blade cuts one sheet of paper at a time, tracing the object outline to create the edges of the part.
(7) Steps 3 to 6 are repeated until the last sheet of paper is added.
(8) The solid object is then removed manually from the surrounding paper.

4.3.4 *Principle*

The principle of Selective Deposition Lamination (SDL) is based on the simple addition of sheets of paper. This process is very similar to Laminated Object Manufacturing (LOM) (see Section 4.4) except that instead of using a roll of paper and a laser, sheets of paper, together with an adhesive dispensing system and tungsten tip knife, are used. Various sheets of papers are accumulated with the adhesive to form the final product, and an adjustable tungsten carbide blade will cut every sheet at a time, tracing the object outline to create the edges of the part. Hence, the final product is produced when all the waste is removed.

4.3.5 *Strengths and Weaknesses*

The main strengths of the SDL technology are as follows:

(1) *Low cost*: Mcor Technologies' 3D printers produce parts at possibly the lowest price in the industry - approximately five percent of other similar technologies' costs because the technology mainly uses papers as its building material.

(2) *High precision*: Mcor Technologies' 3D printers can build models to a precision of 0.00047 in (0.012 mm) and a dimensional accuracy of 0.004 in (0.1 mm).
(3) *Safety*: Throughout the making process, Mcor Technologies Ltd's 3D printers do not use any toxic chemicals, fumes or dust and there is no dangerous heat or light.
(4) *Eco-Friendly*: The products can be easily recycled because they are made of paper.
(5) *High resolution in colour printing*: Mcor Technologies' 3D printers can print more than one million colours simultaneously and its resolution can reach 5760 x 1440 x 508 dpi.

The weaknesses of the SDL technology are as follows:

(1) *Low strength*: The SDL product can be easily damaged because they are made from paper.
(2) *Small build volume*: The largest products printed can only have an area of approximately A4 size.

4.3.6 *Applications*

SDL models can be used in the following general application areas:
(1) Manufacturing uses: SDL helps manufacturers produce models faster, and this helps them enhance the design within a shorter time. Conceptual prototypes can also be produced at a lower cost.
(2) Architectural uses: SDL produces more accurate and precise models or prototypes for architects. This helps to save time and cost for them and it also reduces architects' working period.
(3) Marketing: Models made by SDL are more attractive and powerful in showcasing prospects to customers before the actual products are launched in the market.

4.3.7 *Examples*

4.3.7.1 *Versatile inventor uses Mcor 3D Printing Technology to improve Malta's art, architecture and manufacturing*

Malta, one of the smallest countries in the world, seeks inventors to bring in 3D printing technology to boost its art, architecture and manufacturing. Recently, Malta's furniture manufacturers' association asked Aquilina to invcstigatc the value of 3D printing.

Mcor Technologies' Matrix 300+ was chosen by Aquilina to build models and prototypes. After scanning the clay models using 3D scanners and post-editing via Autodesk 3ds Max, the final CAD file was sent to Matrix 300+. The piece of work is then printed and reproduced with high quality and durability for artists and architects to further study.

Moreover, with Mcor Technologies' 3D printers, it only took one-tenth of the time to produce a 3D printed model. Compared to the handcrafted products, the 3D printed products also had higher quality and accuracy.

4.3.7.2 *Mcor IRIS anchors powerful industrial 3D printing service in Australia*

Williams 3D, a company in Sydney, Australia, provides 3D printing services to clients and customers who require 3D models. Its focus is on high-volume industrial and manufacturing customers.

Williams 3D adopts Mcor IRIS printers for producing the first initial prototypes during the testing phase. After clients check the quality and the design of the printed prototype, Williams 3D will continue to use other plastic-based 3D printers to produce the final product. Hence, Mcor IRIS printers help the company to save time and costs during the testing period.

Furthermore, Williams 3D does not face any problems regarding the deposition of products because all 3D printed products by Mcor IRIS printers can be safely recycled, thus reducing environmental costs.

4.3.8 *Research and Development*

Mcor Technologies Ltd has been working on cutting the printing time on its IRIS and Matrix lines of 3D printers by 50%, resulting in print times nearly twice as fast as they were previously. Additional speed enhancements are being developed by the company to make 3D printing faster, more affordable and accessible to all.

4.4 Cubic Technologies' Laminated Object Manufacturing (LOM™)

4.4.1 *Company*

Cubic Technologies was established in December 2000 by Michael Faygin, the inventor who developed Laminated Object Manufacturing (LOM). In 1985, Faygin set up the original company, Helisys Inc., to market the LOM additive manufacturing machines. However, sales figures did not meet expectations [6] and the company ran into financial difficulties. Helisys Inc. subsequently ceased operation in November 2000. Currently, Cubic Technologies, successor to Helisys Inc., is the exclusive manufacturer of LOM additive manufacturing machines. However, it mostly offers service, support, and parts for LOM systems. The company's address is Cubic Technologies, Inc., 2785 Pacific Coast Highway #295, Torrance, CA 90505, USA.

4.4.2 *Products*

Cubic Technologies offers two models of LOM additive manufacturing systems, the LOM-1015Plus™ and LOM-2030H™ (see Figure 4.10). Each of these systems uses a CO_2 laser, with the LOM-1015Plus™ operating a 25W laser and the LOM-2030H™ operating a 50W laser. The optical system, which delivers a laser beam to the top surface of the work, consists of three mirrors that reflect the CO_2 laser beam and a focal lens that focuses the laser beam to about 0.25 mm (0.010"). The control of the laser during cutting is by means of a XY positioning table that is servo-based as opposed to a galvanometer mirror system. The LOM-2030H™ is a larger machine meant for building larger prototypes. The work volume of the LOM-2030H™ is 810 x 550 x 500 mm (32 x 22 x 20 in.) and that of the LOM-1015Plus™ is 380 x 250 x 350 mm (15 x 10 x 14 in.). Detailed specifications of these two machines are summarised in Table 4.6.

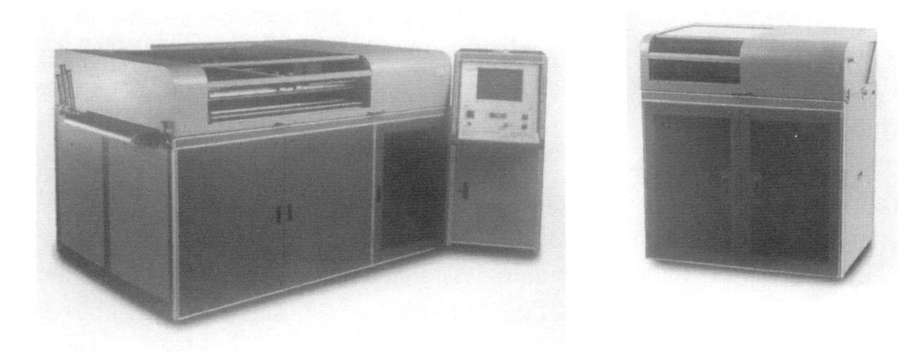

Fig. 4.10. LOM-1015Plus™ and LOM-2030H™.

Table 4.6. Specifications of LOM-1015Plus™ and LOM-2030H™.

Models	LOM-1015Plus™	LOM-2030H™
Max. Part Envelope Size, mm (in.)	381 x 254 x 356 (15 x 10 x 14)	813 x 559 x 508 (32 x 22 x 20)
Max. Part Weight, kg (lbs)	32 (70)	204 (405)
Laser, power and type	Sealed 25W, CO_2 Laser	Sealed 50W, CO_2 Laser

Table 4.6. *(Continued)* Specifications of LOM-1015Plus™ and LOM-2030H™.

Models	LOM-1015Plus™	LOM-2030H™
Laser Beam Diameter, mm (in.)	0.20 – 0.25 (0.008 – 0.010)	0.203 – 0.254 (0.008 – 0.010)
Motion Control	Servo-based X-Y motion systems with a speed up to 457 mm/sec (18"/sec);	Brushless servo-based X-Y motion systems with a speed up to 457 mm/sec (18"/sec);
Part accuracy XYZ directions, mm (in.)	±0.127 mm (±0.005 in.)	±0.127 mm (±0.005 in.)
Material Thickness, mm (in.)	0.08 – 0.25, (0.003 – 0.008)	0.076 – 0.254, (0.003 – 0.008)
Material Size	Up to 356 mm (14") Roll Width and Roll Diameter	Up to 711 mm (28") Roll Width and Roll Diameter
Floor Space, m (ft.)	3.66 x 3.66, (12 x 12)	4.88 x 3.66, (16 x 12)
Power	Two (2) 110VAC or Two (2) 220VAC, 50/60 Hz, 20 Amp, Single Phase	220VAC, 50/60 Hz, 30 Amp, Single Phase
Materials	LOMPaper® LPH series, LPS series LOMPlastics® LPX series	LOMPaper® LPH series, LPS series LOMPlastics® LPX series, LOM Composite® LGF series

4.4.3 *Process*

The patented Laminated Object Manufacturing (LOM) process [7, 8, 9] is an automated fabrication method in which a 3-D object is constructed from a solid CAD representation by sequentially laminating the part cross-sections. The process consists of three phases: preprocessing; building; postprocessing. The overall LOM process is illustrated in Figure 4.11.

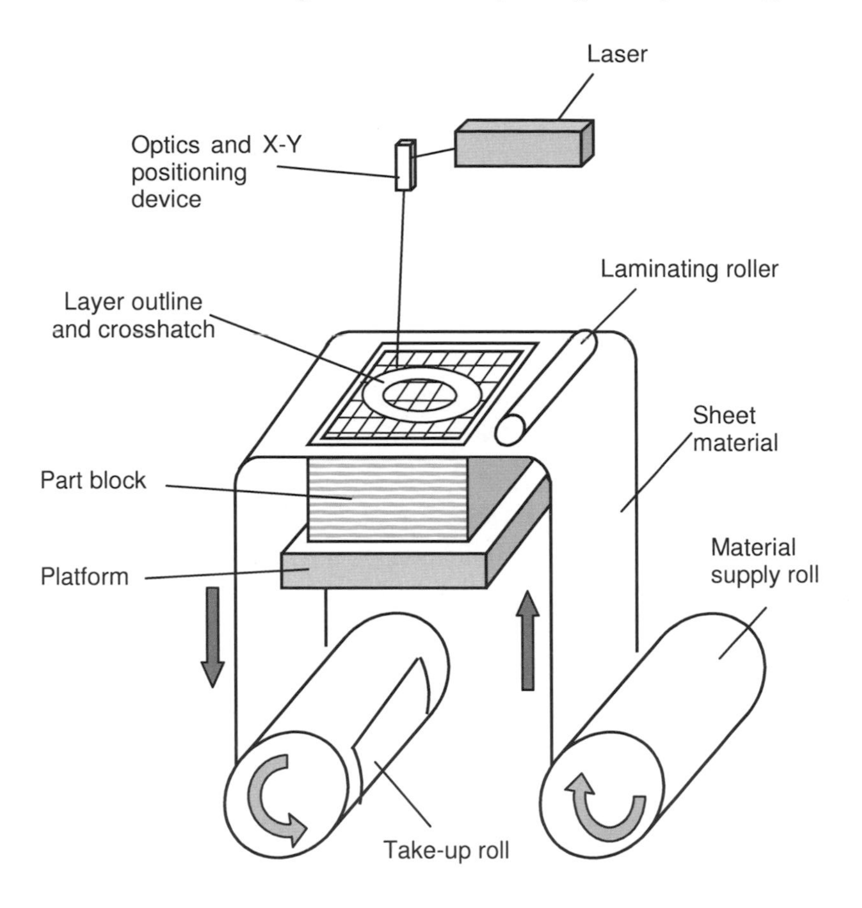

Fig. 4.11. The LOM process.

4.4.3.1 *Preprocessing*

The preprocessing phase comprises several operations. The initial steps include generating an image from a CAD-derived STL file of the part to be manufactured, sorting input data, and creating secondary data structures. These are fully automated by LOMSlice™, the LOM system software which calculates and controls the slicing functions. Orienting and merging the part on the LOM system are done manually. These tasks are aided by LOMSlice™, which provides a menu-driven interface to perform transformations (e.g. translation, scaling, and mirroring) as well as merges.

4.4.3.2 *Building*

In the building phase, thin layers of adhesive-coated material are sequentially bonded to each other and individually cut by a CO_2 laser beam (see Figure 4.12). The build cycle has the following steps:

(1) LOMSlice™ creates a cross-section of the 3-D model measuring the exact height of the model and slicing the horizontal plane accordingly. The software then images crosshatches which define the outer perimeter and converts these excess materials into a support structure.

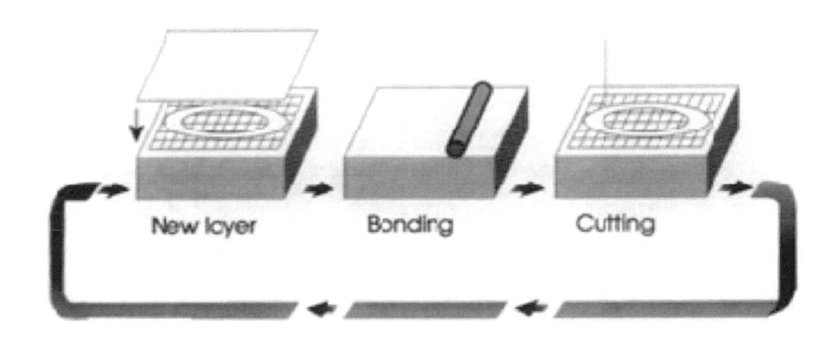

Fig. 4.12. Building process. (Courtesy of Cubic Technologies, Inc.)

(2) The computer generates precise calculations, which guide the focused laser beam to cut the cross-sectional outline, the crosshatches, and the model's perimeter. The laser beam power is designed to cut exactly the thickness of one layer of material at a time. After the perimeter is burned, everything within the model's boundary is 'freed' from the remaining sheet.

(3) The platform with the stack of previously formed layers descends and a new section of material advances. The platform ascends and the heated roller laminates the material to the stack with a single reciprocal motion, thereby bonding it to the previous layer.

(4) The vertical encoder measures the height of the stack and relays the new height to LOMSlice™, which calculates the cross-section for the next layer as the laser cuts the model's current layer.

This sequence continues until all the layers are built. The product emerges from the LOM machine as a completely enclosed rectangular block containing the part.

4.4.3.3 *Postprocessing*

The last phase, postprocessing, includes separating the part from its support material and finishing it. The separation sequence is as follows (see Figure 4.13 (a-d)):

(1) The metal platform, home to the newly created part, is removed from the LOM machine. A forklift may be needed to remove the larger and heavier parts from the LOM-2030H.
(2) Normally a hammer and a putty knife are all that is required to separate the LOM block from the platform. However, a live thin wire may also be used to slice through the double-sided foam tape, which serves as the connecting point between the LOM stack and the platform.
(3) The surrounding wall frame is lifted off the block to expose the crosshatched pieces of the excess material. Crosshatched pieces may then be separated from the part using wood carving tools.

After the part is extracted from surrounding crosshatches, the wood-like LOM part can be finished. Traditional model-making finishing techniques, such as sanding, polishing, painting, etc. can be applied. After the part has been separated, it is recommended that it be sealed immediately with urethane, epoxy, or silicon spray to prevent moisture absorption and expansion of the part. LOM parts can also be machined using conventional machines like those for drilling, milling and turning.

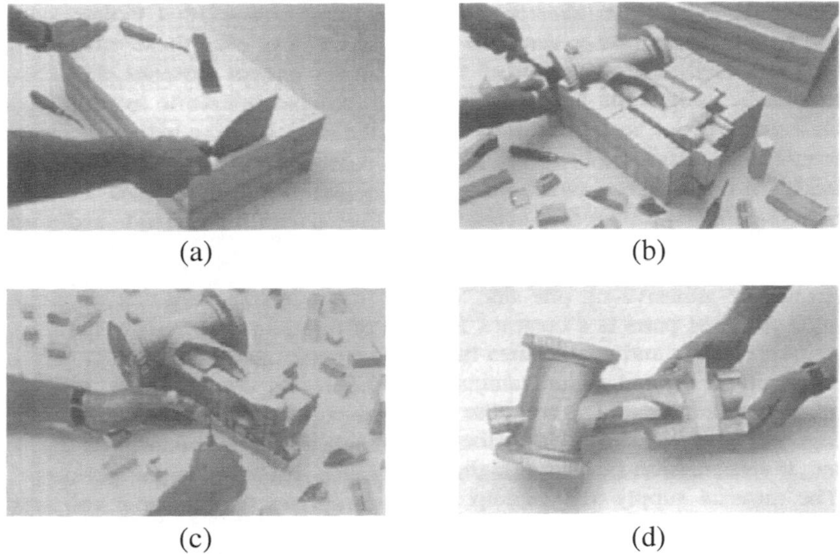

(a) (b) (c) (d)

Fig. 4.13. Separation of the LOM object, (a) The laminated stack is removed from the machine's elevator plate, (b) The surrounding wall is lifted off the object to expose cubes of excess material, (c) Cubes are easily separated from the object's surface, (d) The object's surface can then be sanded, polished or painted, as desired. (Courtesy of Cubic Technologies, Inc.)

4.4.3.4 *System structure*

The LOM-1015Plus™ and LOM-2030H™ have a similar system structure which can be broken down into several subsystems: computer hardware and software, laser and optics, X-Y positioning device, platform and vertical elevator, laminating system, material supply and take-up (see Figure 4.11).

The LOM software, LOMSlice™, is a true 32-bit application and is completely integrated, providing preprocessing, slicing, and machine control within a single program. Z-dimension accuracy is maintained through a closed-loop real-time feedback mechanism and is calculated upon each lamination.

LOMSlice™ can also overcome STL file imperfections that violate facets of normal vector orientation or vertex-to-vertex rules [10], or even those missing facets. In order to facilitate separation of the part from excess material, LOMSlice™ automatically assigns reduced crosshatch sizes to intricate regions.

To make it easier, faster and safer to align the laser beam, a helium-neon visible laser which projects a red beam of light and is collinear with the live CO_2 laser beam is used. The operator can switch on the innocuous red laser beam and the mirrors automatically.

Lamination is accomplished by applying heat and pressure by way of rolling a heated cylinder across the sheet of material that has a thin layer of a thermoplastic adhesive on one side. Studies [11] have indicated that the interlaminate strength of LOM parts is a function of bonding speed, sheet deformation, roller temperature, and contact area between the paper and the roller. By increasing the pressure of the heated roller, lamination is improved with fewer air bubbles. Increased pressure also augments the contact area and bolsters interlaminate strength. Pressure is controlled by the limit switch mounted on the heated roller. Too high a compression can cause distortion in the part.

The material supply and take-up system comprises two material roll supports (supply and rewind), several idle rollers to direct the material, and two rubber-coated nip rollers (driving and idle), which advance or rewind the sheet material during the preprocessing and building phases. To make material flow through the LOM systems more smoothly, mechanical nip rollers are used. The friction resulting from compressing moving material between rubber-coated rollers on both the feed and wind mechanism ensures a clean feed and prevents jamming.

4.4.3.5 *Materials*

Potentially, any sheet material with adhesive backing can be utilised in LOM. Cubic Technologies has demonstrated that plastics, metals, and even ceramic tapes can be used. However, the most popular material has been Kraft paper with a polyethylene-based heat seal adhesive system because it is widely available, cost-effective, and environmentally benign [12].

In order to maintain uniform lamination across the entire working envelope, it is critical that the temperature is kept constant. A temperature control system with closed-loop feedback ensures the system's temperature remains constant, regardless of its surrounding environment.

4.4.4 *Principle*

The LOM process is based on the following principles:

(1) Parts are built layer-by-layer by laminating each layer of paper or other sheet-form materials. The contour of the part of each layer is cut by a CO_2 laser.
(2) Each layer of the building process contains the cross-sections of one or many parts. The next layer is then laminated and built directly on top of the laser-cut layer.
(3) The Z-control is activated by an elevation platform, which lowers when each layer is completed, and the next layer is then laminated and ready for cutting. The Z-height is then measured for the exact height so that the corresponding cross-sectional data can be calculated for that layer.
(4) No additional support structures are necessary as the 'excess' materials, which are crosshatched for later removal, act as the support.

4.4.5 *Strengths and Weaknesses*

The main strengths of using LOM technology are as follows:

(1) *Wide variety of materials.* In principle, any material in sheet-form can be used in LOM systems. These include a wide variety of organic and inorganic materials such as paper, plastics, metals, composites and ceramics. Users can select the type and thickness of

materials that meet their specific functional requirements and applications of the prototype.

(2) *Fast build time*. The laser in the LOM process does not scan the entire surface area of each cross-section; rather, it only outlines its periphery. Therefore, parts with thick sections are produced just as quickly as those with thin sections, making the LOM process especially advantageous for the production of large and bulky parts.

(3) *High precision*. The feature to feature accuracy that can be achieved with LOM machines is usually better than 0.127 mm (0.005 in.). Through design and selection of application specific parameters, higher accuracy levels in the X-Y and Z-dimensions can be achieved. If the layer does shrink horizontally during lamination, there is no actual distortion as the contours are cut post-lamination, and laser cutting itself does not cause shrinkage. If the layers shrink in the transverse direction, a closed-loop feedback system gives the true cumulative part height upon each lamination to the software, which then slices the 3-D model with a horizontal plane at the appropriate location.

(4) The LOM system uses a precise X-Y positioning table to guide the laser beam; it is monitored throughout the build process by the closed-loop, real-time motion control system, resulting in an accuracy of within 0.127 mm regardless of part size. The Z-axis is also controlled using a real-time, closed-loop feedback system. It measures the cumulative part height at every layer and then slices the CAD geometry at the exact Z location. Also, as the laser cuts only the perimeter of a slice, there is no need to translate vector data into raster form. Therefore the accuracy of the cutting only depends on the resolution of CAD model triangulation.

(5) *Support structure*. There is no need for additional support structures as the part is supported by its own material, which is outside the periphery of the part built.

(6) *Post-curing*. The LOM process does not need to convert expensive, and in some cases toxic, liquid polymers to solid plastics or plastic powders into sintered objects. Because sheet materials are not subjected to either physical or chemical phase changes, the finished

LOM parts do not experience warpage, internal residual stress, or other deformations.

The weaknesses of the LOM technology are as follows:

(1) *Precise power adjustment*. The power of the laser used for cutting the perimeter (and the crosshatches) of the prototype needs to be precisely controlled so that the laser cuts only the current layer of lamination and does not penetrate the previously cut layers. Poor control of the cutting laser beam can cause distortion to the entire prototype.
(2) *Fabrication of thin walls*. The LOM process is not well suited for building parts with delicate thin walls, especially in the Z-direction. This is because such walls usually are not sufficiently rigid to withstand postprocessing stresses when the crosshatched portion of the block is being removed. The operator performing the postprocessing task of separating the thin wall of the part from its support must be aware of where such delicate parts are in the model and take sufficient precaution so as not to damage these parts.
(3) *Integrity of prototypes*. The part built by the LOM process is essentially held together by heat-sealed adhesives. The integrity of the part is therefore entirely dependent on the adhesive strength of the glue used, and as such is limited to it. Therefore, parts built may not be able to withstand the vigorous mechanical loading that functional prototypes may require.
(4) *Removal of supports*. The most labour-intensive part of the LOM process is its final phase of postprocessing, when the part has to be separated from its support material within the rectangular block of laminated material. This is usually done with wood carving tools and can be tedious and time consuming.

4.4.6 *Applications*

LOM can be applied across a wide spectrum of industries, including industrial equipment for aerospace or automotive industries, consumer products, and medical devices ranging from instruments to prostheses.

LOM parts are ideal in design applications where it is important to visualise what the final piece will look like, or to test for form, fit and function; as well as in a manufacturing environment to create prototypes, create production tooling or even small volume production.

(1) *Visualisation.* LOM's ability to produce exact dimensions of a potential product lends it well for visualisation purposes. LOM part's wood-like composition allows it to be painted or finished to be a good mockup of the product. As the LOM procedure is inexpensive, several models can be built, giving sales and marketing executives opportunities to utilise these prototypes for consumer testing, product introductions, packaging samples, and samples for vendor quotations.
(2) *Form, fit and function.* LOM parts lend themselves well for design verification and performance evaluation. In low-stress environments, LOM parts can withstand basic tests, giving manufacturers the opportunity to make changes as well as evaluate the aesthetic property of the prototype in its total environment.
(3) *Manufacturing*. LOM part's composition is such that, based on the sealant or finishing products used, it can be further tooled for use as a pattern or mould for many secondary tooling techniques, including investment casting, casting, sanding casting, injection moulding, silicon rubber mould, vacuum forming and spray metal moulding. LOM parts offer several advantages important for the secondary tooling process, namely: predictable level of accuracy across the entire part; stability and resistance to shrinkage, warpage and deformity; and the flexibility to create a master or a mould.
(4) *Rapid tooling*. Two-part negative tooling can be easily created with the LOM. As the material is solid and inexpensive, complicated tools can be cost-effective to produce. These wood-like moulds can be used for the injection of wax, polyurethane, epoxy or other low pressure and low temperature materials. The tooling can also be converted to aluminium or steel via the investment casting process for use in high temperature moulding processes.

4.5 Ultrasonic Consolidation

Ultrasonic consolidation (UC) is a solid-state metal additive manufacturing process in which metal foils are bonded together primarily using ultrasonic welding to produce three dimensional objects [13]. Computer numerically controlled (CNC) milling is sequentially used to introduce features that are designed in the part and for the final finishing of the part. Therefore, the process combines both additive and subtractive manufacturing to obtain a desired part. One of the key features which make UC distinct from other metal additive manufacturing process is that it does not involve melting. The temperatures evolving at the mating interfaces are less than half of the melting point of the materials [14]. This makes the process suitable for some important applications such as embedding electronics, sensors and optical fibres.

4.5.1 *Principle*

The ultrasonic consolidation process utilises a cylindrical sonotrode to produce high frequency (20 kHz) and low amplitude (20 -50 microns) mechanical vibrations to induce shear with the two mating surfaces of two metallic foils held together by normal forces. The high frequency vibration induces localised friction and enables the breakage of oxide films at the contact surfaces of the foils. The oxide layer debris and any other contaminants between the foils are flushed out due to vibration, creating two nascent surfaces. When the two nascent surfaces are brought into atomic level contact by the application of a normal force and by the localised heat generation due to friction, a metallurgical bond is realised. The bonding mechanism at the interface between two foils is due to the localised plastic deformation of the asperities and grain growth across the interface [15].

To fabricate a 3D part, a thin metallic foil (100 -150 μm) is held firmly on a substrate plate either by mechanical clamping or by pneumatically assisted clamping. A sonotrode is then traversed over the foil at a certain amplitude and frequency. The metal foil also vibrates in ultrasonic frequency, resulting in a proper bonding of the metal foil to the substrate. Then a second foil is placed over the already deposited foil and the process of sonotrode traverse is repeated. As the layer-by-layer addition of the foil progresses, a CNC milling head is brought in to create any features designed on the part or to finish the surface or contours. Depending upon the nature of material and to reduce the effect of any residual stresses, the base plate is pre-heated to an elevated temperature. Figure 4.14 shows the schematic of the working principles of the ultrasonic consolidation process.

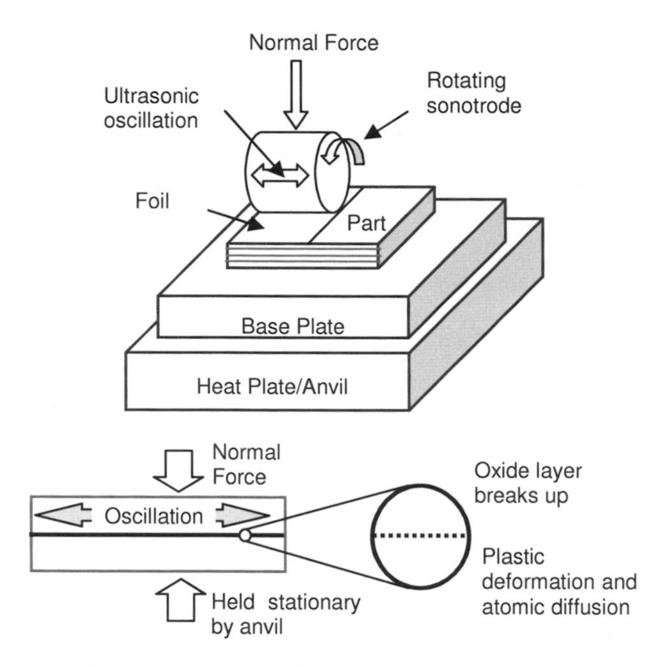

Fig. 4.14. Schematic representation of ultrasonic consolidation process.

4.5.2 *Process Parameters*

Ultrasonic consolidation process parameters have a direct consequence on the quality of the final part. The important parameters in UC are the ultrasonic amplitude (approximately 5–50 μm), the applied normal force (500–2000 N), the travel speed of the sonotrode horn (up to 50 mm/s), the texture of the sonotrode horn (Ra between 4 and 15 μm) and the substrate preheat temperature (up to 150 °C). These parameters vary for different materials and need to be optimised for every material [2]. The placement of foils and its orientation also play a major role in deciding the final properties of the part.

A commonly used parameter to represent the bond quality in UC is linear weld density (LWD), which is given by

LWD = Bonded interface length / Total interface length

4.5.3 *Applications*

Ultrasonic consolidation combines the advantages of both additive manufacturing and subtractive manufacturing. This allows the fabrication of complex three-dimensional parts with good surface finish and high dimensional accuracy. Some of the typical parts include complex objects with internal cooling channels, honeycomb structures, multi-material laminates and objects integrated with fibre optics, sensors and other electronic instruments. These benefits make the process very attractive for a variety of applications in tooling, automotive, aerospace, electronics and defence industries [17].

4.6 Other Notable Solid-Based AM Systems

4.6.1 *Solidimension's Plastic Sheet Lamination (PSL)*

4.6.1.1 *Company*

Solidimension is an Israel-based company that has been developing PVC lamination-based 3D printers since 1999. In 2003, the company established a partnership with Japanese Graphtec Corp. and the following year, Graphtec agreed to distribute 900 units of Solidimension's XD700 3D printer under its own brand name in Japan, Taiwan, Korea, Australia, Germany, France, and the UK. [18]. In 2005, 3D Systems became Solidimension's marketing partner and commercialised its 3D printer as

3D Systems' InVision™ LD 3D Modeller [19]. 3D Systems has since stopped marketing this system, which is now represented by Cubic Technologies in the US. The address of company is Solidimension Ltd., Shraga Katz Building, Be'erot Itzhak, 60905, Israel.

4.6.1.2 *Products*

The main product of Solidimension is the SD300, a low-cost, entry-level desktop 3D printer aimed at the concept modelling process as well as form and fit analysis in an office environment. The corresponding system sold by 3D Systems is known as InVision™ LD 3D Modeller. The SD 300 uses the SolidVC®, whereas the InVision™ LD uses the VisiJet® LD 100 both use thermoplastic (polyvinylchloride, PVC-based) sheets in the form of sheet rolls, together with adhesive and release agents or masking fluid. 3D Systems improved the support removal system by introducing an easy "peel-away" process called EZPeel®, which allows the user to remove the support simply by peeling away the zigzag support. The SDview® software supplied enables the user to manipulate, position, and edit parts, as well as to split large models into multiple components to be glued manually as part of the postprocessing. It supports .STL input files. Solidimension's SD300 and 3D Systems' InVision™ LD are shown in Figure 4.15 and their specifications summarised in Table 4.7.

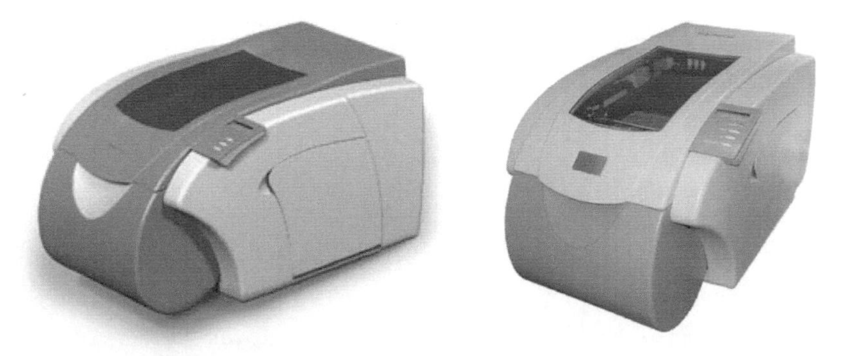

Fig. 4.15. Solidimension's SD300 (left) and 3D Systems' InVision™ LD (right). (Courtesy of Solidimension Ltd. and 3D Systems)

Table 4.7. Specification of SD300 and InVision™ LD.

Model	SD300	InVision™ LD
Technology	Sheet Lamination	
Accuracy	± 0.1 mm	± 0.2 mm
Build Volume, mm (in)	160 x 210 x 135 (6.29 x 8.26 x 5.31)	160 x 210 x 135 (6.29 x 8.26 x 5.31)
Layer Thickness, mm (in)	0.168	0.15 (0.0059)
Electrical	100-120 VAC, 50/60 Hz; 200-240 VAC, 50/60Hz	
Weight, kg (lb)	36 (79.3)	
Overall Dimensions, mm (in.)	465 x 770 x 420 (18.3 x 30.3 x 16.5)	
Input Format	STL, 3DS	
Material	SolidVC® PVC Material	VisiJet LD 100

4.6.1.3 *Process*

The concept modelling process begins when the STL model data is loaded into SDview®software. Within the software, the user is able to edit and orientate the model, as well as perform the splitting of the model if the model is found to be larger than the maximum build size. The software then slices the model and generates the required cutting and gluing action for the model and the support EZPeel®. The processed data is then sent to the machine through an USB connection. The machine then builds the model layer-by-layer from the bottom up, using the layer build-up cycle.

The layer build-up cycle consists of the following steps:

(1) *Preparation*. The machine first prepares a new PVC plastic sheet in the buffer.
(2) *Application of glue*. Glue is then applied over the entire area of the new plastic sheet.

(3) *Ironing*. The ironing unit then irons the plastic sheet onto the model block, joining a new layer to the model.
(4) *Cutting*. The new plastic layer is then cut by the cutting knife unit according to the pattern of the layer. Only the single layer is cut.
(5) *Trimming*. The plastic sheet is then trimmed by the trim knife and the supply is rolled back in preparation for the next layer.
(6) *Masking*. The final step of the layer build-up cycle is the application of masking fluid by anti-glue pens near the model to create the EZPeel® support.

This cycle is repeated until the model is complete. The components and parts of the SD300 used in the layer build-up cycle are illustrated in Figure 4.16. The material near the model is laminated and cut in such a way that it forms a continuous spiral-like form from the top-most layer to the bottom-most layer. During postprocessing, larger blocks of waste material on the outer part can be separated directly because of the straight cut at the edges. The support material near the model can be peeled off by just pulling the upper-most layer. The spiral made from the layers in a zigzag fashion is like a thread on a spool being pulled to the axis of the spool cylinder. This forms the EZPeel® support removal system and it greatly simplifies the removal of the support material.

4.6.1.4 *Principle*

The main principle of the PSL system is similar to that of the LOM, which is building the object layer-by-layer by laminating sheets of plastic into a model block. Each layer is glued to the previous layer and the contour of the part is cut for every layer. The unique feature of the PSL that differentiates it from other solid-based systems is the laminating and cutting of the material around the model to form the EZPeel® support removal that makes postprocessing very easy and simple.

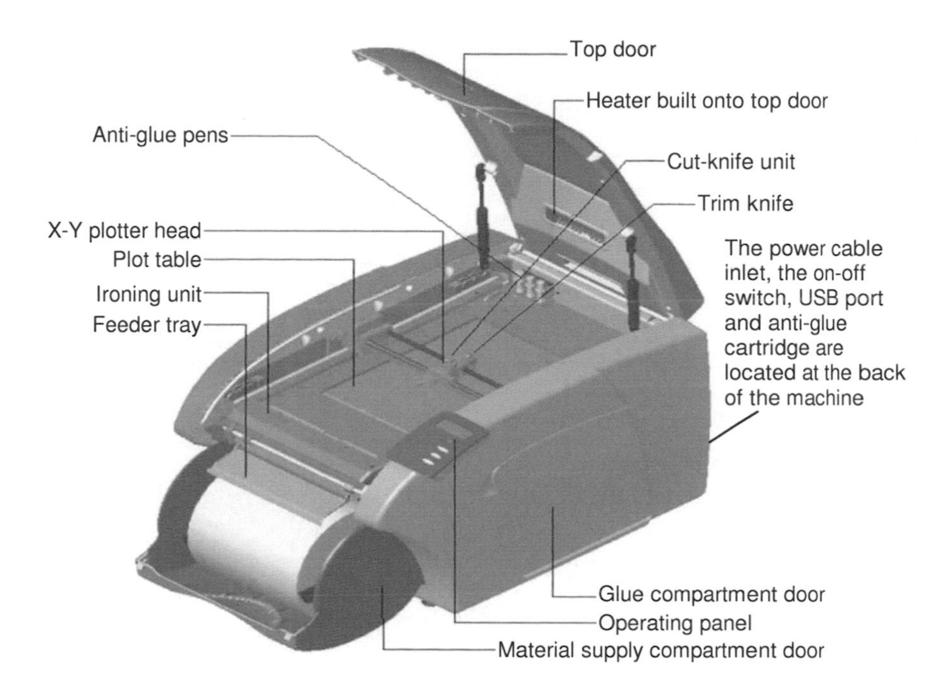

Fig. 4.16. Operating components of the SD300 3D printer. (Courtesy of Solidimension Ltd.)

4.6.1.5 *Strengths and weaknesses*

The strengths of Solidimension's PSL system are as follows:

(1) *Support structures.* There is effectively no need for additional support structures as the part is supported by its own material at the periphery of the part built. The removal of the part is also made easy and simple by the use of EZPeel® support peel-away during postprocessing.
(2) *Cost-effective.* The thermoplastic material used is inexpensive and provides for cost-effective modelling. The machine also has one of the lowest costs in the 3D printer market.

(3) *Office-friendly process.* The system is clean, efficient and safe without the use of toxic chemicals. Postprocessing is relatively clean and the PVC waste can be disposed of safely. This makes the system suitable for an office environment.
(4) *Fast build time.* As the cutting of the patterns of each layer is at the part peripheral and not the entire surface area, each layer can be built very quickly. This greatly shortens the overall process time.

The weaknesses of Solidimension's PSL system are as follows:

(1) *Small build size.* As a compact desktop machine, a small build volume is expected. The ability to split big parts inside SDview® software is a helpful addition, although the durability of the resulting combined parts depends largely on the glue used and the gluing technique employed by the user.
(2) *Internal voids.* Internal voids cannot be produced on a single build since it is impossible to remove the material in the internal voids in this process. The model may need to be split into multiple parts to produce the void.
(3) *Inability to vary layer thickness.* The thickness of the layer is dependent on the thickness of the sheet material supplied, 0.168 for the SD300 and 0.15 for the InVision™ LD.

4.6.1.6 *Applications*

Solidimension's PSL system is mainly for conceptual modelling and visualisation that does not require high accuracy or large build size during the design stage. The system produces tough plastic materials that can also be used for testing fit and form.

4.6.2 *KIRA'S Paper Lamination Technology (PLT)*

4.6.2.1 *Company*

Kira Corporation Ltd. was established in February 1944 and has developed a wide range of industrial products since then. Its line of products includes CNC machines, automatic drilling/tapping machines, recycle units and systems, and folding bicycles. Kira joined the AM industry by developing two additive manufacturing machines, the KSC-50N (PLT-A3) and the PLT-A4. These two systems use a technology known as Paper Lamination Technology (PLT), formerly known as Selective Adhesive and Hot Press (SAHP) [20]. The speed and reliability of both machines have ranked them as one of the better systems. The current PLT technology line continues with the Katana Rapid Mockup System. The address of the company is Kira Corporation Ltd., Tomiyoshishinden, Kira-Cho, Hazu-gun, Aichi 444-0592, Japan.

4.6.2.2 *Products*

Kira Corporation, at the time of writing, only produces the Katana PLT-20 Rapid Mockup System, the name reflecting its speed in producing prototypes (See Figure 4.17). A simple digital camera mockup can be produced in just over an hour. Users may select the paper thicknesses of choice, although for a single build, the paper thickness needs to be the same. Different from its predecessor, the Katana's paper supply is in the form of a paper roll. The specifications of the machine are summarised in Table 4.8.

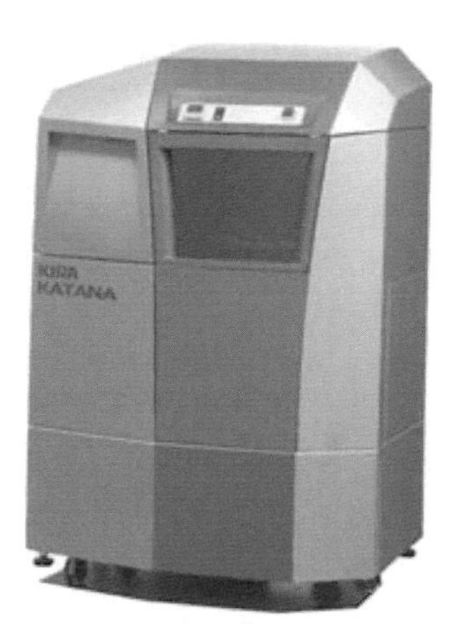

Fig. 4.17. Katana PLT-20 Rapid Mockup System. (Courtesy of of Kira Corporation)

Table 4.8. Specifications of Katana PLT-20. (Courtesy of Kira Corporation)

Model	PLT-20
Technology	Paper Lamination Technology (PLT)
Materials	KATANA paper
Max. build volume, mm	180 x 280 x 150
Resolution, mm	0.025(X,Y), 0.1(Z) or 0.16(Z)
Precision, mm	±0.5(X,Y), ±3.0%(Z)
Power Requirement	AC100/110V ±10% Single phase 15A
System Size, mm	860 x 660 x 1330
Weight, kg	320kg
Data Control Unit	PC/AT compatible PC

4.6.2.3 *Process*

The Katana PLT-20 Rapid Mockup System consists of the following hardware: a PC, a photocopy printer, a paper alignment mechanism, a hot press, and a mechanical cutter plotter [21]. The Katana PLT-20 Rapid Mockup System makes a plain paper layered solid model. The process is somewhat similar to that of the LOM process, except that no laser is used and a flat hot plate press is used instead of rollers. Called Paper Lamination Technology (PLT) and formerly known as Selective Adhesive and Hot Press (SAHP), the process includes six steps: generating a model and printing resin powder, hot pressing, cutting the contour, completing block, removing excess material and postprocessing (see Figure 4.18).

First, the 3D data (STL files) of the model to be built is loaded onto the PC. The model is then oriented within the system with the help of the software for the best orientation for the build. Once achieved, the system software will proceed to slice the model and generate the printing data based on the section data of the model. The plain paper roll is fed onto the machine and trimmed. The resin powder or toner is applied on the paper from the paper roll using a typical laser stream printer and is referred to as the Xerography process (i.e. photocopying). The printed area is the common area of two consecutive sections of the model.

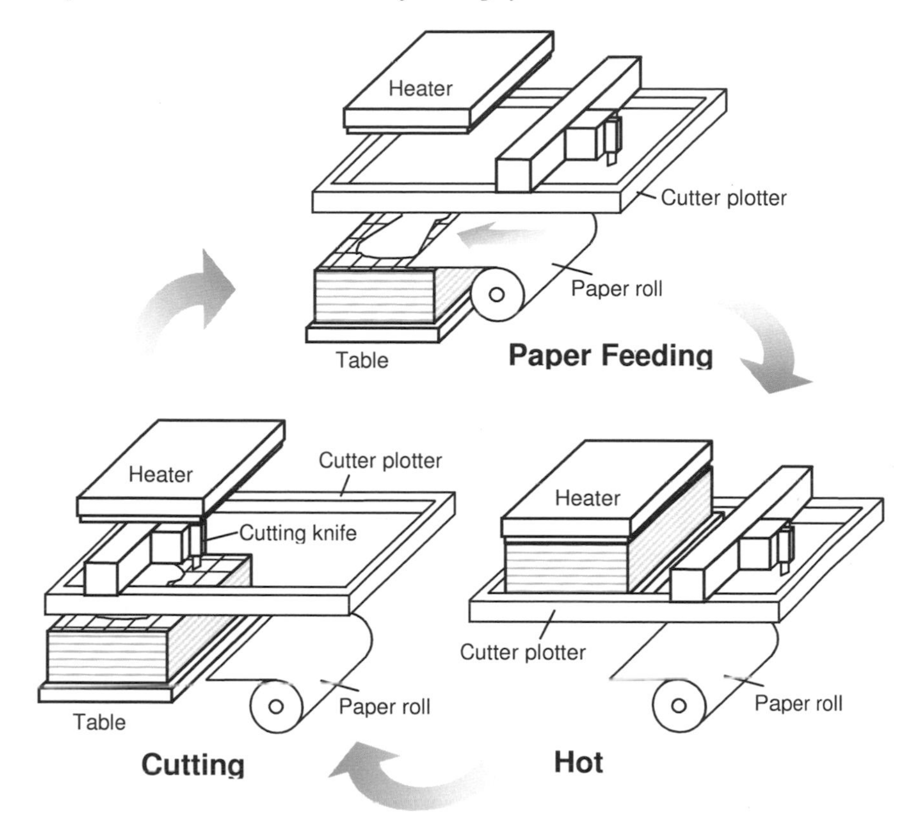

Fig. 4.18. Process diagram of the PLT process.

A paper alignment mechanism then adjusts the printed sheet of paper onto the previous layer on the model. The table then extends to a hot press over model block with the printed sheet pressed to a hot plate at high pressure. The temperature-controlled hot press melts the toner (resin powder), which adheres the sheets together. The hot press also flattens the top surface and prevents the formation of air bubbles between sheets. The PC measures the amount of the movement up to the hot plate to compensate for any deviation of the sheet thickness.

The PC then generates plotting data based on the section data of the model. A precision cutter cuts the top layer of the block along the contour of the section as well as parting lines from which excess paper is removed later. These steps are repeated until the entire model is built.

When printing, hot pressing and cutting are completed, the model block is removed from the machine and unnecessary portions of the paper are disconnected quickly and easily, sheet-by-sheet. When the model is complete, its surface may then be finished by normal woodworking or mechanical means.

The tensile strength and bending strength of the material made by the earlier SAHP process have been shown to be approximately one-half those of wood models. However, enhancement to the PLT process has enabled model hardness to improve significantly and it has been reported that a hardness of up to 25 percent better than the equivalent wood model has been achieved.

4.6.2.4 *Principle*

The principle of the process is based partly on the photocopy principle, conventional mechanical layering and precision cutting techniques. A typical laser stream printer is used for printing and a resin powder instead of print toner is used as a toner, which is applied to the paper in the exact position indicated by the section data, to adhere the two adjacent layers of paper.

Three factors, cutter plotter, temperature and humidity, affect the accuracy of the model being built. The accuracy of the cutter plotter affects the accuracy of the model in the X and Y directions. Shrinkage of the model occurs when the model is cooled down in the hot press unit. Expansion of the model occurs when the model is exposed to varying humidity conditions in the hot press unit.

4.6.2.5 *Strengths and weaknesses*

The key strengths in using PLT, as described by Kira Corporation [22] are as follows:

(1) *Flatness*. The PLT process uses a flat plate and high pressure to bond the layers together. Each layer is pressed with a flat hot plate

and the model remains flat during the entire build process. Since the block is released after cooling, there is minimal internal strain and therefore there is little or no curling in the final model.

(2) *Surface smoothness*. The PLT process uses a computerised knife to cut sheets of paper, which results in a smooth surface for the model built. Better surface finishing can be further attained by simple sanding tools but are seldom necessary. Prototypes can be sanded, cut or coated according to the user's needs.

(3) *Hardness*. High lamination pressure used in the PLT process has resulted in products that are 25 percent harder than the equivalent wood model and this is often strong enough for most prototyping applications.

(4) *Support structures*. Additional support structures are not necessary in the PLT process as the part is supported by its own material that is outside the perimeter of the cut-path. These are not removed during the hot press process and thus act as natural supports for delicate or overhang features.

(5) *Office-friendly process*. Kira's machines can be installed in any environment where electricity is available. There is no need for special facilities or utilities for running the Katana. Moreover, the process is safe as no high-powered laser or hazardous materials are used.

The weaknesses of using PLT are as follows:

(1) *Inability to vary layer thickness*. Fabrication time is slow in the Z-direction as the process builds prototypes layer-by-layer, and the height of each layer is fixed by the thickness of the paper used. Thus, the speed of build cannot be easily increased as the thickness of the paper used cannot be varied.

(2) *Fabrication of thin walls*. Like the LOM process, the PLT process is also not well suited for building parts with delicate thin walls, especially walls that are extended in the Z-direction. This is due to the fact that such walls are joined transversely and may not have sufficient strength to withstand postprocessing. However, the flat

hot plate used in the process may be able to pack the heat seal adhesive resin a little better to limit the problem.

(3) *Internal voids.* Models with internal voids cannot be fabricated within a single build as it is impossible to remove the unwanted support materials from within the 'void.'

(4) *Removal of supports.* Similar to that of the LOM process, the most labour-intensive part of the build comes at the end—separating the part from the support material. However, as the support material is not adhesively sealed, the removal process is simpler, though woodworking tools are sometimes necessary. The person working to separate the part needs to be cautious and aware of the presence of any delicate parts on the prototype so as not to damage the prototype during postprocessing.

4.6.2.6 *Applications*

The PLT has been used in many areas, such as the automobile, electric machine and component, camera and office automation machine industries. Its main application area is in conceptual modelling and visualisation. In Japan, Toyota Motor Corp., NEC Corp. and Mitsubishi Electric Corp. have reportedly used Kira Corporation's Katana to model their products.

4.6.3 *CAM-LEM'S CL 100*

4.6.3.1 *Company*

CAM-LEM (Computer Aided Manufacturing of Laminated Engineering Materials) Inc., a privately owned company in the US, commercialises an additive manufacturing technology akin to Cubic Technology's Laminated Object Manufacturing (LOM). The process uses the 'form-then-bond' laminating principle, where the contour of the cross-section is first cut before being laminated to previous layers. This technology was developed by a team of researchers from Case Western Reserve University. The company's strategy is not to sell the additive manufacturing machine it developed; rather, CAM-LEM intends to act as a service provider to interested customers. The company's address is CAM-LEM, Inc., 1768 E. 25th St., Cleveland, Ohio 44114, USA.

4.6.3.2 *Products*

The additive manufacturing system developed by CAM-LEM Inc. is called CL-100 (see Figure 4.19) and is able to produce parts up to 150 × 150 × 150 mm in size. One distinct feature of CL-100 is that the system is able to build metal parts and ceramic parts with excellent mechanical properties after sintering. The layer thickness ranges from 100 microns to 600 microns and above, and up to five different materials can be incorporated into a single automated build cycle.

Fig. 4.19. CAM-LEM's CL100 additive manufacturing system. (Courtesy of CAM-LEM Inc.)

4.6.3.3 *Process*

The CAM-LEM approach (see Figure 4.20), like other additive manufacturing methods, originates from a CAD model decomposed into boundary contours of thin slices [23]. In the CAM-LEM process, these individual slices are laser cut from sheet stock of engineering material (such as 'green' ceramic tape) as per the computed contours. The resulting part-slice regions are extracted from the sheet stock and stacked to assemble a physical 3-D realisation of the original CAD description. The assembly operation includes a tacking procedure that fixes the position of each sheet relative to the pre-existing stack. After assembly, the layers are laminated by warm isostatic pressing (or other suitable methods) to achieve intimate interlayer contact, promoting high-integrity

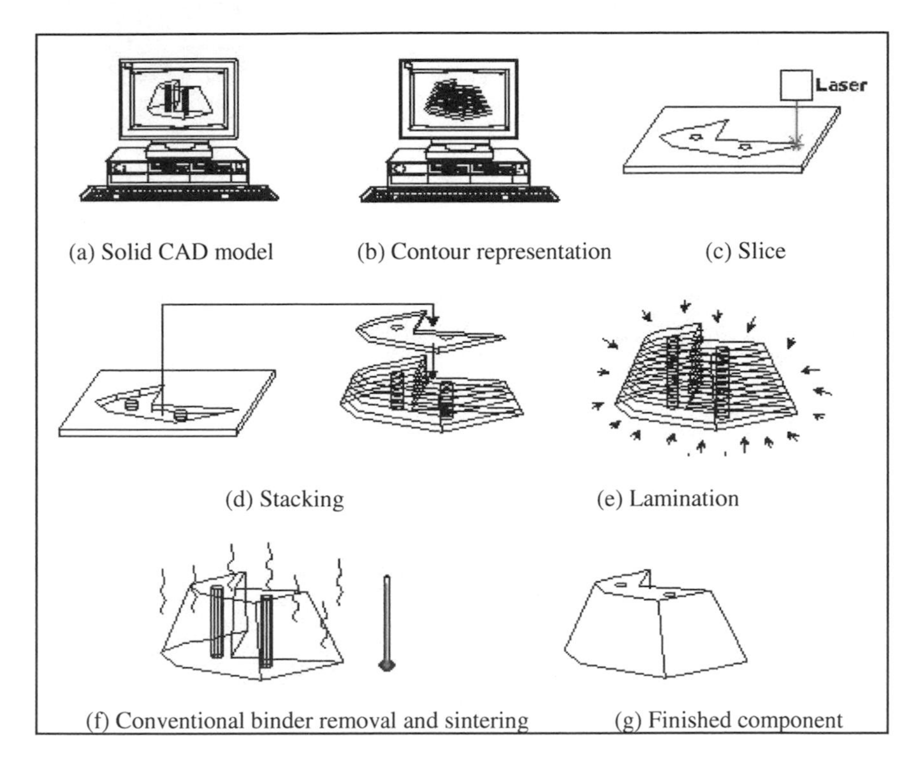

Fig. 4.20. Schematic of CAM-LEM Process. (Courtesy of CAM-LEM, Inc.)

bonding in the subsequent sintering operation. The laminated 'green' object is then fired (with an optimised heating schedule) to densify the object and fuse the layers and particles within the layers into a monolithic structure. The result is a 3-D part, which exhibits not only correct geometric form, but functional structural behaviour as well.

4.6.3.4 *Principle*

The CAM-LEM process is based on the same principle applied to most solid-based AM systems. Objects are built layer-by-layer and the laminated objects can be fabricated from a wide variety of engineering materials. The sections that form the layers are cut separately from sheet stock with a CO_2 laser and are then selectively extracted and stacked precisely. Multiple material types can be used within a single build. This process allows for the formation of interior voids and channels without manual waste removal, thus overcoming the problem of entrapped volume that plagues most other AM systems. The distinct characteristic of this process is the separation of the geometric formation process from the material process, thus providing users with more flexibility. The crucial factors affecting the quality of the model built are laser cutting, indexing and tacking, the alignment of the stacking process, the binding process and the sintering process.

4.6.3.5 *Strengths and weaknesses*

The key strengths of using CAM-LEM technology are as follows:

(1) *Allow for the formation of interior voids and channels.* The CAM-LEM process separates the laser cutting process from the stacking and lamination process, thus allowing for the formation of interior voids and channels, thereby eliminating the problem of entrapped volumes that troubles many other AM systems.
(2) *Laser power adjustment.* The cutting laser power of CAM-LEM technology does not need to be precisely adjusted because the process uses the 'form-then-bond' laminating principle, where the contour of the cross-section is first cut before being laminated to the previous layers. Thus, it eliminates the problem of laser burning into the previous layers.
(3) *High quality prototypes.* As the technology uses the 'form-then-bond' principle, it ensures that the layers are free from fine grains of unwanted materials before bonding them to previous layers. This is highly desirable as the unwanted materials trapped in between layers would affect the mechanical property of the final product, and such a situation should be eliminated so as to fabricate high quality prototypes.
(4) *Adjustable build layers.* The CAM-LEM process allows prototypes to be built using different material thicknesses, which could

effectively speed up the process. Regions with large volumes are built with thicker sheets of paper, and surfaces that require a smooth surface finish are built with thinner sheets of paper.

The weaknesses of using CAM-LEM technology are as follows:

(1) *Significant shrinkage*. The main disadvantage of using CAM-LEM's technology is that the prototype will shrink in size by around 12-18%, which makes the dimensional and geometric control of the final prototype difficult.
(2) *Precise alignment*. The process requires high system accuracy to align the new bonding layer to the previous layer before bonding them. Any slight deviation in alignment with previous layers will not only affect the accuracy of the model, but also its overall shape.
(3) *Lacks natural supports*. While the process eliminates the problem of entrapped volume, it does require users to identify the locations of supports for the prototype, especially for overhanging features. As this process only transfers the desired layers to bond with previous layers, all unwanted materials, which could have been used as supports, are thus left behind.

4.6.3.6 *Applications*

The CAM-LEM process has been used mainly to create rapid tools for manufacturing. Functional prototypes and even ceramic and metal components have been built.

4.6.4 *ENNEX Corporation's Offset Fabbers*

4.6.4.1 *Introduction*

Ennex Corporation started as Ennex Fabrication Technologies in 1991 by a successful physicist and computer entrepreneur, Marshall Burns. Burns was then joined by a team of experienced and enthusiastic professionals to develop a range of state-of-the-art additive manufacturing machines, which focused mainly on digital manufacturing technology. The company named these machines "digital fabricators" or simply "fabbers." A "fabber," as defined by the company, is a *"factory in a box" that makes things automatically from digital data* [24]. Besides developing this technology, the company also provides valuable consulting advice to related companies that deal with "fabbers." Ennex Corporation is located at 549 Landfair Ave., Los Angeles, California 90024-2172, USA.

4.6.4.2 *The Genie® Studio Fabber*

One of the "fabbers" which the company has developed is the Genie® Studio Fabber, based on a technology known as "Offset™ Fabbing." The technology is somewhat similar to that of Kira Corporation's Paper Laminating Technology (PLT). The main difference between the two technologies is that 'Offset™ Fabbing' uses the "form-then-bond" fabrication [25] method. The working principle of this fabrication method is simple. It uses a mechanical knife to cut the outline of the layer and after the sheet of material is cut to the required shape, it is then transferred and laminated onto the previous layer. The process of cutting and laminating of the materials is repeated until the whole model is built layer-by-layer. Although Ennex Corp first announced "Offset™ Fabbing" technology in 1996, the company has yet to commercialise its first Genie® Studio Fabber. However, a working prototype has been built in the company's development laboratory. Ennex has claimed that its output will be fast, up to ten times more than other leading systems.

4.6.4.3 *Process*

The process of "Offset™ Fabbing" can be described as follows (see Figure 4.21):

(1) Thin fabrication material (any thin film that can be cut and bonded to itself by means of adhesive) is rested on a carrier.
(2) A horizontal plotting knife is used to cut the outlines of cross-sections of the desired object into the fabrication material without cutting through the carrier.

(3) The plotter can also cut parting lines and outlines for support structures.
(4) The film is then "weeded" to remove some or all of the "negative" material.
(5) The film is inverted so that the carrier is facing up and the cut pattern is brought into contact with the top of the growing object and bonded to it.
(6) The carrier is then peeled off to reveal the new layer just added and a fresh surface ready to bond to the next layer.

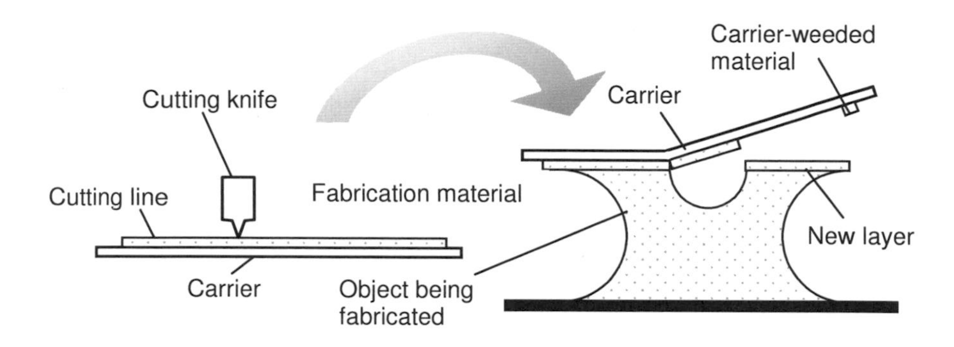

Fig. 4.21. Schematic of Ennex "Offset™ Fabbing" process. [24]

4.6.4.4 *Strengths and weaknesses*

The strengths of "Offset™ Fabbing" technology are as follows:

(1) *Wide variety of materials*. The process is not restricted to a few materials. Basically, any thin film that can be cut and bonded to itself can be used for the process. Paper and thermoplastic sheets are typical examples.
(2) *Minimal shrinkage*. As the bonding layers do not experience any significant change in temperature during the fabrication process, the shrinkage of the parts built is minimal.
(3) *Office-friendly process*. The process is clean and easy to use in ordinary office environment. Furthermore, the process is safe as it does not involve the use of laser or hazardous materials.

The weaknesses of Offset™ Fabbing technology used on the Genie Studio Fabber are as follows:

(1) *Need for precise cutting force.* The cutting force of the cutting knife must be precise, as a deep cut will penetrate the carrier while a shallow cut will not give a clean cut to the layer, resulting in tearing and sticking.
(2) *Need for precise alignment.* The process requires high accuracy from the system to align the new bonding layer to the previous layer before bonding it. Precise position is critical as any slight deviation in alignment from previous layers will not only affect the accuracy of the model, but also the overall shape of the model.
(3) *Wastage of materials.* The weeded (unwanted) materials cannot be re-used after the layers have been bonded with the previous layer and hence a significant amount of materials are wasted during the process.

4.6.4.5 *Applications*

The "Offset™ Fabbing" process can be used mainly to create concept models for visualisation and proofing during the early design process and is intended to function in an office environment near CAD workstations. Functional prototypes and tooling patterns intended for rapid tooling can also be built.

4.6.5 *Shape Deposition Manufacturing Process*

4.6.5.1 *Introduction*

While most additive manufacturing processes based on the discretised layer-by-layer process are able to build almost any complex shape and form, they suffer from the very process of discretisation in terms of surface finish as well as geometric accuracies. The Shaped Deposition Manufacturing (SDM) process, first pioneered by Professor Fritz Prinz and his group at Carnegie Mellon University and later Stanford University, is an additive manufacturing process that overcomes these difficulties by combining the flexibility of the additive layer manufacturing process with the precision and accuracy attained with the subtractive CNC machining process. Shot peening, microcasting, and a weld-based material deposition process can be further combined within a CAD-CAM environment using robotic automation to enhance the capability of the process [26]. The SDM process, though well researched and full of capabilities, has yet to be commercialised.

4.6.5.2 *Process*

The SDM process is an additive manufacturing process that systematically combines the advantages of layer-by-layer manufacturing with the advantages of precise material removal [27]. The process is illustrated by Figure 4.22.

Materials for the individual segments of the part are first deposited at the deposition station to form the layer of the part (See Figure 4.22(a)). One of the several deposition processes is a weld-based deposition process called microcasting [26], and the result is a near-net shape deposition of the part for that layer. The part is then transferred to the shaping station, usually a 5-axis CNC machining centre, where material is removed to form the desired or net shape of the part (Figure 4.22(b)). After the shaping station, the part is transferred to a stress relief station, such as shot peening, to control and relieve residual stress build up due to the thermal process during deposition and machining. The part is then transferred back to the deposition station where complementary-shaped, sacrificial support material is deposited to support the part. The sequence of depositing part or support material is dependent on the geometry of the part, as explained in the following paragraph. The process is repeated until the part is complete, after which the sacrificial support material is removed and the final part is revealed (Fig. 4.22(c)).

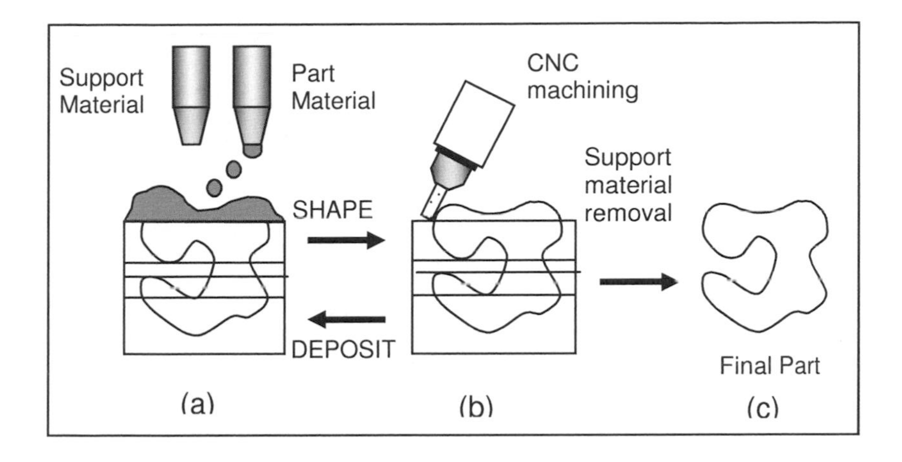

Fig. 4.22. Shape Deposition Manufacturing building process.

One major difference of SDM is that unlike most additive manufacturing processes, the CAD model of the part is decomposed into slices or segments that maintain the full 3D geometry of the outer surface. The major advantage of this strategy of decomposing shapes into segments, or "compacts" as called by the inventors, is that it eliminates the need to machine undercut features. Instead, such features are formed by depositing either support or part material, as appropriate, onto previously deposited and shaped segments (See Figure 4.22(a)). For example, if the part overhangs the support (like in a 'U'-shape), then the support material for that layer will be deposited first. Conversely, if the part is clear of the support in the upward direction (as in Figure 4.22(b)), then the part material for the layer will be deposited first. As such, this flexibility of sequencing part-support material deposition totally eliminates the need to machine difficult undercut features. The total layer thickness is not set arbitrarily but is dependent on the local geometry of the part and the deposition process constraints. Because of the alternating deposition-shaping sequences, discretisation steps (which plague many other additive manufacturing systems) are also eliminated.

Microcasting is a non-transferred MIG (Metal Inert Gas) welding process, which deposits discrete, super-heated molten metal droplets to form dense metallurgical-bonded structures. Apart from the microcasting deposition process, several other alternative processes are also available to deposit a variety of materials in SDM. These processes and their processing materials are summarised in Table 4.9 [28].

Table 4.9. Deposition process in SDM [28].

Deposition Process	Description	Part Materials	Support Materials
Microcasting	An arc is established between a tungsten electrode of a MIG system and a feedstock wire that is fed from a charged contact tip. The wire melts in the arc, forming a molten droplet which, having accumulated sufficient molten material, falls from the wire to the part. The droplets remain super-heated and have enough energy to locally remelt the part to form metallurgical bonding on solidification. A laminar curtain of shielding gas prevents oxidation.	Stainless steel	Copper
Extrusion	Materials are deposited with an extruder with a single screw drive	Thermoplastics, ceramics	Water-soluble thermoplastics
Two-part resin system	Polyurethanes and epoxy resins are deposited as 2-part resin-activator systems	Polyurethanes, epoxy resins	Wax
Hot wax dispensation	Waxes are deposited with a hot-melt extrusion system. Waxes can be used as either part or support material	Wax	Wax, water-soluble photocurable resins
Photocurable resin dispensation	Photocurable resins are deposited with a simple syringe pumping system. This is usually used as support material for wax parts		Water-soluble photocurable resins

Table 4.9. *(Continued)* Deposition process on the SDM [28].

Deposition Process	Description	Part Materials	Support Materials
MIG Welding	MIG welding is used to deposit directly onto the part for fast deposition	Steel alloys	Copper
Thermal Spraying	Thermal spraying is used to spray thin layers of 'high performance' materials. Plasma sprayers are typically used	Metals, plastics, and ceramics	

4.6.5.3 *Strengths and weaknesses*

The strengths of Shape Deposition Manufacturing are as follows:

(1) *Wide variety of materials*. The process is capable of handling a wide variety of materials, including stainless steel, steel alloys, metals, thermoplastics, photocurable plastics, waxes, ceramics, etc. There is very little limitation or constraint on the type of materials that this process can handle.
(2) *Ability to build heterogeneous structures*. In addition to the additive manufacturing of complex shapes, the SDM process is also able to fabricate multi-material structures and it permits pre-fabricated components to be embedded within the built shapes. These provide the process with the unique capability of fabricating heterogeneous structures, which enables the manufacture of novel product designs.
(3) *Variable layer thickness*. As the CAD model of the part is decomposed into slices or segments that maintain the full 3D geometry of the outer surface, the layer thickness is varied. The actual layer thickness will depend on the local geometry of the part and the deposition process constraints.
(4) *Direct creation of functional metal shapes*. The process is one of only few that are able to directly create or fabricate functional metal shapes without the secondary processes needed in other AM systems.
(5) *Ease of creating undercut features*. The decomposition strategy used in the SDM process overcomes the many difficulties of

building undercuts that constrains many other AM systems. As such, highly complex shapes and forms can be built with little problem.

The weaknesses of Shape Deposition Manufacturing are as follows:

(1) *Need for precise control of automated robotic system.* The process requires moving the part from station to station (deposition, shaping, shot peening, etc.). Such a process demands precise control in placing the part accurately for each layer built. Invariably, because of movement, errors can accumulate over several thicknesses built and thus affect the overall dimensional accuracy of the part. Also, the control of the deposition process has to be precise to prevent the creation of voids, which causes excess remelting.
(2) *Thermal stresses due to temperature gradient.* The microcasting process of melting metals and precisely depositing them on the part to be solidified with the previous layer results in a steep temperature gradient. That introduces thermal stresses into the part and can result in distortion. Controlled shot peening can alleviate the problem but cannot entirely eliminate it. A careful balance of the temperature gradient, the associated internal stress build up, the thermal energy accumulation and their relief has to be achieved in order for SDM to produce quality precision parts.
(3) *Need for a controlled environment.* As several deposition and shaping processes can be incorporated into SDM, the environment within which the system functions has to be controlled. For example, MIG welding and other hot work can result in waste gas discharges, which have to be dealt with.
(4) *A large area is required for the system.* Due to the multiple capabilities of the SDM process and the several stations (deposition, shaping, etc) needed to accomplish the building of the parts, a relatively large area is necessary for the entire system. This means that the system is not suitable for any office environment.

4.6.5.4 *Applications*

With its capability to handle multiple materials and deposition processes, SDM can be applied to many areas in many industries, especially building finished parts and heterogeneous products. One example of finished parts is the direct building of ceramic silicone nitrite components for aircraft engines, which have been tested and remained unscathed on a jet engine rig at temperatures of up to 1250°C. The SDM process has also successfully fabricated heterogeneous products like an electronic device by building a non-conductive housing package and simultaneously embedding and interconnecting electronic components within the housing. This is especially useful in fabricating purpose-built devices for special applications like wearable computers. Many interesting heterogeneous products with multiple materials embedded within the product have also been tested [28].

References

[1] Crump, S. (1992). *The extrusion of fused deposition modeling, Proc. 3rd Int. Conf.* Rapid Prototyping, pp. 91 - 100.

[2] Stratasys Inc., *Transatlantic Copying from Xerox.* (2008). Retrieved from http://www.dimensionprinting.com/successstories/xerox.aspx

[3] Stratasys Inc., *Stratasys FDM Case Studies (Automotive)—Hyundai Mobis.* (2008). Retrieved from http://intl.stratasys.com/media.aspx?id=416

[4] Solidscape Inc., *3D Printing Advances Cerebral Aneurysm Research Case Study.* Retrieved from http://www.solid-scape.com/case-studies/research-medical/asu-advanced-research-case-study

[5] Solidscape Inc., *Miniature Masterpiece Design Powered by Solidscape Case Study.* Retrieved from http://www.solid-scape.com/miniature-masterpiece-design-powered-solidscape

[6] Wohlers, T. (2000). *Wohlers Report 2000: Rapid Prototyping and Tooling, State of the Industry.* Wohlers Association, Inc.

[7] Feygin, M. *Apparatus and Method for Forming an Integral Object from Laminations.* U.S. Patent No.4,752,352, 6/21/1988.

[8] Feygin, M. *Apparatus and Method for Forming an Integral Object from Laminations.* European Patent No. 0,272,305, 3/2/1994.

[9] Feygin, M. *Apparatus and Method for Forming an Integral Object from Laminations*. U.S. Patent No. 5,354,414, 10/11/1994.

[10] Industrial Laser Solutions. (2001). *Application Report, NASA Builds Large Parts Layer by Layer.* Retrieved from http://www.industrial-lasers.com/ archive/1999/05/0599fea3.html

[11] Wohlers, T. (2001). *Wohlers Report 2001: Industrial Growth, Rapid Prototyping & Tooling State of the Industry*. Wohlers Association, Inc.

[12] CAD/CAM Publishing, Inc. (2001). *Rapid Prototyping Report. Stratasys' New Insight Software*, 11(3).

[13] White, D. R., (2003). *Ultrasonic consolidation of aluminium tooling, Advanced Materials and Process,* Vol. 161, pp. 64-65.

[14] Pal, D., & Stucker, B. (2013). *A study of sub-grain formation in Al 3003 H-18 foils undergoing ultrasonic additive manufacturing using a dislocation density based crystal plasticity finite element frame work.* Journal of Applied Physics 113, 203517; doi: 10.1063/1.4807831.

[15] Dehoff, R. R. & Babu, S. S. (2010). *Characterization of interfacial microstructures in 3003 aluminum alloy blocks fabricated by ultrasonic additive manufacturing*. Acta Materialia, 58, pp. 4305 – 4315.

[16] Janaki Ram, G. D., Yang, Y., & Stucker, B. E. (2006). *Effect of process parameters on bond formation during ultrasonic consolidation of aluminum alloy 3003*. Journal of Manufacturing Systems, 25(3), pp. 221-238.

[17] Wohlers, T. (2004). *Wohlers Talk, Solidimension Finally on a Roll.* Retrieved from http://wohlersassociates.com/blog/2004/08/solidimension-finally-on-a-roll/

[18] BNET Business Network. (2005). *3D Systems Unveils its Plan to Launch the InVision LD 3-D Printer — a New Affordable Desk-Top 3-D Printer*. Retrieved from http://findarticles.com/p/articles/mi_m0EIN/is_2005_April_6/ai_n13559041

[19] Inui, E., Morita, S., Sugiyama, K., & Kawaguchi, N. (1994). SHAP - A Plain 3-D Printer/Plotter Process. In *Proceedings of the Fifth International Conference on Rapid Prototyping*, Dayton, Ohio, USA, 17-26.

[20] Kira Corporation. (2007). *The Katana Process and Feature*. Retrieved from http://www.rapidmockup.com/eg/menu2_5_e.htm

[21] Kira Solid Centre. (2001). *Characteristics: Why PLT Is So Good*. Retrieved from http://www.kiracorp.co.jp/EG/pro/rp/RP_characteristic.html

[22] CAM-LEM, Inc. (2005). *The CAM-LEM process*. Retrieved from http://www.camlem.com/camlemprocess.html

[23] Ennex Corporation. (2001). *All About Fabbers*. Retrieved from http://www.ennex.com/fabbers/index.sht

[24] Burns, M., Hayworth, K. J., & Thomas, C. L. (1996). Offset® Fabbing. In *Solid Freeform Fabrication Symposium*. University of Texas at Austin, USA.

[25] Merz, R., Prinz, F. B., Ramaswami, K., Terk, M., & Weiss, L. E. (1994). Shape Deposition Manufacturing. In *Solid Freeform Fabrication Symposium*, University of Texas at Austin, USA.

[26] *The Shape Deposition Manufacturing Process: Methodology*. (2001). Retrieved from http://www-2.cs.cmu.edu/~sdm/methodology.htm

[27] *The Shape Deposition Manufacturing Process*. (2001). Retrieved from http://www-2.cs.cmu.edu/~sdm

Problems

1. Describe the process flow of Cubic's Laminated Object Manufacturing.

2. Describe the process flow of Stratasys' Fused Deposition Modelling.

3. Compare and contrast the laser-based LOM™ process and the FDM systems. What are the advantages (and disadvantages) of each system?

4. Describe the critical factors that will influence the performance and functions of:

   (i) Stratasys' FDM,

   (ii) Solidscape's Benchtop System,

   (iii) Mcor Technologies' SDL,

   (iv) Cubic's LOM™,

   (v) Solidimension's PSL.

5. Compare and contrast Stratasys' FDM with Solidscape's Benchtop System. What are the advantages (and disadvantages) of each system?

6. Compare and contrast Stratasys' FDM with 3D Systems' liquid-based SLA® machines. What are the advantages (and disadvantages) of each system?

7. Compare and contrast Cubic's LOM™ process with Mcor Technologies' SDL. What are the advantages (and disadvantages) of each system?

8. What are the advantages and disadvantages of solid-based systems compared with liquid-based systems?

9. What do you think are the factors that limit the work volume of LOM™ systems?

Chapter 5

POWDER-BASED ADDITIVE MANUFACTURING SYSTEMS

This chapter describes the special group of solid-based additive manufacturing systems which primarily use powder as the basic medium for prototyping. Some of the systems in this group, such as Selective Laser Sintering, bear similarities to the liquid-based additive manufacturing systems described in Chapter 3, i.e. they generally have a laser to "draw" the part layer-by-layer, but the medium used for building the model is a powder instead of photocurable resin. Others, like colorjet printing has similarities with the solid-based additive manufacturing systems described in Chapter 4. The common feature among these systems described in this chapter is that the material used for building the part or prototype is invariably powder-based.

5.1 3D Systems' Selective Laser Sintering (SLS)

5.1.1 *Company*

3D Systems was founded by Charles W. Hull and Raymond S. Freed in 1986 commercialising the SLA systems. 3D Systems acquired DTM Corporation, the original company that first introduced the SLS® technology, in August 2001. DTM was first established in 1987. With financial support from BFGoodrich Company, and based on the technology that was developed and patented at the University of Texas, Austin, USA, DTM shipped its first commercial machine in 1992. It had worldwide exclusive license to commercialise the SLS® technology till its acquisition by 3D Systems. The address of 3D Systems' head office is 333 Three D Systems Circle Rock Hill, SC 29730 USA.

5.1.2 *Products*

3D Systems has introduced several generations of the SLS system. The current generation consists of the ProX 500 and sPro series printers. The ProX 500 is the latest printer developed by 3D Systems with SLS technology. It is designed to increase the productivity and precision of the machine. It uses DuraForm® ProX™ materials to produce high quality prototypes and parts in various areas. Furthermore, the sPro™ SLS systems can be used in several different models to achieve different quality and print speed. Table 5.1(a) summarises the specifications of the ProX 500 and sPro™ 60 SD, HD Base and HD-HS. Table 5.1(b) summarises the specifications of the sPro™ 140 Base, sPro™ 140 HS, sPro™ 230 Base, and sPro™ 230 HS. Fig. 5.1 shows the sPro™ 60 HD printer.

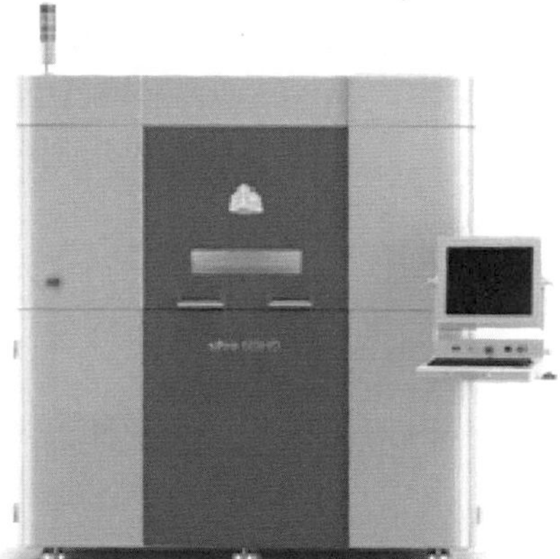

Fig. 5.1. sPro™ 60 HD printer. (Courtesy of 3D Systems)

Table 5.1. (a) Specifications of the ProX 500 and sPro™ 60 SD, HD Base and HD-HS. (Courtesy of 3D Systems)

Models	ProX 500	sPro™ 60 SD	sPro™ 60 HD Base	sPro™ 60 HD-HS
Build envelope capacity (XYZ)	15 x 13 x 18 in (381 x 330 x 457 mm)	15 x 13 x 18 in (381 x 330 x 437 mm), 15.2 U.S. gal (57.5 l)		
Powder layout	Variable Speed Counter Rotating Roller	Precision Counter Rotating Roller		

Table 5.1. (a) *(Continued)* Specifications of the ProX 500 and sPro™ 60 SD, HD Base and HD-HS. (Courtesy of 3D Systems)

Models	ProX 500	sPro™ 60 SD	sPro™ 60 HD Base	sPro™ 60 HD-HS
Layer thickness range (typical)	0.003 - 0.006 in (0.08 - 0.15 mm) (0.004 in, 0.10 mm)	0.003 in (Min 0.08 mm); Max 0.006 in (0.15 mm), (0.004 in, 0.10 mm)		Min 0.003 in (0.08 mm); Max 0.06 in (0.15 mm), (0.0047 in; 0.1 & 0.12 mm)
Imaging System	ProScan DX Digital High Speed	High Torque Scanning Motors (analogue)	ProScan™ CX (digital)	ProScan™ DX Dual Mode High Speed (digital)
Scanning speed	Fill - 500 in/s (12.7 m/s) Outline - 200 in/s (5 m/s)	240 in/s (6 m/s)	200 in/s (5 m/s)	240 in/s and 480 in/s (6 m/s and 12 m/s)
Laser power / type	100 W / CO_2	30 W / CO_2	30 W / CO_2	70 W / CO_2
Volume build rate	2 L/hr	0.9 L/hr	1.0 L/hr (60 cu in/hr)	1.8 L/hr (110 cu in/hr)
Electrical Requirements	208 VAC/7.5 kVA, 50/60 Hz, 1PH	240 V/12.5 kVA, 50/60 Hz AC 50/60 Hz, 3-phase (System)		
System Warranty	-	One-year warranty, under 3D Systems' purchase terms and conditions		

Table 5.1. (b) Specifications of the sPro 140 Base, sPro 140 HS, sPro 230 Base, and sPro 230 HS. (Courtesy of 3D Systems)

Models	sPro™ 140 Base	sPro™ 140 HS	sPro™ 230 Base	sPro™ 230 HS
Build envelope capacity (XYZ)	22 x 22 x 18 in (550 x 550 x 460 mm), 8,500 cu in (139 l)		22 x 22 x 30 in (550 x 550 x 750 mm), 13,900 cu in (227 l)	
Powder layout	Precision Counter Rotating Roller			
Layer thickness range (typical)	0.003 in (Min 0.08 mm); Max 0.006 in (0.15 mm), (0.004 in, 0.10 mm)			
Imaging System	ProScan™ Standard Digital Imaging System	ProScan™ GX Dual Mode High Speed Digital Imaging System	ProScan™ Standard Digital Imaging System	ProScan™ GX Dual Mode High Speed Digital Imaging System
Scanning speed	400 in/s (10 m/s)	400 in/s (15 m/s) (600 and 400 in/s)	400 in/s (10 m/s)	400 in/s (15 m/s) (600 and 400 in/s)

Table 5.1. (b) *(Continued)* Specifications of the sPro 140 Base, sPro 140 HS, sPro 230 Base, and sPro 230 HS. (Courtesy of 3D Systems)

Models	sPro™ 140 Base	sPro™ 140 HS	sPro™ 230 Base	sPro™ 230 HS
Laser power / type	70 W / CO_2	200 W / CO_2	70 W / CO_2	200 W / CO_2
Volume build rate	3.0 L/hr (185 cu in/hr)	5.0 L/hr (300 cu in/hr)	3.0 L/hr (185 cu in/hr)	5.0 L/hr (300 cu in/hr)
Electrical Requirements	208 V/ 17kVA, 50/60 Hz AC 50/60 Hz, 3-phase (System)			
System Warranty	One-year warranty, under 3D Systems' purchase terms and conditions			

5.1.3 *Process*

5.1.3.1 *The SLS® process*

The SLS® process creates 3D objects, layer-by-layer, from CAD data using powdered materials with heat generated by a CO_2 laser within the SLS machine. CAD data files in the .STL file format are first transferred to the SLS machine systems where they are sliced. From this point, the SLS® process (see Fig. 5.2) begins and operates as follows:

(1) A thin layer of heat-fusible powder is deposited onto the part-building chamber.
(2) The bottom-most cross-sectional slice of the CAD part to be fabricated is selectively "drawn" (or scanned) on the layer of powder by a CO_2 laser. The interaction of the laser beam with the powder elevates the temperature to the point of melting, fusing the powder particles to form a solid mass. The intensity of the laser beam is modulated to melt the powder only in areas defined by the part's geometry. Surrounding powder remains a loose compact and serves as natural supports.
(3) When the cross-section is completely "drawn," an additional layer of powder is deposited via a roller mechanism on top of the previously scanned layer. This prepares the next layer for scanning.

(4) Steps 2 and 3 are repeated, with each layer fusing to the layer below it. Successive layers of powder are deposited and the process is repeated until the part is complete.

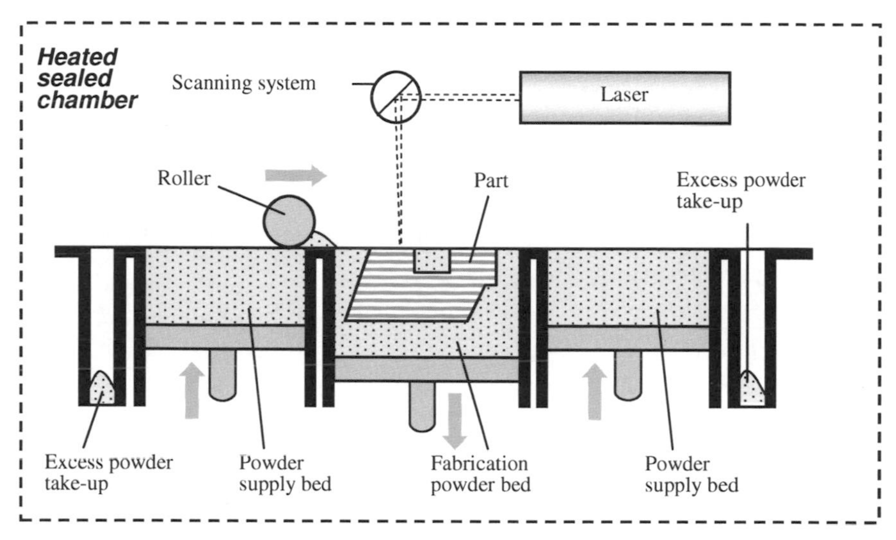

Fig. 5.2. Schematic of the Selective Laser Sintering (SLS) process.

As SLS® materials are in powdered form, the powder that is not melted or fused during processing serves as a customised, inherent built-in support structure. Thus, there is no need to create additional support structures within the CAD design and therefore no post-build removal of these supports is required. After the SLS® process, the part is removed from the build chamber and the loose powder simply falls away. SLS® parts may then require some postprocessing or secondary finishing, such as sanding, lacquering and painting, depending on the application of the prototype built.

The sPro SLS® system contains these hardware components [1]:

(1) *Sinterstation Pro SLS System* - Manufactures part(s) from 3-D CAD data

(2) *Rapid Change Module (RCM)* – Build module mounted on wheels for quick and easy transfer between the Sinterstation, the Offline Thermal Station (OTS) and the Break Out Station (BOS).
(3) *Nitrogen Generator* – Delivers a continuous supply of nitrogen to the SLS system to keep the fabrication inert and prevents oxidation.
(4) *Offline Thermal Station (OTS)* – Pre-heats the RCM before it is loaded into the SLS system and controls the RCM cool down process after a build has been completed.
(5) *Break Out Station (BOS)* – The built parts are extracted from the powder cake here. The non-sintered powder automatically gets sifted and transferred to the IRS.
(6) *Integrated Recycling Station (IRS)* – The IRS automatically blends recycled & new powder. The mixed powder is automatically transferred to the SLS system.
(7) *Intelligent Powder Cartridge (IPC)* – New powder is loaded into the IRS from a returnable powder cartridge. When the IPC is connected to the IRS, electronic material information is automatically transferred to the SLS system.

The software and system controller for Sinterstation® HiQ™ Series SLS® System includes the proprietary SLS system software working on Microsoft's Windows XP operating system. The software that comes with Sinterstation® Pro SLS® system includes the following:

- Build Setup and Sinter (included)
- Sinterscan™ (optional) software provides more uniform properties in X- and Y-directions and improved surface finish
- RealMonitor™ (optional) software provides advanced monitoring and tracking capabilities

5.1.3.2 *Materials*

The materials used in SLS® system can be broadly classified into two groups: DuraForm® materials and CastForm™ materials [2].

The DuraForm® group consists of eight types of materials: DuraForm® GF material, DuraForm® PA material, DuraForm® EX material, DuraForm® Flex plastic DuraForm® FR 100 material, DuraForm® HST Composite material, DuraForm® ProX™ material and DuraForm® AF plastic. These are discussed in the following:

(1) *DuraForm® GF plastic:* These are glass-filled polyamide (nylon) material for tough real-world physical testing and functional applications. The features of the material are as follows: excellent mechanical stiffness, elevated temperature resistance, dimensional stability, easy-to-process, and relatively good surface finish. Applications for the material include housings and enclosures, consumer sporting goods, low to medium batch size manufacturing, functional prototypes, parts requiring stiffness, and thermally stressed parts.

(2) *DuraForm® PA plastic:* These are durable polyamide (nylon) material for general physical testing and functional applications. The features of the material are as follows: excellent surface resolution and feature detail, easy-to-process, compliant with USP Class VI testing, compatible with autoclave sterilisation, and good chemical resistance and low moisture absorption. Applications for the material include producing complex, thin wall ductwork e.g. motorsports, aerospace, impellers and connectors, consumer sporting goods, vehicle dashboards and grilles, snap-fit designs, functional prototypes that approach end-use performance properties, medical applications requiring USP Class VI compliance, parts requiring machining or joining with adhesives.

(3) *DuraForm® EX plastic:* These are impact-resistant plastic offering the toughness of injection-moulded thermoplastics and are suitable for rapid manufacturing. They are available in either natural (white) or black colours. Features of the material are: it offers the toughness and impact resistance of injection-moulded ABS and polypropylene. Applications for the material include complex, thin-walled ductwork, motorsports, aerospace and unmanned air vehicles (UAV's), snap-fit designs, hinges, vehicle dashboards, grilles and bumpers.

(4) *DuraForm® Flex plastic:* This is a thermoplastic elastomer material with rubber-like flexibility and functionality. Features of the material are as follows: flexible, durable with good tear resistance, variability of Shore A hardness using the same material, good powder recycle characteristics, good surface finish and feature detail. The applications for the material include athletic footwear and equipment, gaskets, hoses and seals, simulated thermoplastic elastomer, cast urethane, silicone and rubber parts.
(5) *DuraForm® AF plastic:* These are polyamide (nylon) material with metallic appearance for real-world physical testing and functional use. Features of the material are as follows: metallic appearance with nice surface finish, good powder recycle characteristics, excellent mechanical stiffness, easy-to-process, and dimensional stability. Applications for the material include housings and enclosures, consumer products, thermally stressed parts, and plastic parts requiring a metallic appearance.
(6) *DuraForm® FR 100 material:* This is a halogen and antimony-free, flame retardant engineering plastic. It is suitable for AM of aerospace parts and parts requiring UL 94-V-0 compliance.
(7) *DuraForm® HST Composite material:* This is a fibre-reinforced engineering plastic with stiffness, strength and temperature resistance.
(8) *DuraForm® ProX™ material:* This is an extra-strong engineered production plastic. It is used to produce durable functional prototypes with superior mechanical properties. It was developed in tandem with ProX 500 printer to print smoother wall surfaces of injection moulding-like part quality.

Lastly, the CastForm™ PS material: This directly produces complex investment casting patterns without tooling. Features of this 'foundry friendly' material includes: functions like foundry wax, low residual ash content (less than 0.02%), short burnout cycle, easy-to-process plastic, and good plastic powder recycle characteristics. Applications of the material include creating complex investment casting patterns, indirectly producing reactive metals like titanium and magnesium, near net-shaped components, low melt-temperature metals such as aluminium, magnesium and zinc, ferrous and non-ferrous metals. Smaller parts can be joined to create very large patterns, and sacrificial, expendable patterns.

5.1.4 *Principle*

The SLS® process is based on the following two principles:

(1) Parts are built by sintering when a CO_2 laser beam hit a thin layer of powdered material. The interaction of the laser beam with the powder raises the temperature of the powder to the point of melting, resulting in particle bonding, fusing the particles to themselves and the previous layer to form a solid. This is the basic principle of sinter bonding.
(2) The building of the part is done layer-by-layer. Each layer of the building process contains the cross-sections of one or many parts. The next layer is then built directly on top of the sintered layer after an additional layer of powder is deposited via a roller mechanism.

The packing density of particles during sintering affects the part density. In studies of particle packing with uniform sized particles [3] and particles used in commercial sinter bonding [4], packing densities are found to range typically from 50% to 62%. Generally, the higher the packing density, the better will be the expected mechanical properties. However, it must be noted that scan pattern and exposure parameters are also major factors in determining the part's mechanical properties.

5.1.4.1 *Sinter bonding*

In the process of sinter bonding, particles in each successive layer are fused to each other and to the previous layer by raising their temperature with the laser beam to above the glass-transition temperature. The glass-transition temperature is the temperature at which the material begins to soften from a solid state to a jelly-like condition. This often occurs just prior to the melting temperature at which the material will be in a molten or liquid state. As a result, the particles begin to soften and deform owing to its weight and cause the surfaces in contact with other particles or solid to deform and fuse together at these contact surfaces. One major advantage of sintering over melting and fusing is that it joins powder particles into a solid part without going into the liquid phase, thus avoiding the distortions caused by the flow of molten material during fusing. After cooling, the powder particles are connected in a matrix that has approximately the density of the particle material.

As the sintering process requires the machine to bring the temperature of the particles to the glass-transition temperature, the energy required is considerable. The energy required to sinter bond a similar layer thickness of material is approximately between 300 to 500 times higher than that required for photopolymerisation [5, 6]. This high power requirement can be reduced by using auxiliary heaters to raise the powder temperature to just below the sintering temperature during the sintering process. However, an inert gas environment is needed to prevent oxidation or explosion of the fine powder particles. Cooling is also necessary for the chamber gas.

The parameters which affect the performance and functionalities are the properties of the powdered materials and its mechanical properties after sintering, the accuracy of the laser beam, the scanning pattern, the exposure parameters and the resolution of the machine.

5.1.5 *Strengths and Weaknesses*

The strengths of the SLS® process include:

(1) *Good part stability.* Parts are created within a precise controlled environment. The process and materials provide for functional parts to be built directly.
(2) *Wide range of processing materials.* In general, any material in powder form can be sintered on the SLS. A wide range of materials including nylon, polycarbonates, metals and ceramics are available directly from 3D Systems, thus providing flexibility and a wide scope of functional applications.

(3) *No part supports required.* The system does not require CAD-developed support structures. This saves the time required for support structure building and removal.
(4) *Little postprocessing required.* The finish of the part is reasonably fine and requires only minimal postprocessing such as particle blasting and sanding.
(5) *No post-curing required.* The completed laser sintered part is generally solid enough and does not require further curing.
(6) *Advanced software support.* The New Version 2.0 software uses a Windows NT-style graphical user interface (GUI). Apart from the basic features, it allows for streamlined parts scaling, advanced non-linear parts scaling, in-progress part changes, build report utilities and is available in foreign languages [7].

The weaknesses of the SLS® process include:

(1) *Large physical size of the unit.* The system requires a relatively large space to house it. Apart from this, additional storage space is required to house the inert gas tanks that are required for each build.
(2) *High power consumption.* The system requires high power consumption due to the high wattage of the laser required to sinter the powder particles together.
(3) *Poor surface finish.* The as-produced parts tend to have poorer surface finish due to the relatively large particle sizes of the powders used.

5.1.6 *Applications*

The SLS® system can produce a wide range of parts in a broad variety of applications, including the following:

(1) *Concept models.* Physical representations of designs used to review design ideas, form and style.
(2) *Functional models and working prototypes.* Parts that can withstand limited functional testing, or fit and operate within an assembly.
(3) *Polycarbonate (RapidCasting™) patterns.* Patterns produced using polycarbonate, and then cast in the metal of choice through the

standard investment casting process. These build faster than wax patterns and are ideally suited for designs with thin walls and fine features. These patterns are also durable and heat resistant.

(4) *Metal Tools (RapidTool™).* Direct rapid prototype of tools of moulds for small or short production runs.

(5) *Aerospace ducting.* With parts and components with high precision and high strength produced by SLS® in ducting system, additive manufacturing's products are already been used in many different aircrafts.

5.1.7 *Examples*

5.1.7.1 *SLS® manufactures custom surgical platforms*

FHC is the global leader and manufacturer of neuroscience. In 2012, neurosurgeons depended on FHC equipment for intraoperative micro-electrode recoding and stimulation, mapping the electrical activity of the human brain in order to precisely locate dysfunctional areas prior to removing or modifying them. Since then, FHC's Stereotactic platform is the only product of its type and kind in the market worldwide.

Usually, FHC's customers in the US require a 72-hour turnaround time for each customised model. Hence, FHC uses 3D Systems SLS® System to produce each model quickly and efficiently within the turnaround time. Therefore, FHC can be guaranteed to deliver the products to consumers within 72 hours with the help of 3D Systems' SLS® Technology. SLS® technology also helps FHC to save costs when producing platforms and models compared to traditional injection moulding or urethane casting techniques.

5.1.7.2 *Los Angeles-based TESTA Architecture/Design utilises SLS for large-scale models of Carbon Tower Prototype*

In the "Extreme Textiles: Designing for High Performance" exhibition, the design team of TESTA Architecture/Design utilised SLS® systems to build the prototype of the Carbon Tower Prototype.

The Carbon Tower Prototype demonstrates a new construction system for an all-carbon and glass-fibre high-rise office building that is woven, knitted and braided. This 40-story prototype, engineered with ARUP software, is several times lighter and stronger than conventional steel and concrete structures. Given the intricate nature of the tower design, it was not possible to construct architectural models with traditional modelling materials and techniques. Using 3D Systems' SLS® technology, the very large model measuring 152.4 cm (60 in.) by 35.6 cm (14 in.) in diameter was fabricated in only five interlocking pieces from digital files prepared by TESTA [8].

It has been proven that using SLS to fabricate complex and intricate structures rapidly is possible. Fig. 5.3 shows such a prototype part of the Carbon Tower built with micro-detailed parts.

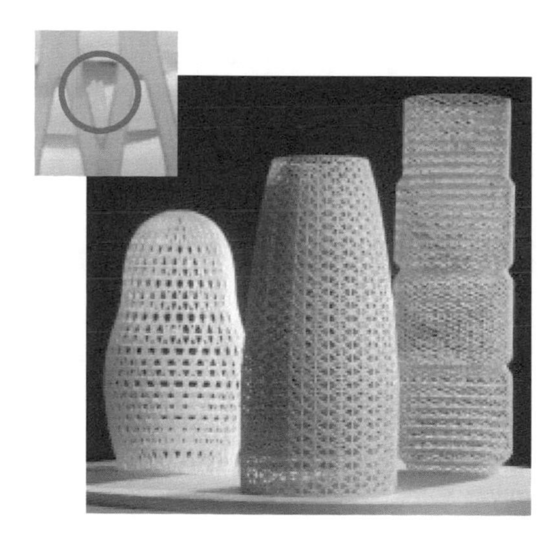

Fig. 5.3. Carbon Tower Prototype with micro-detailed parts and features. Inset shows the human figures built into the model to indicate the scale of model. (Courtesy of 3D Systems)

5.1.7.3 *Frontloading — The Crossblade outside mirror*

General contractors Bertrandt developed the Smart Crossblade for Smart GmbH in just six months, and built an exclusive series of 2000 with

Binz, their production partners [9]. The Smart Crossblade was designed based on the concept of a Smart convertible platform which resulted in a final car design that had no roof, no convertible top, no screens and no doors (see Fig. 5.4). The new concept raised a number of issues, one of which regarded the mirror system. The designer decided to develop a new mirror system to match this new concept. The SLS® system and DuraForm™ PA material were used in the building the model for mounting the upper and lower shells (holding the rear-view mirrors) and attaching them to the door pillars. This helped shorted the lead time required for the injection moulding tools.

As the development of the entire car was tight, it was necessary to perform functional testing, even before the injection moulding tool was made. A pendulum impact test ECE R 46 was required to investigate where the mirrors would fold when a collision occurred with a pedestrian in order to minimise injuries. A laser sintered part was built and positioned on the test rig. After a few minor modifications, the prototype passed the tests for the driver's side. Thus, actual injection-moulded parts were fabricated for the final ECE R 46 testing.

Fig. 5.4. The Smart Crossblade. The inserts show the model of the door pillar with SLS parts (white portion) as the mounting of the mirror base and the new mirror system of the Smart Crossblade. (Courtesy of 3D Systems)

5.1.7.4 *4D concepts for Hekatron*

Hekatron, a German-based company in Sulzburg, are makers of safety systems and OEM supplier for a string of renowned firms. The company developed a 220-volt socket for a smoke alarms and needed a several thousand units for initial sampling and market surveys. The real challenge turned out to be the plastic casing (90 mm diameter) and its intricate interiors with many openings, contra-rotating cores, thin-walled ribs — in short, a component with a highly complex geometry that needed to be made within the shortest possible time.

4D Concepts in Gross-Gerau was appointed by Hekatron to do the task. 4D Concepts used the Vanguard HS SLS system with LaserForm steel material, allowing them to make complicated and difficult tool geometry within a few weeks and at competitive terms. Just three weeks after receiving the 3-D CAD data (IGES) from Hekatron's development department, 4D Concepts delivered the first injection-moulded components. Using SolidWorks software, the provider first delivered the mould design drawn from the original data and then used their 3D Systems SLS system to make both mould halves in the dimensions 120 x 120 x 120 mm (see Fig. 5.5). The actual build time was less than 24 hours.

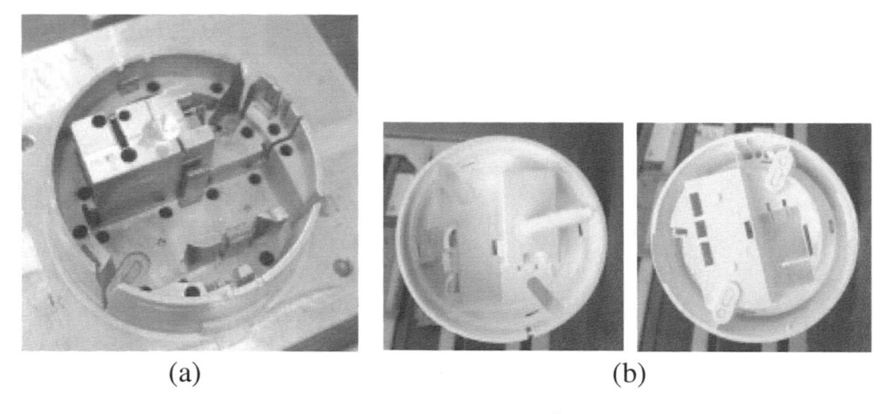

(a) (b)

Fig. 5.5. (a) Metal injection mould produced by SLS® and (b) Parts produced by the metal mould. (Courtesy of 3D Systems)

The Vanguard HS SLS system was used to build the required geometry layer-by-layer using a CO_2 laser with the LaserForm ST-200 steel powder. The sintered inserts were then infiltrated with bronze to obtain a dense mould. When compared directly with a conventionally made solid steel mould, the cost was lowered by a staggering 40 percent. The benefits of this method were particularly evident when it came to geometries requiring a great deal of erosion. [10].

5.1.8 *Research and Development*

Primary research continues to focus on new and advanced materials while further improving and refining SLS® process, software and system. Currently, ProX 500 is one of the latest SLS® machines while 3D Systems is still working to improve the build volume as well as to reduce the cost.

5.2 3D Systems' Colorjet Printing (CJP) Technology

5.2.1 *Company*

Originally invented, patented and developed at the Massachusetts Institute of Technology (MIT) in 1993, Three-Dimensional Printing technology (3DP™) forms the basis of Z Corporation's licensed prototyping process. Z Corp. pioneered the commercial use of 3DP technology, developing 3D printers that leading manufacturers used to produce early concept models and product prototypes for a broad range of applications.

Z Corporation was incorporated in 1994 by Hatsopoulos, Walter Bornhost, Tim Anderson and Jim Brett. It commercialised its first 3D printer, the Z™402 System, based on the 3DP technology in 1997. In 2000, Z Corp. launched its first colour 3D printer and subsequently introduced High Definition 3D Printing (HD3DP) in 2005. It was acquired by 3D Systems in April 2013 and 3D Systems' ProJet x60 series replaced the ZPrinter from Z Corp. The technology was also renamed as ColorJet Printing (CJP). 3D Systems' details are found in Section 3.1.1.

5.2.2 *Products*

3D Systems' current products are ProJet® 360, ProJet® 460*Plus,* ProJet® 660*Pro,* ProJet® 860*Pro* (see Fig. 5.6) and ProJet® 4500. ProJet® 360 is the enhancement of the previous ProJet® 160 and ProJet® 260*C*. It expands the build volume to 203 x 254 x 203 mm and reduces the cost of printing. Hence, it is often used to produce architectural modelling and medium-sized prototypes. ProJet® 460*Plus* further improves the machines with safer build materials, active dust control and zero liquid waste. Furthermore, ProJet® 660*Pro* has a larger building volume of 254 x 381 x 203 mm, incorporating 3D System's 4-channel CMYK full colour 3D printing to print high resolution prototypes. It is mainly used by designers and researchers to print professional models, art pieces and more. Meanwhile, ProJet® 860*Pro* is made to produce larger models. With ColorJet Printing (CJP) technology and the use of VisiJet® C4 Spectrum™ plastic material, ProJet® 4500 is developed to print at a faster speed and to save on materials. Table 5.2 summarises the specifications of ProJet® 360, ProJet® 460*Plus,* ProJet® 660*Pro,* ProJet® 860*Pro* and ProJet® 4500.

Fig. 5.6. ProJet® 860*Pro.* (Courtesy of 3D Systems)

Table 5.2. Specifications of ProJet® 360, ProJet® 460*Plus*, ProJet® 660*Pro*, ProJet® 860*Pro* and ProJet® 4500.

Models	ProJet® 360	ProJet® 460*Plus*	ProJet® 660*Pro*	ProJet® 860*Pro*	ProJet® 4500
Resolution	300 x 450 dpi	300 x 450 dpi	600 x 540 dpi	600 x 540 dpi	600 x 600 dpi
Colour	White (monochrome)	Full CMY	Full CMYK	Full CMYK	Continuous CMY
Pastel or vibrant colour options	No	No	Yes	Yes	-
Minimum Feature Size	0.006 in (0.15 mm)	0.006 in (0.15 mm)	0.004 in (0.1 mm)	0.004 in (0.1 mm)	0.004 in (0.1 mm)
Layer Thickness	0.004 in (0.1 mm)	0.004 in (0.1 mm)	0.004 in (0.1 mm)	0.004 in (0.1 mm)	0.004 in (0.1 mm)
Vertical Build Speed	0.8 in/hr (20 mm/hour)	0.9 in/hr (23 mm/hour)	1.1 in/hr (28 mm/hour)	0.2 - 0.6 in/hr (5 - 15 mm/hour); Speed increases with volume of prototypes	0.3 in/hr (8 mm/hr)
Prototypes per Build	18	18	36	96	18
Draft Printing Mode (Monochrome)	No	No	Yes	Yes	-
Net Build Volume (XYZ)	8 x 10 x 8 in (203 x 254 x 203 mm)	8 x 10 x 8 in (203 x 254 x 203 mm)	10 x 15 x 8 in (254 x 381 x 203 mm)	20 x 15 x 9 in (508 x 381 x 229 mm)	8 x 10 x 8 in (203 x 254 x 203 mm)
Build Materials	VisiJet® PXL™	VisiJet® PXL™	VisiJet® PXL™	VisiJet® PXL™	VisiJet® C4 Spectrum™
Number of Jets	304	604	1520	1520	
Number of Print Heads	1	2	5	5	
Automated Setup and Self-Monitoring	Yes	Yes	Yes	Yes	Yes
Core Recycling	Yes	Yes	Yes	Yes	Yes
Automatic Build Platform Clearing	Yes	Yes	Yes	Yes	Yes

Table 5.2. *(Continued)* Specifications of ProJet® 360, ProJet® 460*Plus,* ProJet® 660*Pro,* ProJet® 860*Pro* and ProJet® 4500.

Models	ProJet® 360	ProJet® 460*Plus*	ProJet® 660*Pro*	ProJet® 860*Pro*	ProJet® 4500
Fine Core Removal	Yes	Yes	Yes	Yes	Yes
Integrated Materials	Yes	Yes	Yes	Yes	Yes
Intuitive Control Panel	Yes	Yes	Yes	Yes	Yes
E-mail Notice Capability	Yes	Yes	Yes	Yes	Yes
Tablet/Smartphone Connectivity	Yes	Yes	Yes	Yes	Yes
Print3D App	Remote monitoring and control from tablet, computers and smartphones				
Input Data File Formats Supported	STL, VRML, PLY, 3DS, FRX, ZPR				STL, VRML, PLY, ZPR
Client Operating System	Windows® 7 and Vista®				
Operating Temperature Range	55 - 75°F (13 - 24 °C)	55 - 75°F (13 - 24 °C)	55 - 75°F (13 - 24 °C)	55 - 75°F (13 - 24 °C)	55 - 75°F (13 - 24 °C)
Operating Humidity Range	20 - 50% -non-cond.	20 - 50% -non-cond.	20 - 50% - non-cond.	20 - 50% - non-cond.	20 - 50% - non-cond.
Printer Dimensions	48 x 31 x 55 in (122 x 79 x 140 cm)	48 x 31 x 55 in (122 x 79 x 140 cm)	74 x 29 x 57 in (188 x 74 x 145 cm)	47 x 46 x 68 in (119 x 116 x 162 cm)	Printer Crated: 75 x 48 x 68 in (190 x 122 x 172 cm) Printer Uncrated: 64 x 60 x 31.5 in (162 x 80 x 152 cm)
Printer Weight	395 lbs (179 kg)	425 lbs (193 kg)	750 lbs (340 kg)	800 lbs (363 kg)	Printer Crated: 750 lbs (340 kg) Printer Uncrated: 600 lbs (272 kg)

Table 5.2. *(Continued)* Specifications of ProJet® 360, ProJet® 460*Plus*, ProJet® 660*Pro*, ProJet® 860*Pro* and ProJet® 4500.

Models	ProJet® 360	ProJet® 460*Plus*	ProJet® 660*Pro*	ProJet® 860*Pro*	ProJet® 4500
Electrical	90 - 100 V, 7.5 A 110 - 120 V, 5.5 A 208 - 240 V, 4.0 A	90 - 100 V, 7.5 A 110 - 120 V, 5.5 A 208 - 240 V, 4.0 A	100 - 240 V, 15 - 7.5 A	100 - 240 V, 15 - 7.5 A	100 - 240 V, 15 - 7.5 A
Noise					
Building	57 dB	57 dB	57 dB	57 dB	-
Core Recovery	66 dB	66 dB	66 dB	66 dB	-
Vacuum (open)	86 dB	86 dB	86 dB	86 dB	-
Fine Decoring	80dB	80dB	80dB	-	-
Office Compatibility	Yes	Yes	Yes	No	Yes

5.2.3 *Process*

3D Systems' ColorJet Printing (CJP) technology, formerly known as the 3D Printing (3DP) technology, creates 3D physical prototypes by solidifying layers of deposited powder using a liquid binder. The CJP process is shown in Fig. 5.7.

(1) The machine spreads a layer of powder from the feed box to cover the surface of the build piston. The printer then prints binder solution onto the loose powder, forming the first cross-section. For multi-coloured parts, each of the four print heads deposits a different colour binder, mixing the four colour binders to produce a spectrum of colours that can be applied to different regions of a part.

(2) The powder is glued together by the binder at where it is printed. The remaining powder remains loose and supports the following layers that are spread and printed above it.

(3) When the cross-section is complete, the build piston is lowered, a new layer of powder is spread over its surface, and the process is repeated. The part grows layer-by-layer in the build piston until the part is complete, completely surrounded and covered by loose

powder. Finally the build piston is raised and the loose powder is vacuumed away, revealing the complete part.

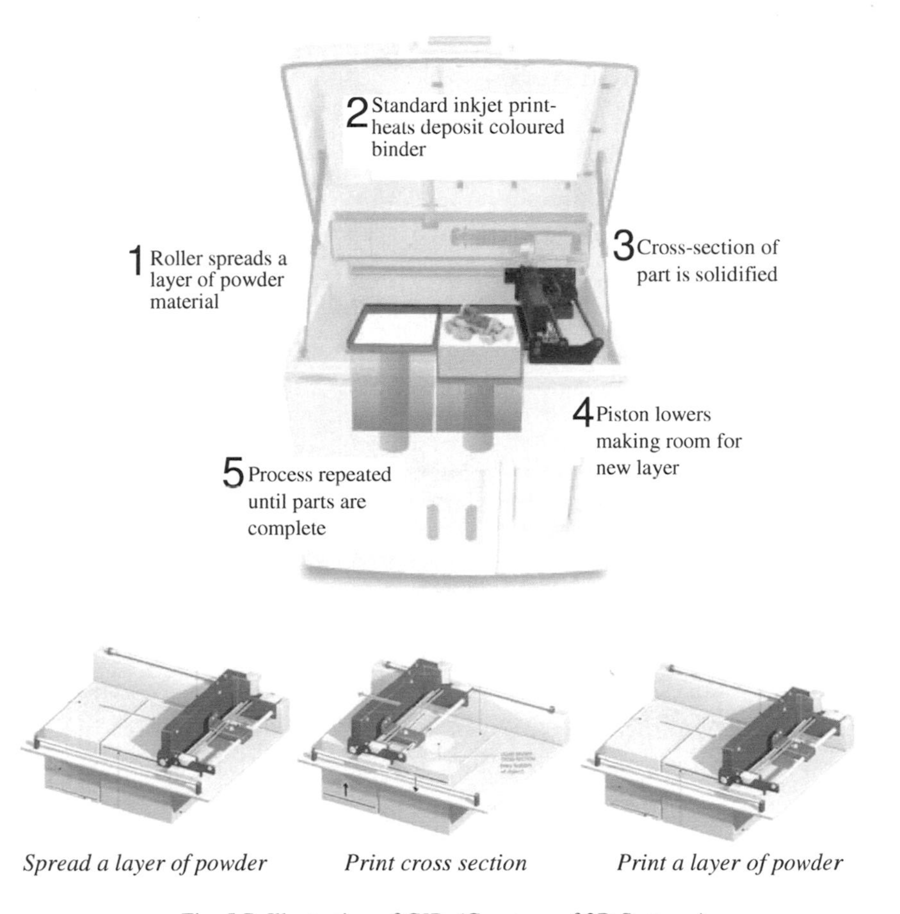

Fig. 5.7. Illustration of CJP. (Courtesy of 3D Systems)

(4) Once a build is complete, the excess powder is vacuumed away and the parts are lifted from the bed. Once removed, parts can be finished in a variety of ways to suit your needs. For a quick design review, parts can be left raw or “green.” To quickly produce a more robust model, parts can be dipped in wax. For a robust model that can be sanded, finished and painted, the part can be infiltrated with a resin or urethane.

5.2.3.1 *Materials*

With 3D Systems' CJP, VisiJet® PXL™ material is used by ProJet® x60 series 3D printers to produce strong, high performance and full colour models and prototypes.

5.2.4 *Principle*

3D Systems' CJP creates parts by a layered printing process and adhesive bonding, based on sliced cross-sectional data. A layer is created by adding another layer of powder. The powder layer is selectively joined where the part is to be formed by "inkjet" printing of a binder material. The process is repeated layer-by-layer until the part is complete.

As described in Section 5.1.4, the packing density of the powder particle has a profound impact on the results of the adhesive bonding which in turn affects the mechanical properties of the model. Like powders used on the SLS, packing densities are from 50% to 62% [4]. When the ink droplet impinges on the powder layer, it forms a spherical aggregate of binder and powder particles. Capillary forces will cause adjacent aggregates, including that of the previous layer, to merge. This will form the solid network which will result in the solid model. The binding energy for forming the solid comes from the liquid adhesive droplets. This energy is composed of two components, its surface energy and kinetic energy. As this binding energy is low, it is about 10 times [4] more efficient than sinter binding in converting powder to a solid object [3].

Parameters which influence the performance and functionalities of the process are the properties of the powder, the binder material, and the accuracy of the XY table and Z-axis control.

5.2.5 *Strengths and Weaknesses*

The key strengths of the CJP systems are:

(1) *High speed.* CJP are high speed printers. Each layer is printed in seconds, reducing the prototyping time of a hand-held part to 1 to 2 hours.
(2) *Versatile.* Parts are currently used for automotive, packaging, education, footwear, medical, aerospace, and telecommunications industries. Parts are used in every step of the design process for communication, design review and limited functional testing. Parts can be infiltrated if necessary, offering the opportunity to produce parts with a variety of material properties to serve a range of modelling requirements.
(3) *Simple to operate.* The office system is straightforward to operate and does not require a designated technician to build a part. The system is based on the standard, off the shelf components developed for the inkjet printer industry, resulting in a reliable and dependable 3D printer.
(4) *Minimal wastage of materials.* Powder that is not printed during the cycle can be reused.
(5) *Colour.* It enables complex colour schemes for AM parts from a full 24-bit palette of colours to be made possible.

The limitations of the CJP systems include:

(1) *Limited functional parts.* Relative to the SLS, parts built are much weaker, thereby limiting the functional testing capabilities.
(2) *Poor surface finish.* Parts built by CJP have relatively poorer surface finish and postprocessing is frequently required.

5.2.6 *Applications*

The CJP process can be used in the following areas:

(1) *Concept and functional models*. Creating physical representations of designs used to review design ideas, form and style. With the infiltration of appropriate materials, it can also create parts that are used for functional testing, fit and performance evaluation.
(2) *CAD-Casting metal parts*. CAD-Casting is a term used to connote a casting process where the mould is fabricated directly from a computer model with no intermediate steps. In this method, a ceramic shell with integral cores may be fabricated directly from a computer model. This results in tremendous streamlining of the casting process.
(3) *Direct metal parts*. Metal parts in a range of material including stainless steel, tungsten and tungsten carbide can be created from metal powder with CJP process. Printed parts are post-processed using techniques borrowed from metal injection moulding.
(4) *Structural ceramics*. CJP can be used to prepare dense alumina parts by spreading submicron alumina powder and printing a latex binder. The green parts are then isostatically pressed and sintered to densify the component. The polymeric binder is then removed by thermal decomposition.
(5) *Functionally gradient materials*. CJP can create composite materials as well. For example, ceramic mould can be 3D printed, filled with particulate matter, and then pressure infiltrated with a molten material. Silicon carbide reinforced aluminium alloys can be produced directly by 3D printing a complex SiC substrate and infiltrating it with aluminium, allowing localised control of toughness.

5.2.7 *Examples*

5.2.7.1 *3D Systems' ProJet® 660 the ultimate solution for Hankook Tire concept design*

Korea's Hankook Tire, founded in 1941, is one of the largest tire manufacturers in the world. Its products are sold in 185 countries around the world. Intense competition has made the company put more emphasis on their products design and development [11].

Hence, the company decided to invest in 3D Systems ProJet® 660 to create perfect full colour prototypes for its products to allow for full form and function testing. Designers are able to print the tire design overnight in the ProJet 660 before leaving work and the final model will be ready the next morning. Waiting times are significantly reduced with the ability to produce the models overnight. Communication between design and engineering departments has improved with the use of AM technology. With detailed, realistic 3D prints on-hand to touch, review and observe, errors are minimised and decision-making process are greatly enhanced. The company reported that meeting times are about 70% shorter than the past.

Prior to using AM, Hankook Tire was outsourcing part of their design mockups. These mockups were handcrafted and very rarely matched the original CAD design perfectly. Myungjoong Lee, CAD designer in Hankook Tire commented, "Now, the 3D print takes the exact dimensions of the CAD data and reproduces it perfectly."

Finally, the company can be assured that new design innovations are securely kept in-house without having to worry about intellectual property theft and abuse.

5.2.7.2 *Sports shoe industry*

The 3D printer has been used by designers, marketers, manufacturers, and managers in the footwear industry. Leading athletic shoe companies, such as adidas, New Balance and Wolverine, have used this AM system to radically reduce prototype development time and communicate in new ways. With the introduction of full colour printing, appearance of prototypes can be close to the actual product, so that more comprehensive communication on design can be done (see Fig. 5.8) [12]. Shoe industries are faced with constantly changing consumer preferences and have to react quickly to stay ahead of the business. With the 3D printer, lead times are drastically reduced, beating the competition to the shelves with the latest design trends while avoiding an excess inventory of unwanted designs.

Fig. 5.8. Actual production shoe with full colour prototype created with the CJP. (Courtesy of 3D Systems)

5.2.8 *Research and Development*

3D Systems is launching a consumer 3D printer that produces ceramic parts ready for firing and glazing. Using the CJP technology, the CeraJet is capable of producing intricate and detailed ceramic objects that many artists and designers would like to work with.

5.3 EOS' EOSINT Systems

5.3.1 *Company*

EOS was founded in 1989 and is the world's leading manufacturer of laser sintering systems today. Laser sintering is the key technology for e-Manufacturing Solutions which enabled the fast, flexible and cost-effective production of products, patterns or tools. While accelerating product development and optimising production processes, e-Manufacturing Solutions is able to fulfil the increasing demand for customised products, shifting from static methods to generative methods. EOS has sold more than 1000 laser sintering systems worldwide. It gained market leadership by assisting numerous industries in making use of laser sintering and setting up branches in Europe, North America and Asia. Its address is at EOS GmbH Electro Optical Systems, Robert-Stiring-Ring 1, D-82152 Krailling, Germany. Their e-mail is available at info@eos.info.

5.3.2 *Products*

5.3.2.1 *Models and specifications*

EOSINT systems share identical software but offer a different spectrum of building materials. The systems EOSINT P and FORMIGA P are specifically designed for fabrication of plastic parts. Another system, called EOSINT M, has been developed for sintering of metal powder. The latest EOSINT M 280 incorporates the leading-edge Direct Metal Laser Sintering (DMLS) technology. The EOSINT S system is used to build sand cores and moulds for metal casting. Table 5.3 shows the specifications of the EOS Additive Manufacturing systems.

Table 5.3. Specifications of EOSINT machines. (Courtesy of EOS GmbH)

Models	FORMIGA P 110	EOSINT P 396	EOSINT P 760	EOSINT M 280	EOSINT S 750
Effective building volume (mm)	200 x 250 x 330	340 x 340 x 600	700 x 380 x 580	250 x 250 x 325	720 x 380 x 380
Building speed (material dependent)	Up to 20 mm/h	Up to 48 mm/h	Up to 32 mm/h	-	Up to 2500 cm^3/h
Layer thickness (material dependent)	0.06 mm, 0.1mm, 0.12 mm	Typically 0.06 - 0.18 mm	Typically 0.06 - 0.18mm	-	0.2 mm
Laser type	CO_2, 30W	CO_2, 70W	CO_2, 2 x 50W	Yb-fibre laser, 200W or 400W (optional)	CO_2, 2 x 100W
Support Structure	Not required	Not required	Not required	Required	Required

Table 5.3. *(Continued)* Specifications of EOSINT machines. (Courtesy of EOS GmbH)

Models	FORMIGA P 110	EOSINT P 396	EOSINT P 760	EOSINT M 280	EOSINT S 750
Precision optics	F-theta lens	F-theta lens	F-theta lens	F-theta lens, high speed scanner	2 x F-theta lens, 2 x high speed scanner
Scan speed during build process	Up to 5 m/s	Up to 6 m/s	Up to 2 x 6 m/s	Up to 7.0 m/s	Up to 3.0 m/s
Power supply	16 A	32 A	32 A	32A	32 A
Power consumption	2 kW (nominal)	2.4 kW (nominal)/ 10 kW (max)	3.1 kW (nominal)/ 12 kW (max)	3.2 kW (nominal)/ 8.5 kW (max)	6 kW (avg), 12 kW (max)
Nitrogen generator	Integrated	Integrated (optional)	Integrated	Integrated	-
Compressed air supply	Min 6000 hPa; 10 m^3/h	Min 5000 hPa; 10 m^3/h	Min 6000 hPa; 20 m^3/h	7000 hPa; 20 m^3/h	Min 6000 hPa; 15 m^3/h
Recom-mended approximate installation space (mm)	3200 x 3500 x 3000	4300 x 3900 x 3000	4800 x 4800 x 3000	4800 x 3600 x 2900	4500 x 4600 x 2700
PC	Current Windows operating system				
Software	EOS RP Tools; Magics RP (Materialise); Desktop PSW; EOSTATE				
CAD interface	STL. Optional: converter to all common formats				
Network	Ethernet				

The FORMIGA P 110 (see Fig. 5.8 (a)) represents laser sintering in the compact class. With a build envelope of 200 mm x 250 mm x 330 mm, the FORMIGA P 110 produces plastic products from a wide range of plastic materials within a few hours and directly from CAD data. The machine is ideally suited for the economic production of small series and individualised products with complex geometry – requirements which apply among others to the medical device industry as well as for high value consumer goods [13]. At the same time, it provides capacity for the quick and flexible production of fully functional prototypes and patterns for plaster, investment and vacuum casting. With turnover times of less than 24 hours the FORMIGA P 110 integrates itself perfectly in a production environment that requires the highest level of flexibility. The system distinguishes itself also by comparatively low investment costs and being environmentally friendly.

EOSINT P 396 (see Fig. 5.8 (b)) offers economical solutions for a broad range of applications. It is a highly productive system for building serial components, spare parts, prototypes and models from digital 3D data within a few hours. The system processes the widest range of EOS materials within all P system series. EOSINT P 396 encompasses the innovative Blade Cartridge Concept from EOS which allows layer thickness to be varied swiftly during the building process. This ensures high productivity of the parts with maximal precision. It creates products without support structures which is both time and cost saving. The build volume of the machine is 340 mm x 340 mm x 600 mm which covers medium construction area.

The EOSINT P 760 (see Fig. 5.8 (c)) is one of the largest plastic laser sintering systems available in global market. The system uses premier EOS plastic to manufacture workpieces up to 700 mm x 380 mm x 580 mm with top-quality surface finishing well above the market standard. It achieves sustainability by using new material such as EOS material PrimePart PLUS (PA 2221) with maximum recyclability. The machine capacity is optimally utilised via the installation of the Cool Down station which permits the finished parts to cool down outside the machine while commencing a new building job immediately. Fig. 5.8 shows the (a) FORMIGA P 110, (b) EOSINT P 395 and (c) EOSINT P 760 machine.

EOSINT M 280 (see Fig. 5.9) is a state-of-art system which builds metal parts using Direct Metal Laser Sintering (DMLS). The technology fuses metal powder into a solid part by melting it locally using a focussed laser beam. The objects are built up additively layer-by-layer. Even parts with highly complex geometries including freeform surfaces, deep slots and coolant ducts could be created directly from CAD design data, fully automatically, in a few hours and without any tooling. It is a net-shape process, producing parts with high accuracy and detail resolution, good surface quality and excellent mechanical properties. The system is flexible in terms of materials that can be processed, ranging from light alloys to high-grade steels, tool steels and super alloys. Moreover, it is equipped with Laser Measurement Kit to monitor the quality of the whole building process and Comfort Power Module to allow convenient handling of metal powder in the enclosed process volume.

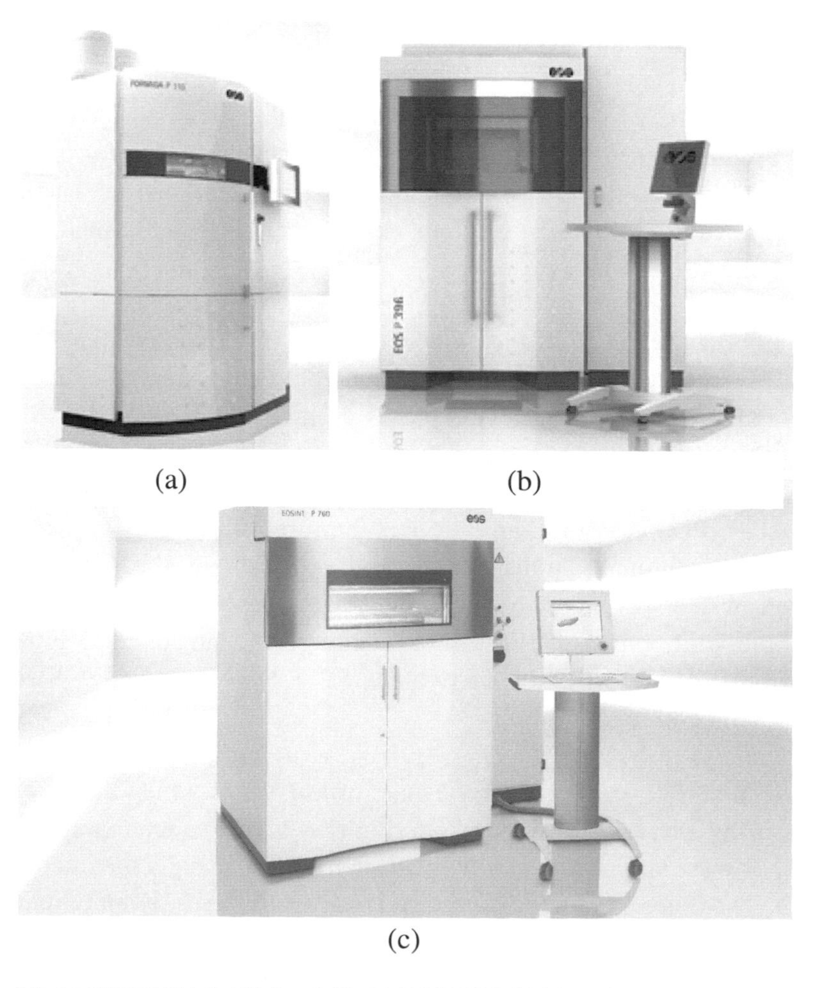

Fig. 5.8. (a) FORMIGA P 110 (top left), (b) EOSINT P 396 (top right) and (c) EOSINT P 760 (bottom). (Courtesy of EOS GmbH)

Fig. 5.9. EOSINT M 280 machine. (Courtesy of EOS GmbH)

The EOSINT S 750 (see Fig. 5.10) is the only double laser sintering system worldwide for the processing of croning moulding material. Using the DirectCast method, the system builds cores and moulds for sand casting directly from CAD data, fully automatically, with a building speed of up to 2,500 cm3/h (0.09 ft3/h.) and without any tooling. Sand parts of any complexity are built layer-by-layer, with high accuracy, detail resolution and surface quality. The maximum part size adds up to 720 mm x 380 mm x 380 mm. The resulting cores or core packages are realised with significant savings in time and costs compared to conventional technologies. They usually consist of fewer parts which are thus assembled faster and more precisely.

DirectCast with EOSINT S 750 enables the production of castings in batch sizes that would be extremely labourious, economically unviable or even impossible to manufacture with conventional techniques. In this way, high quality castings are produced for the engine development, for pumps or hydraulic applications. These castings can be used as fast, cost-effective prototypes or as final products in small series. The technology allows foundries to cater for new trends such as spare parts on demand. EOSINT S 750 uses different croning sands which are commonly used in foundries. These sands have been optimised by EOS for the DirectCast application. The system is ideally suited for building highly complex, detailed sand cores and moulds for premium castings in series quality. Laser sintering of foundry sand achieves excellent results for lightweight constructions using aluminium or magnesium. The technology also opens up new applications for cast iron and steel.

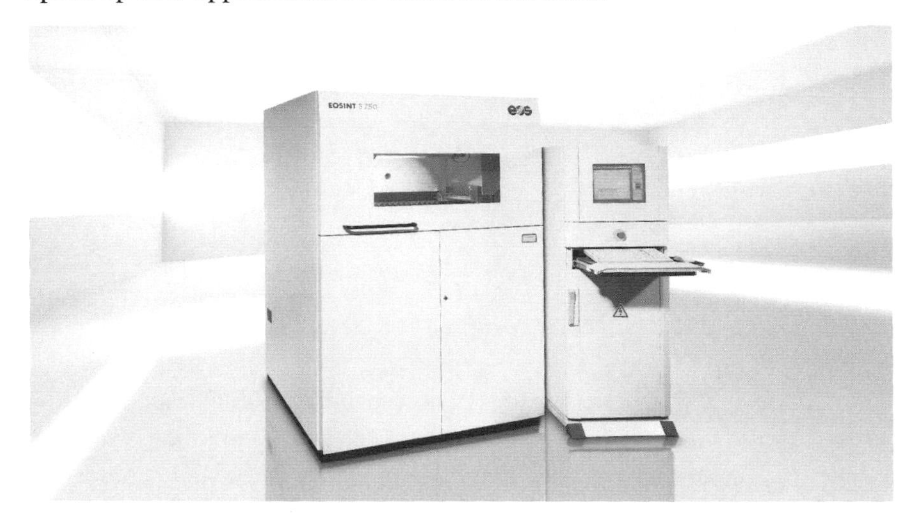

Fig. 5.10. EOSINT S 750 machine. (Courtesy of EOS GmbH)

5.3.2.2 *Materials*

•*Plastic*

(1) PA 2200/ 2201: Fine polyamide. Both are fine powders on the basis of polyamide 12. They share similar material properties such as high strength and stiffness, good chemical resistance, bio compatible and excellent long term constant behaviour etc. Typical applications of the materials include fully functional plastic parts with high end finish, medical applications such as prostheses and substitute typical injection moulding plastics etc.
(2) PrimePart PLUS (PA 2221): Fine polyamide based on polyamide 12. It is optimised for low warpage and minimum refresh rate which makes it an economical and ecological material. It allows products

to have a balanced relationship between mechanical strength and elasticity over a wide temperature range. Typical applications include fully functional parts of premier quality and substitute for typical injection moulding plastics.

(3) PA 2210 FR: Fine polyamide. It contains a chemical flame retardant free of halogens. In case of fire, a carbonating coating develops at the surface of the part, isolating the plastic below. Typical applications extend to aerospace and electric & electronics industry. It is best for parts with increased requirements on flame protection.

(4) PA 3200 GF: Glass-filled fine polyamide with excellent stiffness, wear resistance and thermal loadability. Typical applications of the material are stiff housings, thermally stressed parts and parts under frequent abrasion.

(5) EOS PEEK HP3: semi-crystalline Polyaryletherketone (PEAK). It is the world-first high performance polymer for the laser sintering process. With a tensile strength up to 95 MPa, a Young's modulus up to 4400 MPa and excellent temperature performance up to 260 °C, this thermoplastic material could be used for metal replacement, aerospace, automotive and motorsports and medical (ideal replacement for stainless steel and titanium) industries.

(6) Alumide®: Aluminium-filled fine polyamide characterised by its high stiffness, metallic appearance and good postprocessing possibilities. Typical applications for Alumide® are the manufacture of stiff parts of metallic appearance for applications in automotive manufacture (e.g. wind tunnel tests or parts that are not safety relevant), for tool inserts for injecting and moulding small production runs and for illustrative models (metallic appearance), etc. Surfaces of parts made of Alumide® can be finished by grinding, polishing or coating. An additional advantage is that low tool-wear machining is possible.

(7) CarbonMide®: Carbon-fibre reinforced polyamide. CarbonMide® has outstanding mechanical properties characterised by excellent stiffness and weight-strength-ratio. Typical applications of the material are fully functional prototypes with high end finish for wind tunnel tests or other aerodynamic applications.

(8) PrimeCast® 101: Polystyrene with good dimensional accuracy and a low melting point. One typical application of the material is the production of lost patterns for investment casting process. It is also suitable for plaster and ceramic shell casting and master patterns for vacuum casting.

•***Metal***

(1) Aluminium AlSi10Mg: Typical casting alloy. Due to good casting properties, it is used for cast parts with thin walls and complex geometry. The alloying elements silicon and magnesium promise high strength and hardness which ensure the cast parts are of good loading capability. Typical applications include direct manufacturing of functional prototypes, customised products or spare parts, aerospace and automotive industries and producing prototypes for aluminium die casting.
(2) EOS CobaltChrome MP1: Cobalt-chrome-molybdenum-based superalloy. It is characterised by excellent mechanical properties, corrosion resistance and temperature resistance. Typical applications include biomedical implants and high temperature engineering applications such as in aero engines.
(3) EOS CobaltChrome SP2: Cobalt-chrome-molybdenum-based superalloy. It is CE-certified metal material for series production by EOSINT M 270 system. With a biocompatible property combined with high tensile strength, high temperature and corrosion resistance, such superalloys are used for series production in dental industry.
(4) EOS MaragingSteel MS1: Martensite-hardenable steel. Its chemical composition corresponds to US classification 18% Ni Maraging 300, European 1.2709 and German X3NiCoMoTi 18-9-5. Characterised by excellent strength and high toughness, parts made of such steel are easily machinable and can be easily post-hardened to more than 50 HRC. It is commonly used for high performance series tooling.
(5) EOS NickelAlloy IN625: Nickel-chromium alloy. Its chemical composition corresponds to UNS N06625, AMS 5666F, AMS

5599G, W.Nr 2.4856, DIN NiCr22Mo9Nb. Characterised by high tensile, creep and rupture strength, it is suitable for building complex parts for high temperature and high strength applications. Typical applications include aerospace, motor sport and maritime industries.

(6) EOS NickelAlloy IN718: Nickel-based heat resistant alloy. Its chemical composition corresponds to UNS N07718, AMS 5662, AMS 5664, W.Nr 2.4668 and DIN NiCr19Fe19NbMo3. This kind of precipitation-hardening nickel-chromium alloy has good tensile, fatigue, creep and rupture strength at temperatures up to 700°C. The alloy also has outstanding corrosion resistance in various corrosive environments. Hence it is ideal for applications such as aero and land based turbine engine parts, rocket and space application components and chemical and process industry parts.

(7) EOS StainlessSteel GP1: Stainless steel. Its chemical composition corresponds to US classification 17-4, European 1.4542 and German X5CrNiCuNb16-4. It has good mechanical properties, especially excellent ductility in laser processed state. Typical applications include building parts such as functional metal prototypes, small series products, individualised products or spare parts.

(8) EOS StainlessSteel PH1: Stainless steel. Its chemical composition corresponds to 15-5 PH, DIN 1.4540 and UNS S15500. It possesses excellent mechanical properties, especially in the precipitation hardened state. It is widely used in a variety of medical, aerospace and other engineering applications requiring high hardness and strength.

(9) EOS Titanium Ti64: Ti6AlV4 alloy. This well-known light alloy is characterised by having excellent mechanical properties and corrosion resistance combined with low specific weight and biocompatibility. The ELI (extra-low interstitial) version has particularly low levels of impurities. Typical uses include those for aerospace and engineering applications and biomedical implants.

•*Sand*

(1) Foundry Sand: Ceramics 5.2 is a phenolic resin-coated aluminium silicate sand (synthetic mullite). The material is suited for generative fabrication of complex sand cores and sand moulds for all casting applications. Due to its high heat capacity and low temperature expansion this ceramic sand can especially be used for high temperature casting. Quartz 4.2 / Quartz 5.7 is a phenolic resin-coated quartz sand. The material is suited for generative fabrication of complex sand cores and sand moulds for all casting applications.

5.3.3 *Process*

The EOSINT process apply i.e. the following steps to creating parts, processing data, preparing new layer, scanning and removal of unsintered powder [14]:

(1) First, the part is created in a CAD system on a workstation. Then, the CAD data are processed by various software to control the sintering process. Specifically, orientation and positioning is tackled by Magics RP (from Materialise) while EOS RP tools convert the data into the cross-section format. The desktop process software (PSW) then combines the data in a construction order which is sent to laser sintering plant. To have an overview of the building process of multiple products, EOSTATE is employed for quality control.
(2) At the build stage, a new layer powder covers the platform. The laser scans the new powder layer and sinters the powder together according to the cross-sectional data. Simultaneously, the new layer is joined to the previous layer.
(3) When the sintering of the cross-section is complete, the platform lowers and another new layer is prepared for the next step. The processes are repeated till finally, the part is finished.
(4) After this, the powder around the part is removed.

The EOSINT system typically contains a Silicon Graphics workstation and an EOSINT machine including working platform, a laser, and an optical scanner calibration system.

5.3.4 *Principle*

The principle of the EOSINT systems is based on the laser sintering principle and layer manufacture principle similar to that of the SLS® described in Section 5.1.4.

The parameters that influence the performances and functionalities of the EOSINT systems are properties of powder materials, the laser and the optical scanning system, the precision of the working platform and working temperature.

5.3.5 *Strengths and Weaknesses*

The key strengths of the EOSINT systems are as follows:

(1) *Good part stability.* Parts are created in a precisely monitored environment. The process and materials provide for directly produced functional parts to be built.
(2) *Wide range of processing materials.* A wide range of materials including polyamide, glass-filled polyamide composite, polystyrene, metals and foundry sands are available thus providing flexibility and wide scope of functional applications.
(3) *Support structures not required.* The FORMIGA P and EOSINT P systems do not require support structures and the EOSINT S uses only simplified support structures as in the case of the Direct Croning Process®. This enhances efficiency by reducing processing time of the build.
(4) *Little postprocessing required.* The surface finishing of the built part is very satisfactory thus requiring only minimal postprocessing.
(5) *High accuracy.* For EOSINT P system, the polystyrene it uses can be laser sintered at a relatively low temperature, thereby causing low shrinkage and high inherent building accuracy.

(6) *Parts of various sizes can be built.* Machines with large build volume allow for relatively larger and taller parts to be built which saves the trouble of building smaller parts then joining together whereas systems with smaller build volume offer possibilities to build small parts as required.
(7) *High performance plastic products can be produced.* EOSINT P 800 offers the world's first laser sintering system for the A of high performance plastic products which exploits the possibilities of thermoplastics with laser sintering production to the full.

The limitations of the EOSINT systems are as follows:

(1) *Dedicated systems.* Only separate dedicated systems for plastic, metal and sand are available respectively.
(2) *High power consumption.* The EOSINT systems require relatively high laser power in order to directly sinter the metal powders.
(3) *Large physical size of the unit.* A relatively large space is required to house the system.

5.3.6 *Applications*

(1) *Concept models.* Physical representations are used to visualise design ideas, form and style.
(2) *Functional models and working prototypes.* Parts that can withstand limited functional testing, or fit and operate within an assembly. The system is suitable for the automotive, aerospace, machine tools and other industries for consumer products.
(3) *Wax and styrene cast patterns.* Patterns are produced in wax, and then cast in metal of choice using the investment casting process. Styrene patterns, on the other hand, can be evaporated pattern cast.
(4) *Metal tools.* The EOSINT M system is ideal for manufacturing rapid tooling [15]. It is used primarily for creating tools for investment casting, injection moulding and other similar manufacturing processes.
(5) *Biomedical implants.* Some of the EOSINT metal materials are certified by CE to be processed for biomedical implants.

5.3.7 *Examples*

5.3.7.1 *BEGO USA's use of Direct Metal Laser Sintering to manufacture dentures*

BEGO USA is a company in dental industry with a long history. In recent years, it succeeded in significantly increasing its productivity by employing laser sintering system from EOS. Shifting from the traditional lost-wax technique to Direct Metal Laser Sintering (DMLS), BEGO scaled up the production from 20 units per day to 450. In addition, DMLS is extremely precise with an accuracy of +/-20μm. The dentures produced by embracing this technology are durable, strong and of a consistently high quality [16].

5.3.7.2 *Bionic Gripper by Festo*

Festo, a world leading supplier of automation technology, employs the laser sintering process for the production of some of its bionic industrial applications. One example would be the DHDG adaptive gripper (See Fig. 5.11) which is manufactured on a FORMIGA P 100 provided by

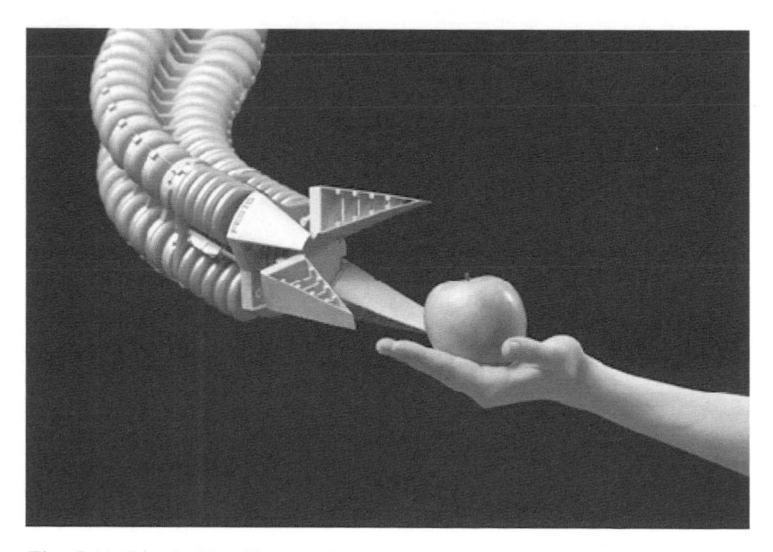

Fig. 5.11. Bionic Handling Assistant. (Courtesy of Festo AG & Co. KG)

EOS. The AM technology allows the gripper to be flexible yet stable enough to precisely adjust to the shape of the object including those with different contours. Laser sintering enables production in small batch while being cost-effective. Klaus Müller-Lohmeier, Head of Advanced Prototyping Technology at Festo AG & Co. KG was amazed that "its functionality and its structure that incorporates complex plastic components make it impossible to produce the high-tech arm without laser sintering [17]."

5.3.8 *Research and Development*

EOS is working on Micro Laser Sintering (MLS) which extends Additive Manufacturing to the Micro World. Since the global markets for miniaturised parts are growing, EOS has entered the "MicroForm" project to develop its technology on MLS [18]. The soon-to-be-launched EOSINT µ60 is believed to have the potential in offering new solutions in various fields including medical devices, automotive, jewellery, watchmakers and space research, etc.

5.4 Optomec's Laser Engineered Net Shaping (LENS®) and Aerosol Jet Systems

5.4.1 *Company*

Optomec is a privately held company founded in 1982, with corporate headquarters located in Albuquerque, New Mexico, and with an Advanced Applications Lab and New Product Development Centre located in Saint Paul, Minnesota. Since 1997, Optomec has focussed on commercialising LENS® additive manufacturing technology for 3D Printed Metals, which was originally developed by Sandia National Laboratories. Optomec delivered its first commercial LENS® system to Ohio State University in 1998. In 2004, Optomec released their first commercial Aerosol Jet® system for printed electronics. Optomec Inc. is located at 3911 Singer Boulevard, N.E., Albuquerque, NM. USA 87109.

5.4.2 *Products*

Optomec offers two product lines for additive manufacturing, LENS® systems for 3D Printed Metals and Aerosol Jet systems for printed electronics. LENS® systems use proprietary blown powder technology (see Section 5.4.2.2), to fabricate and repair high value metal components such as aircraft engine and industrial machinery parts. Aerosol Jet systems use proprietary gas based deposition technology (see Section 5.4.3.2) to print high resolution circuitry and components at the micron to centimetre scale on 2D and 3D substrates. LENS® and Aerosol Jet systems are used commercially today at over 150 companies and institutions. Fig. 5.12 provides an overview of Optomec LENS® and Aerosol Jet systems and the market applications they serve.

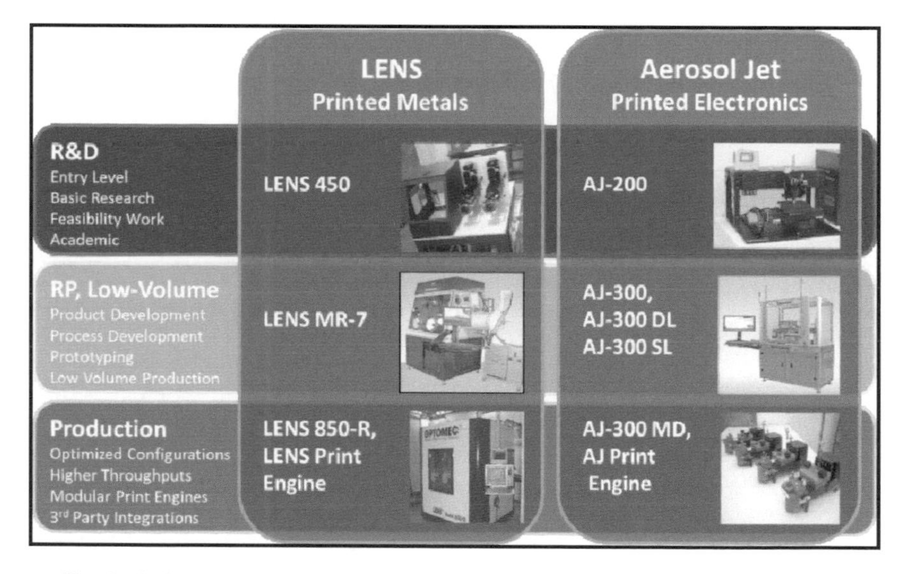

Fig. 5.12. Optomec Printers for R&D, RP and Production. (Courtesy of Optomec)

5.4.2.1 *LENS® systems*

Optomec offers three models of standard LENS® systems for 3D Printed Metals each targeted to meet specific market needs. The LENS® 450 system is ideally suited for R&D application such as basic research or educating students on metal additive manufacturing technology. The LENS® MR-7, with its larger build envelope and advanced processing features, provides an ideal platform for 3D Printing applications such as additive manufacturing, new alloy development and low volume production of metal parts. The LENS® 850-R is a state of the art additive manufacturing system ideally suited for the repair, rework, and modification of large, high value industrial components. In addition to these three standard turnkey systems, Optomec offers LENS® core hardware and software technology packaged as a Print Engine for integration with any machine tool or automation platform. The LENS® Engine and standard LENS® turnkey systems utilise the same industry proven LENS® additive manufacturing technology for printing and repairing 3D metal components. Table 5.4 compares the standard features available in each LENS® System model.

Table 5.4. LENS® System Models Comparison.

LENS® Features	Standard LENS® System Models		
	LENS® 450	LENS® MR-7	LENS® 850 R
Process	Blown Powder	Blown Powder	Blown Powder
Build Volume (mm)	100x100x100	300x300x300	900x1500x900
Isolated Pass Through Chamber	NA	Yes	Yes
Integrated Gas Purification System	NA*	Yes	Yes
Gas Recirculation System	NA	Yes	Yes
IPG Fibre Laser Power	400 W	500 W to 2kW	1 kW to 4kW
CNC Motion Control System	3 axis	3 axis	5 axis
Powder Feeders	1	2	2
System Control and Tool Path Software	Yes	Yes	Yes
Materials	See Table 5.5 for LENS® Process Supported Materials		

* LENS® 450 uses direct purge to achieve low oxygen levels

5.4.2.2 *LENS® process*

The LENS® process is housed in a hermetically-sealed chamber which is purged with argon so that the oxygen and moisture levels are controlled to stay below 10 parts per million. This keeps the part clean during the build process, preventing oxidation. The metal powder feedstock (see Table 5.5 for LENS® supported powdered metals) is delivered to the material deposition head by a proprietary powder-feed system which is able to precisely regulate mass flow. The LENS® process builds components in an additive manner from powdered metals using a fibre laser to fuse powder into a solid, as shown in Figure 5.13. It is a blown powder (aka freeform metal fabrication) process in which a fully dense metal component is formed. The LENS® process comprises the following steps:

(1) A deposition head supplies metal powder to the focus of a high-powered laser beam to be melted. This laser is typically directed by fibre optics or precision angled mirrors.

Table 5.5. LENS Powdered Metals.

LENS Materials Used Commercially			
Alloy Class	**Alloy**	**Alloy Class**	**Alloy**
Titanium	CP Ti	Tool Steel	H13, S7
	Ti 6-4	Stainless Steel	13-8, 17-4
	Ti 6-2-4-2		304, 316
Nickel	IN625		410, 420
	IN718	Aluminum	4047
	Waspalloy	Cobalt	Stellite 6, 21
	Rene 41	Carbide	Ni-WC
			Co-WC
LENS Materials Used in R&D			
Alloy Class	**Alloy**	**Alloy Class**	**Alloy**
Titanium	Ti 6-2-4-6	Tool Steel	A-2
	Ti 48-2-2	Stainless Steel	15-5PH
	Ti 22Al-23Nb		AM355
Nickel	IN690		309, 416
	Hastelloy X	Copper	GRCop-84
	MarM 247		Cu-Ni
	Rene 142	Refractories	W, Mo, Nb
Ceramics	Alumina	Composites	TiC, CrC

(2) The laser is focussed on a particular spot by a series of lenses, and a motion system underneath the platform moves horizontally, laterally and vertically as the laser beam traces the cross-section of the part being produced. LENS® 850R systems include two additional axis for tilting and rotating the part, especially useful for repair applications.

(3) When a layer is completed, the deposition head moves up and continues with the next layer. The process is repeated layer-by-layer until the part is completed. Generally the prototypes need additional finishing but are fully dense products with excellent metallurgical properties.

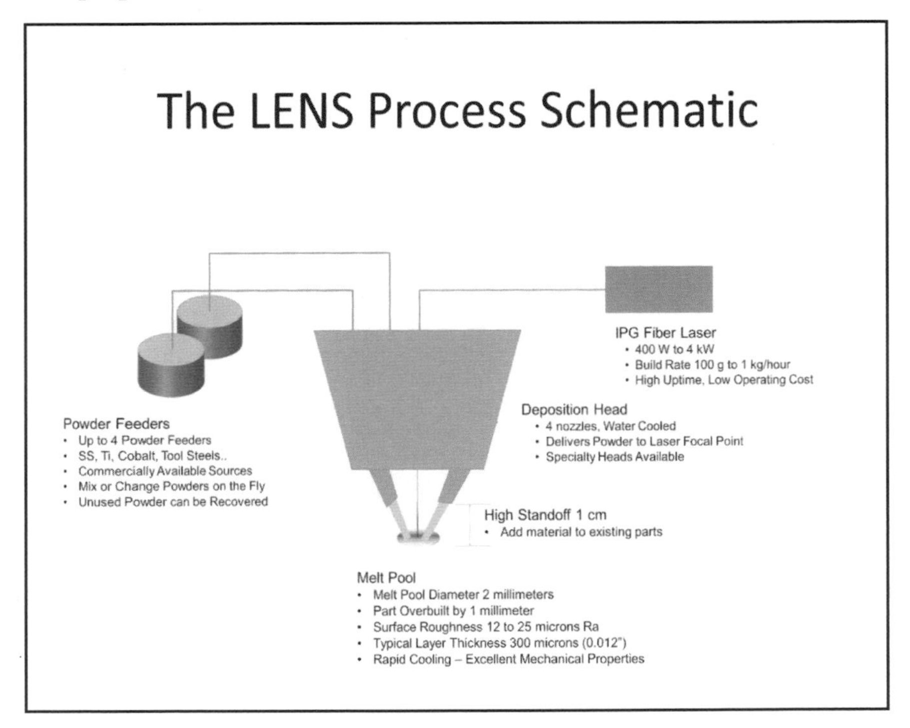

Fig. 5.13. LENS® Process.

5.4.2.3 *LENS® Principles*

The LENS® process is based on the following two principles:

(1) A high-powered fibre laser focussed onto a metal substrate creates a molten puddle on the substrate surface. Powder is continuously injected into the molten puddle to increase material volume.
(2) A "printing" motion system moves a platform horizontally and laterally as the laser beam traces the cross-section of the part being produced. After formation of a layer of the part, the machine's powder delivery nozzle moves upwards prior to building next layer.

5.4.2.4 *LENS® Strengths and weaknesses*

The key strengths of the LENS® systems are:

(1) *Superior material properties.* The LENS® process is capable of producing fully dense metal parts. Metal parts produced can also include embedded structures and superior material properties. Microstructure produced is excellent.
(2) *Complex parts.* Functional metal parts with complex features can be produced on the LENS® system.
(3) *Add material onto existing parts.* LENS® systems can not only fully build 3D metal parts, but also can add materials to existing parts. For example wear coatings, or to repair worn/damaged components, or to combine LENS® with conventional manufacturing methods to create unique hybrid manufacturing solutions.
(4) *Blend materials.* LENS® systems can blend multiple powders together during the build process to construct parts with functional gradients or new alloys tailored to meet specific requirements.
(5) *Rapid deposition rate.* Materials can be deposited at up to 1kg/hr.
(6) *Low powder cost.* The LENS® process uses a wide powder mesh range, and thus powder costs are minimised.

The primary limitations of the LENS® system are:

(1) *Geometrical complexity.* The LENS® process can manufacture relatively complex shapes, but is limited in complexity compared to casting or to powder bed processes.
(2) *Surface finish.* The surface finish of LENS®-produced materials tends to be rougher than powder bed processes

(3) *Dimensional accuracy.* Due to the large laser spot size, the melt pool is larger, and thus the dimensional accuracy not as good as powder bed techniques.

5.4.2.5 *LENS® applications:*

The LENS® technology can be used in the following areas:

(1) Building mould and die inserts
(2) Producing titanium parts in the race car industry
(3) Producing functionally gradient structures
(4) Component repair
(5) Producing titanium components for the aerospace industry
(6) Manufacturing cermets and carbon nanotube-reinforced nickel [19, 20]

Figure 5.14 highlights metallic automobile, medical and hybrid parts produced using LENS®. The LENS® system can be integrated with conventional processes to create unique hybrid manufacturing solutions. For example, LENS® can be used for feature enhancement to an existing component by adding layers of wear-resistant material or other surface treatments. LENS® can also produce value-added features on existing parts, such as adding a boss or flange to a large casting component.

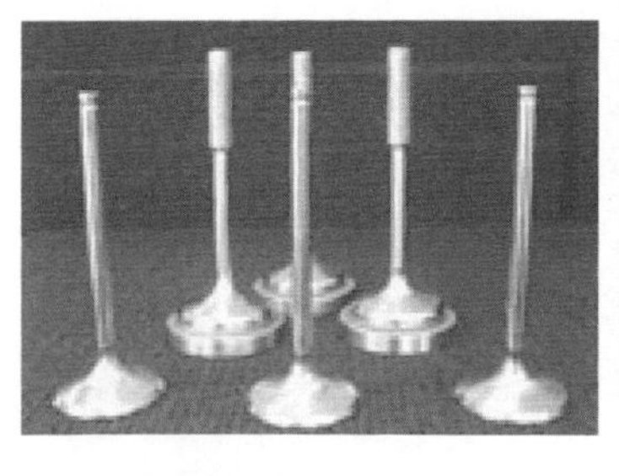

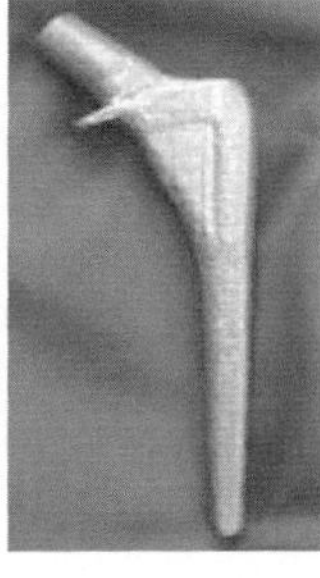

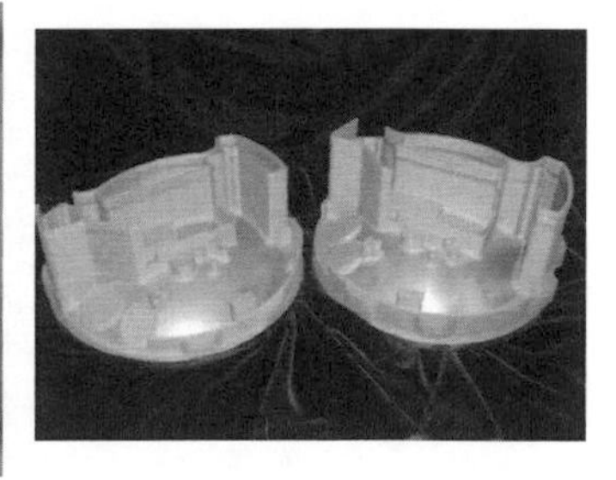

Fig. 5.14. Piston (left), medical (centre) and hybrid (right) metal parts produced with LENS®. (Courtesy of Optomec)

LENS® offers unique flexibility in the geometries and material range it supports, enabling the use of innovative design concepts such as hollow and internal structures, and functional material gradients that incorporate mechanical property transitions within a single part.

5.4.3 *Aerosol Jet Systems*

Optomec offers three models of standard Aerosol Jet systems for printed electronics each targeted to meet specific market needs. The Aerosol Jet 200 Series is a professional grade benchtop system ideally suited for universities, ink developers, and others exploring the benefits of additive manufacturing technology for electronic applications. The Aerosol Jet 300 System Series, with its higher print accuracy motion system and larger work area, is ideally suited for developing next generation printed electronics and biologics processes/prototypes as well as low volume production. For production level printed electronics, Optomec offers the Aerosol Jet Micro Dispense system for applications such as die & component attach, underfill and component encapsulation, and 3D printed interconnects. An Aerosol Jet Print Engine is also available for integration with industrial automation platforms for high volume production applications. The Aerosol Jet Print Engine and standard systems all utilise the same industry proven Aerosol Jet core deposition technology which enables processing of wide variety of electronic materials and printing of a wide range of feature sizes on both 2D and 3D substrates. Table 5.6 compares the standard features available in each Aerosol Jet System model.

Table 5.6. Aerosol Jet Systems Features Comparison.

	Aerosol Jet System Models		
Standard Features	AJ 200	AJ 300	AJ MD
Process	Aerosol	Aerosol	Aerosol
Build Volume (mm)	200x200	300x300	370x470
Atomisers	Ultrasonic, w/ Pneumatic option	Ultrasonic & Pneumatic	Ultrasonic & Pneumatic
Fine Feature Print Head	10 to 250 microns	10 – 250 microns	10 to 250 microns
Wide Feature Print Head	Option	Option	0.5mm to 2.0mm

Table 5.6. *(Continued)* Aerosol Jet Systems Features Comparison.

	Aerosol Jet System Models		
Standard Features	AJ 200	AJ 300	AJ MD
X,Y Motion Accuracy	+/- 25 microns	+/- 6 microns/axis	+/- 6 microns/axis
Safety Enclosure	No	Yes	Yes
Heated Platen	Vacuum & Heating	Vacuum & Temp Control	Vacuum & Temp Control
Lighting	LED	LED	Red, white or blue LED, with dynamic Control
Camera	1288 x 964 pixels	1288 x 964 pixels	1288 x 964 pixels
Laser for In Situ Sintering	No	Option	Option
System Control &Tool Path Software	Yes	Yes	Yes
Materials	See Figure 5.7 for Materials Supported by the AJ Process		

5.4.3.1 *Aerosol Jet process*

Aerosol Jet Systems atomises a source material into a dense aerosol mist composed of droplets with diameters between approximately 1 and 5 microns. The resulting aerosol stream contains particles that are further treated on the fly to provide optimum process flexibility. The material stream is then focussed using a flow guidance deposition head, which creates an annular flow of sheath gas to collimate the aerosol. The co-axial flow serves to focus the material stream. The resulting high velocity material stream is deposited onto planar and 3D substrates, creating features ranging from 10 microns to centimetres in size as shown in Fig. 5.15. A motion control system allows the creation of complex patterns on the substrate. For low temperature substrates, deposited material can be laser sintered to achieve properties near those of the bulk material without damage to the surrounding substrate. The Aerosol Jet process can print a wide variety of materials on a wide variety of substrates. Typically if a source material can be suspended in a solution and atomised, it can be printed with the Aerosol Jet process. Table 5.7 provides a list of inks that have been printed on various substrates using Aerosol Jet technology.

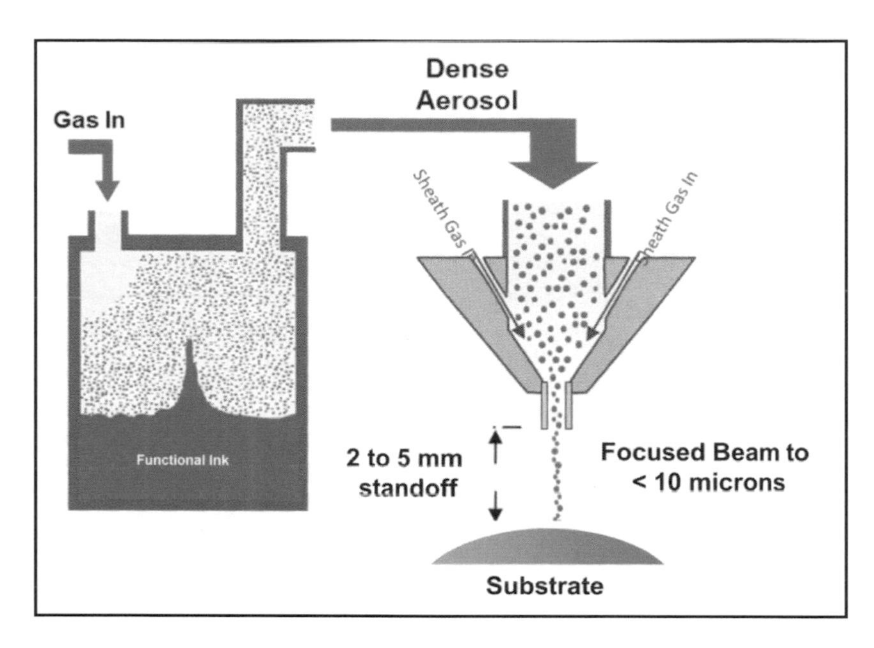

Fig. 5.15. Aerosol Jet Process with Ultrasonic Atomiser.

Table 5.7. Supported Materials using the Aerosol Jet Process.

<table>
<tr><th colspan="2" rowspan="2">Aerosol Jet Materials Matrix</th><th colspan="6">Substrates</th></tr>
<tr><th>Polyester</th><th>Polyimide</th><th>Glass</th><th>C-Si</th><th>ITO</th><th>Metals</th></tr>
<tr><td rowspan="2">Inks</td><td>Conductor</td><td>Ag
AU
PEDOT:P
SS
SWCNT</td><td>Ag
AU
Cu
SWCNT</td><td>Ag
Cu
PEDOT:P
SS
SWCNT</td><td>Fritted
Ag</td><td>Ag
AU
Al
PEDOT:
PSS</td><td>Ag
AU
Al
PEDOT:P
SS</td></tr>
<tr><td>Resistor</td><td>FTU
series</td><td>TU series
SWCNT
M-2031-
pol
PC11223</td><td>TU series
SWCNT
M-2031-
pol
PC11223</td><td></td><td></td><td>TU series</td></tr>
</table>

Table 5.7. *(Continued)* Supported Materials using the Aerosol Jet Process.

Aerosol Jet Materials Matrix		Substrates					
		Polyester	Polyimide	Glass	C-Si	ITO	Metals
Inks	Dielectric Adhesive	Loctite 3492 Norland 65 SU-8 GM 1010 Polyimide (various) Teflon AF Luvitec PVP Ablestik	Loctite 3492 Norland 65 Teflon AF Luvitec PVP	Norland 65 SU-8 GM 1010 Summers Optical CX-16 Teflon dispersion Luvitec PVP		Norland 65 Summer s Luvitec PVP	Norland 65 Loctite SU-8 GM 1010 Teflon AF Luvitec PVP
	Biological			Water based solvent based PLGA 5%			Water based solvent based PLGA 5%
	Etch Resists Catalysts Etch Chemicals				Enlight		SU-8 GM1010 Acids (various)
	Semi-conductor	P3HT SWCNT	P3HT SWCNT	Merck P3HT SWCNT	Enlight		SU-8 GM1010 Acids (various)

5.4.3.2 *Aerosol Jet's strengths and weaknesses*

The key strengths of Aerosol Jet systems are:

(1) *Planar and non-planar (3D) printing capabilities.* Enables functionalised circuitry on 3D substrates
(2) *Printed feature sizes ranging from 10 microns to centimetres.* Addresses shrinking electronic package size challenges and enables denser component spacing than previously possible

(3) *Thin layer deposits from 100 nanometres.* Targeted at precise coating and multi-material applications such as SOFC using less material than traditional manufacturing methods.
(4) *A wide range of inks and substrates.* From nanomaterials such as silver and gold, to polymers, epoxies, etchants, dopants, and biological materials have all been dispensed with Aerosol Jet onto substrates such as Kapton, polycarbonate, glass, silicon, fabrics, etc.
(5) *Low temperature processing.* Enable Aerosol Jet dispensing at ambient temperatures without the need for specialised chambers or temperature controlled environments.

The limitations of Aerosol Jet systems are:

(1) Requirement for particle sizes of <1 micron will limit use of some current conductive inks.
(2) Current nozzle spacing of ~2.5mm will limit use for certain applications.

5.4.3.3 *Aerosol Jet applications*

The Aerosol Jet system allows "high-tech" printed electronics applications such as:

- Fine Feature (10 micron) to Wide Feature Circuits (1 cm)
- Printed passive components (resistors, capacitors)
- Printed active components (antennas, sensors, thin film transistors)
- Printed batteries and fuel cells
- 3D Printed Electronics (stacked dies interconnects, cell phone antennas, sensors (See Figure 5.16)
- Printed biologics

Fully printed components and circuits eliminate the need for mounting and soldering physical devices to a carrier board. Using Aerosol Jet technology, components can even be printed directly onto or inside 3D structures, such as the wing of an aircraft or a cell phone case, thereby eliminating the need for separate circuit boards. This capability to print electronics on 3D surfaces opens up new packaging opportunities for lighter, thinner electronic devices with increased functionality.

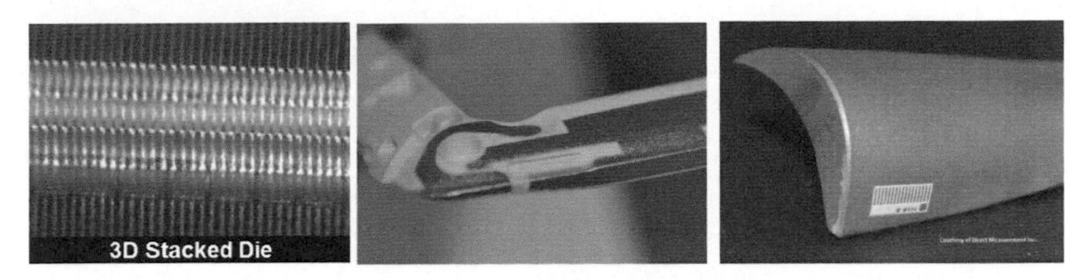

Fig. 5.16. 3D Interconnects (left), 3D Printed antenna (centre), 3D Creep Sensor on Ti Blade (right). (Courtesy of Optomec)

5.4.4 *Research and Development*

Optomec's R&D focus is on optimising LENS® and Aerosol Jet technology for high volume repair and manufacturing application. LENS® and Aerosol Jet Print Engine technologies are ideally suited for high volume production. For example, the Aerosol Jet Print Engine can be scaled to enable simultaneous printing with print heads on a single automation platform. Optomec users are currently qualifying Aerosol Jet Print Engines in high volume production. Similar activity is progressing with LENS® Print Engine technology.

5.5 Arcam's Electron Beam Melting (EBM)

5.5.1 *Company*

Arcam AB, a Swedish technology development company was founded in 1997. The company's main activity is concentrated in the development of the Electron Beam Melting (EBM) technique for the production of solid metal parts directly from metal powder based on a 3D CAD model. The fundamental development work for Arcam's technology began in 1995 in collaboration with Chalmers University of Technology in Gothenburg, Sweden. The Arcam EBM technology was commercialised in 2001. The company's address is Krokslätts Fabriker 27A, SE-431 37 Mölndal, Sweden.

5.5.2 *Products*

Arcam has introduced three main series of products, Arcam Q10 (see Figure 5.17), Arcam Q20 and Arcam A2X. The Arcam Q10 system is designed to produce orthopaedic implants with its high productivity and high resolution features. With the Arcam Q20, they targeted the aerospace market, racing industry and general industry with a machine fulfilling these industries' need for the production of larger components. The Arcam A2X system on the other hand, process materials that require high process temperatures.

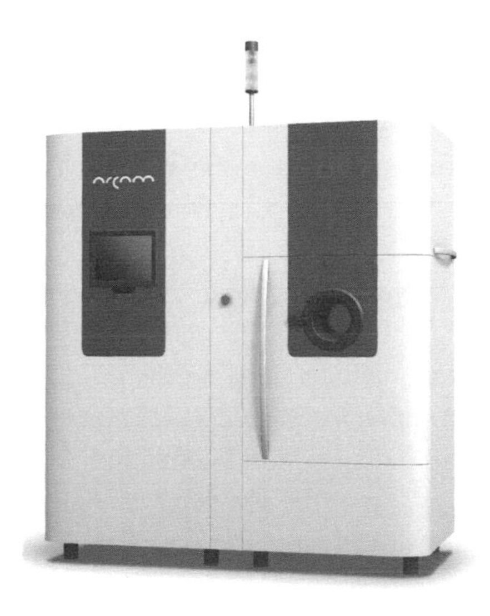

Fig. 5.17. Arcam Q10. (Courtesy of Arcam AB)

Table 5.8 shows the specifications for the Arcam Q10, Arcam Q20 and Arcam A2X. Currently, 4 metal powders are available, namely Titanium alloy Ti6Al4V (Grade 5), Titanium alloy Ti6Al4V ELI (Grade 23), Titanium CP (Grade 2) and CoCr Alloy (ASTM F75).

Table 5.8. Specifications of Arcam Q10, Arcam Q20 and Arcam A2X.

Models	Arcam Q10	Arcam Q20	Arcam A2X
Build tank volume	250 x 250 x 200 mm (W x D x H)	420 x 420 mm (Θ x H)	250 x 250 x 400 mm (W x D x H)
Actual build envelope	200 x 200 x 180 mm (W x D x H)	350 x 380 mm (Θ x H)	200 x 200 x 380 mm (W x D x H)
Power supply	3 x 400 V, 32 A, 7kW		
Size (W x D x H)	Approx. 1850 x 900 x 2200 mm	Approx. 2300 x 1300 x 2600 mm	Approx. 2000 x 1060 x 2370 mm
Weight	1420 kg	2900 kg	1570 kg
Process computer	PC	PC, Windows XP	
CAD interface	Standard: STL		
Network	Ethernet 10/100/1000		
Certification	CE		

5.5.3 *Process*

The Arcam EBM process consists of the following steps:

(1) The part to be produced is first designed in a 3D CAD program. The model is then sliced into thin layers, approximately a tenth of a millimetre thick.
(2) An equally thin layer of powder is scraped onto a vertically adjustable surface. The first layer's geometry is then created through the layer of powder melting together at those points directed from CAD file with a computer-controlled electron beam.
(3) Thereafter, the building surface is lowered and the next layer of powder is placed on top of the previous layer. The procedure is then repeated so that the object from the CAD model is shaped layer-by-layer until a finished metal part is completed.

5.5.4 *Principle*

The EBM process is based on the following two principles [21]:

(1) Parts are built up when an electron beam is fired at metal powder. The computer-controlled electron beam in vacuum melts the layer

of powder precisely as indicated by CAD model with the gain in electron kinetic energy.

(2) The building of the part is accomplished layer-by-layer. A layer is added once the previous layer has melted. In this way, the solid details are built up as thin metal slices melted together.

The basis for the Arcam Technology is essentially Electron Beam Melting (EBM). During the EBM process, the electron beam melts metal powder in a layer-by-layer process to build the physical part in a vacuum chamber. The Arcam EBM machines use a powder bed configuration and are capable of producing multiple parts in the same build. The vacuum environment in the EBM machine maintains the chemical composition of the material and provides an excellent environment for building parts with reactive materials such as titanium alloys. The electron beam's high power ensures a high rate of deposition and an even temperature distribution within the part, which gives a fully melted metal with excellent mechanical and physical properties.

5.5.5 *Strengths and Weaknesses*

The key strengths of the Arcam EBM systems are:

(1) *Superior material properties.* The EMB process produces fully dense metal parts that are void free and have excellent strength and material properties.

(2) *Excellent accuracy.* The vacuum provides a good thermal environment that results in good form stability and controlled thermal balance in the part, greatly reducing shrinkage and thermal stresses. The vacuum environment also eliminates impurities such as oxides and nitrides.

(3) *Excellent finishing*. The high-energy density melting process results in parts that have excellent surface finishing.

(4) *Good build speed.* High-energy density melting with a deflecting electron beam also results in a high speed build and good power efficiency.

The limitations of the Arcam EBM system are:

(5) *Need to maintain the vacuum chamber.* The process requires a vacuum chamber that has to be maintained as it has direct impact on the quality of the part built.
(6) *High power consumption.* The power consumed for using the electron beam is relatively high.
(7) *Gamma rays.* The electron beam used in the process can produce gamma rays. The vacuum chamber acts as a shield to the gamma rays, thus it is imperative that the vacuum chamber has to be properly maintained.

5.5.6 *Applications*

(1) *Rapid manufacturing.* Rapid manufacturing includes the fast fabrication of the tools required for mass production, such as specially-shaped moulds, dies, and jigs [22].
(2) *Medical implants.* The technology offers production of small lots as well as customisation of designs, adding important capabilities to the implant industry. A unique feature is the possibility of building parts with designed porosity and scaffolds, which enables the building of implants with a solid core and a porous surface to facilitate bone ingrowth. The ability to directly build complex geometries makes this technology viable for the manufacture of fully functional implants. The process uses standard biocompatible materials such as Ti6Al4V ELI, Ti Grade 2 and cobalt-chrome.
(3) *Aerospace.* This AM system offers significant potential cost savings for the aerospace industry and enables designers to create completely new and innovative systems, applications and vehicles.

5.5.7 *Examples*

5.5.7.1 *Arcam systems produce new orthopaedic implants* [23]

With longer life expectancies and more active lifestyles, more and more people are undergoing joint replacement surgeries. Ideally, an implant should be able to promote for long term fixation by enabling the bone to grow into it making the implant an integral part of the body. This is also called osseointegration. Using the Arcam's EBM technology, Adler Ortho Group, an Italian manufacturer of orthopaedic implants, came up with a ground breaking implant called Fixa Ti-Por acetabular cup to improve osseointegration.

The Arcam EBM S12 machine was installed in Adler Ortho's manufacturing facilities in Milan. During its first year on the market, more than 1000 acetabular cups produced by Arcam machines were implanted at several reference centres in Italy. The surgeons' post-op feedback were positive showing the primary fixation granted by the hemispherical press-fit is supported by the strong surface grip of the cup design. This proves Arcam machines' capability to manufacture products that can be customised to meet the market's demand.

5.5.7.2 *An implant manufactured with Arcam EBM was used to reconstruct a young girl's skull*

A young girl suffered a tragic car accident and needed to remove a large part of the skull bone to relieve the pressure on the brain. A titanium alloy implant was fabricated within 12 hours using the Arcam EBM machine (See Figure 5.18). The mechanical properties were excellent with 100% density of the material. The implant produced with EBM featured a chemical composition within stipulated standards, fully solid material with fine microstructure, high ductility and good fatigue characteristics. The entire surgical operation took only 2 hours as the implant fitted perfectly without complications. From this case, it was shown that EBM implants can be more rapidly and cost effectively manufactured compared to CNC machined ones. The design aspect is smoother and less complicated with an EBM manufactured part and the surface is almost perfect for the implant.

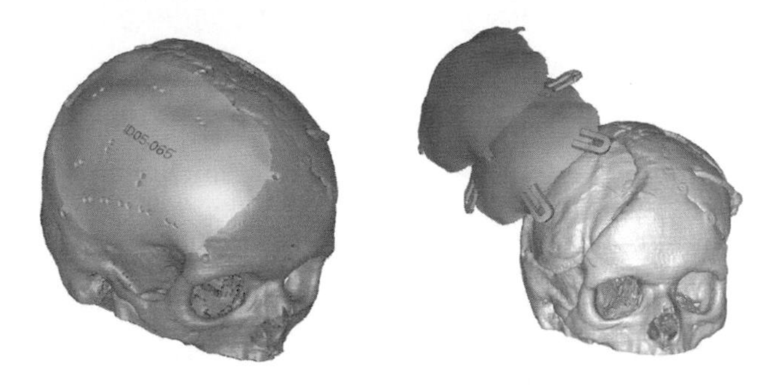

Fig. 5.18. CAD image of the young girls' skull and implant. (Courtesy of Arcam AB)

5.5.8 *Research and Development*

In 2013, Arcam AB received a European Union research grant to develop a "fast EBM" technology for 3D printing. The company is committed to enhance and improve the productivity of the existing EBM technology process by a factor of 5 with the introduction of new models and new hardware [24].

5.6 Concept Laser's LaserCUSING®

5.6.1 *Company*

Concept Laser GmbH (Hofmann Innovation Group) was founded in Lichtenfels, Germany in 2000 with a vision to optimise the SLS process. In pursuit of this vision, Frank Herzog, the firm's Managing Director and his team of engineers developed a completely new concept strategy to perform laser processing. Currently, Concept Laser GmbH has 50 employees and it is the leading manufacturer in the field of machines for metal laser melting process. Its address is An der Zeil 8, 96215 Lichtenfels, Germany.

5.6.2 *Products*

Figure 5.19, Figure 5.20 and Figure 5.21 show Concept Laser's modular laser processing systems. The machine consists of the laser station containing the laser and the axes that are powered by high-powered dynamic linear motors. Three technology modules allow effective time and cost-saving work in addition to the flexibility of choosing and exchanging of technologies. The system specifications are presented in Table 5.9.

Fig. 5.19. Mlab Cusing.

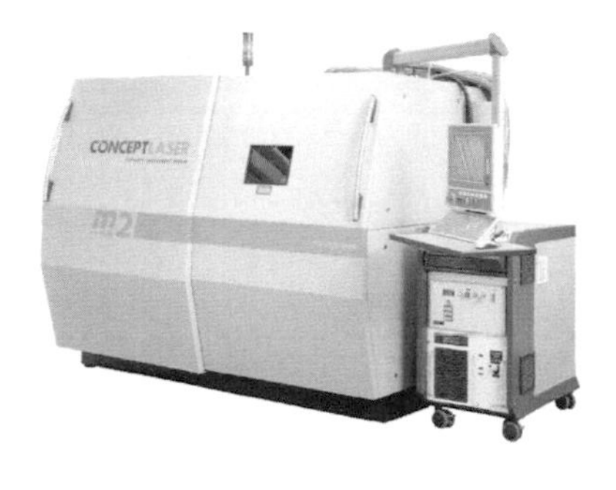

Fig. 5.20. M2 Cusing.

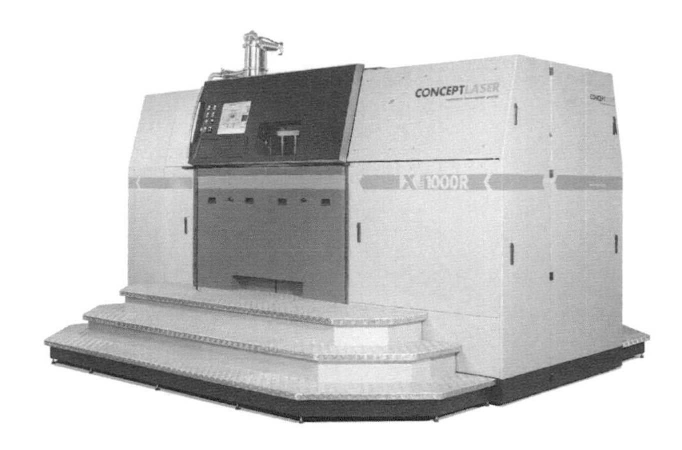

Fig. 5.21. X line 1000R. (Courtesy of Concept Laser GmbH)

Table 5.9. Specifications of Concept Laser's LaserCUSING systems.

Models	MLab Cusing	M1 Cusing	M2 Cusing	X line 1000R cusing
Building envelope	50 x 50 mm (x, y) 70 x 70 mm (x, y) 90 x 90 mm (x, y) z = 80 mm	250 x 250 x 250 mm (x,y,z)	250 x 250 x 280 mm (x,y,z)	630 x 400 x 500 mm (x,y,z)
Layer thickness	20 - 50 μm	20 - 50 μm	20 - 80 μm	30 - 200 μm
Production speed	1 - 5 cm^3/h	2 - 20 cm^3/h	2 - 20 cm^3/h	10 - 100 cm^3/h
Laser system	Fibre laser 100 W (cw)	Fibre laser 200 W (cw)	Fibre laser 200 W (cw), optional 400 W (cw)	Fibre laser 1 kW (cw)
Max. scanning speed	7 m/s	7 m/s	7 m/s	7 m/s
Focus diameter	20 - 80 μm	70 - 200 μm	50 - 200 μm	100 - 500 μm
Connected loads Power consumption	1.5 kW	7.4 kW	7.4 kW	13 kW
Power supply	1/N/PE AC 230 V, 16 A	3/N/PE AC 400 V, 32 A	3/N/PE AC 400 V, 32 A	3/N/PE AC 400V, 63 A
Compressed air	-	5 bar	5 bar	
Inert gas supply	1 gas connection provided / Nitrogen or Argon	Nitrogen; external generator	Nitrogen; 2 gas connections provided	2 gas connections provided
Inert gas consumption	Approx. 0.6 - 0.8 l/min	< 1 m^3/h	< 1 m^3/h	Approx. 17 - 34 l/min
Dimensions	705 x 1220 x 1848 mm (W x D x H)	2362 x 1535 x 2308 mm (W x D x H)	2440 x 1630 x 2354 mm (W x D x H)	4415 x 3070 x 3900 - 4500 mm (W x H x D)
Weight	-	1500 kg	2000 kg	8000 kg (net weight)
Operating conditions	-	15 - 35 °C	15 - 35 °C	15 - 30 °C

Table 5.9. *(Continued)* Specifications of Concept Laser's LaserCUSING systems.

Models	MLab Cusing	M1 Cusing	M2 Cusing	X line 1000R cusing
Materials	CL 91 RW Stainless hot work steel CL 100NB Nickel-based alloy (Inconel 718) CL 110CoCr Cobalt-chromium alloy (F75) remanium® star CL Cobalt-chromium alloy (by Dentaurum)	CL 20ES Stainless steel CL 50WS Hot work steel (1.2709) CL 100NB Nickel-based alloy (Inconel 718)	CL 20ES Stainless steel CL 30AL Aluminium (AlSi12) CL 40TI Titanium alloy (TiAl6V4) CL 40TI EFL Titanium alloy (TiAl6V4 ELI) CL 50WS Hot work steel (1.2709) CL 91RW Stainless hot work steel CL 100NB Nickel-based alloy (Inconel 718) CL 110CoCr Cobalt-chromium alloy (F75) remanium® star CL Cobalt-chromium alloy (by Dentaurum) rematitan® CL Titanium alloy (by Dentaurum)	CL 31AL Aluminium alloy (AlSi10Mg) CL 41TI ELI Titanium alloy (TiAl6V4 ELI) CL 100NB Nickel-based alloy (Inconel 718)

With the Concept Laser technology, the tool-less manufacturing of components with very complex geometries for one-off metal products or small batches is now a real alternative to the conventional machining or casting process. Components manufactured by LaserCUSING systems have the same properties as cast or machined products.

The Mlab cusing is an excellent choice for manufacturing components with elaborate structures. It provides high surface quality and fine component structures. The M1 cusing is a low-cost version relative to the larger laser processing machine such as the M3 linear system.

The M2 cusing is based on the Concept Laser technology and in addition meets the special requirements of processing the reactive powder materials out of aluminium and titanium alloys. Powder materials of titanium alloys are characterised by their high oxygen affinity and thus have to be stored and processed without atmospheric oxygen, i.e., in an inert atmosphere. This makes their processing, in a gas atmosphere indispensable. With the M2 cusing handling station, up to two materials can be stored. Inert atmosphere is filled in M2 cusing handling station; this station also has an integrated lock system and an inert powder extraction unit with integrated screening process. The powder is thus extracted from the build chamber and returned to the storage chamber within a matter of minutes automatically, removing the risk of the machine operator coming into contact with the powder.

The X line 1000R is the improved version of the M3 linear. It is a high performance production machine with LaserCUSING technology in XXL format. It has huge build envelopes of 630 x 400 x 500 mm and it can save processing of reactive materials.

5.6.3 *Process*

LaserCUSING® involves fine metallic powder being locally fused by a fibre laser. Metallic powder is fully fused layer-by-layer to produce a 100% component density with a high-energy laser. A specially developed exposure strategy allows the generation of solid and large-volume components without any deformation. The detailed process is shown in Figure 5.22.

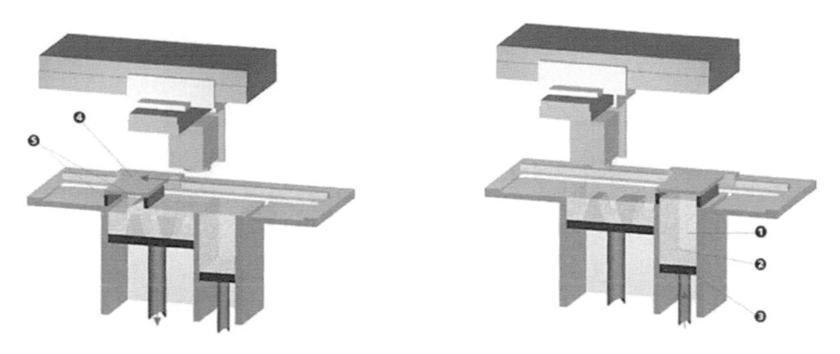

Step 1: Pushing powder into the coating system

Step 2: Lowering of the building platform and applying a layer of powder

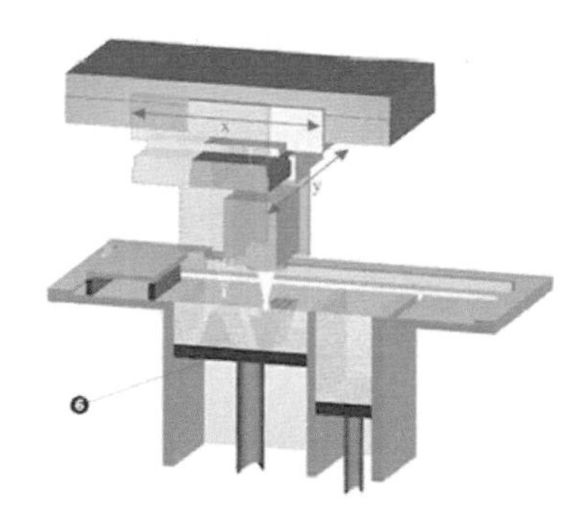

Step 3: Melting of the part boundary, followed by the internal cross-sectional area

Fig. 5.22. Schematic illustrations of LaserCUSING's process. (Courtesy of Concept Laser GmbH)

5.6.3.1 *Materials*

Concept Laser materials are produced in a powder form and have been developed specifically for the LaserCUSING® process. The compositions of the materials as well as the nature of the powder and the distribution of the powder fraction have been optimised for this method and the process control. All Concept Laser powder materials are 100% compatible for re-use in subsequent construction processes. No fresh material has to be added. Typical layer thicknesses for all materials are

20 – 50μm. Some of the materials used are introduced in more detail as follows:

(1) CL 60DG – Hot work steel: This is a material suitable for the production of tool inserts, in pressure die cast moulds for light metal alloys. This material is a hot work steel with tensile strengths up to 1.800 N/mm2 and hardness up to 52HRC.
(2) CL 50WS – Hot work steel: This material is very tough with excellent mechanical properties. Its main field of use is series injection moulding of plastic parts. This material can be used to produce large components that are only limited by the dimensions of the machine's built envelope. Components made with this material have tensile strengths of approximately 1,100 N/mm2 and a hardness of 40-42 HRC directly after the process. Subsequent annealing can increase these parameters to > 1,600 N/mm2 and up to 54 HRC.
(3) CL 20ES – Stainless steel: This is a powder material to produce acid and rust-resistant assemblies or tool components for pre-production tools. The tensile strength is approx. 650 N/mm2 and the hardness approx. 220 HB 30.
(4) CL 90RW and 91 RW – Stainless Hot work steel: these two materials have the characteristics of a hard stainless steel with high chrome content. The powder materials are for the production of tool components for the serial injection moulding of packaging and medical products. Components made with these materials have tensile strengths of approx. 850 N / mm2 and a hardness of 35-40 HRC directly after the process. Subsequent annealing can increase these parameters to > 1,100 N / mm^2 and up to 45-48 HRC for CL 90RW and to > 1,700 N / mm^2 and up to 48-50 HRC for CL 91RW.
(5) CL 30AL, CL 31AL (AlSi10Mg) – Aluminium-based alloy: These are aluminium powder materials that can be employed for high mechanical and dynamic load. The powder material can be used optimally for the production of technical prototypes or small series.
(6) CL 40TI (TiAl6V4) – Titanium-based alloy: CL 40TI is a titanium powder material for the production of technical lightweight components and medical implants.

(7) CL 100NB (Inconel 718) – Nickel-based alloy: This is a powder material made out of a nickel-based alloy for the production of heat resistant components in the automotive and aerospace industry.

5.6.4 *Principles*

Based on new laser technologies and a completely new process, it became possible to overcome the weaknesses of laser sintering. Using the patented exposure strategy and original materials, solid and large-volume components, such as mould inserts can be produced rapidly. The material properties are identical to those of the original material and allow these components to be employed under production conditions. The term "cusing" was coined from the words "concept" and "fusing". LaserCUSING® is based on the fusion of single-component metallic powder materials using a laser. This "generative" method makes it possible to assemble components layer-by-layer from virtually all weldable materials (e.g. stainless steel and hot work steel), by completely fusing the metal powder layer-by-layer. It has shown that it can overcome the internal stress and deformation problems almost entirely and to achieve 100% component density. The typical layer thickness is between 20 and 50 μm. The LaserCUSING® method is considered an excellent link in the process chain between rapid tooling and traditional tool and mould making. The patented exposure strategy also allows the low-deformation generation of solid and large-volume components.

Generating cores and inserts with cooling ducts has been built in practice. This ability allows for the production of highly complicated 3D moulds with such cooling ducted cores/inserts. In other words, the cooling ducts, which conventionally could only be introduced to a limited extent or at great expense, are now adapted exactly to the contours of the mould insert during the process. Thus considerably shorter cycle times are achievable. The compactness of the components ensures that no cooling water can escape. The deformations on the injection-moulded part are minimised due to optimised mould cooling.

5.6.5 *Strengths and Weaknesses*

The strengths of the LaserCUSING® system are as follows:

(1) *Unattended manufacturing of components within only a few days*: Prototypes can be manufactured overnight. This saves waiting time and hence reduces the development time.
(2) *High quality dentures*: Standardised manufacturing process produces very high quality and consistent work in customised dental restorations and implants.
(3) *Reduction of the post-treatment procedures*: All roughing work and pre-finishing is no longer required. Hence, this helps to reduce the product development time.
(4) *Less distortion and blow holes in the injection moulding or aluminium die casting*: Reduce the scrap rate and produce high quality plastic and aluminium parts. Hence, it can gain competitive advantage and has shorter cycle times.

The weaknesses of the LaserCUSING® system are:

(1) *Large physical size of the unit*: The system requires a large space to house.
(2) *High power consumption*: The system consumes high power due to the high wattage of the laser required to perform direct metal powder sintering.

5.6.6 *Applications*

(1) Functional models and working prototypes: Parts that can be used for fit and full functional testing. The system is suitable for the automotive, aerospace, machine tools and other industries for industrial products.

(2) Metal tools and inserts: The main application of the LaserCUSING® system is in rapid tooling. It is used primarily for creating superior tools and inserts with or without optimised cooling channels and ducts for injection moulding and other similar tooling and manufacturing processes.

5.6.7 *Examples*

5.6.7.1 *Digital dental technology laboratory*

Unicim, a dental restoration manufacturer based in Berschis, Switzerland adopted AM with laser melting to ensure the uniformity and accuracy of ceramic-veneered, non-precious metal restorations. The company uses a Concept Laser Mlab Cusing R System with a 100W continuous-wave (CW) fibre laser to produce metal dental implants that are of superior quality. The LaserCUSING process allows the economical production of caps, crowns and bridges, cast parts, primary and secondary structures (See Figure. 5.23). Combining CAD/CAM and metal laser melting, dental components are fabricated layer-by-layer with powder using a laser. This represents a significant transition from manual casting and milling to high precision, high accuracy industrial CAD/CAM production in dental technology, With Concept Laser systems, processed data is transmitted to the machine and the production can be started and left fully automated overnight. This additive manufacturing technology offers many advantages such as less material consumption, lower production costs and greater fit and shape retention thus reducing dental productions risk significantly. New material such as titanium can be used as they are the ideal material for allergy sufferers. Traditional milling method would have created high amount of material waste rendering the process too expensive and casting would be impractical. Thus, AM processes offered by Concept Laser systems are creating and establishing new alternatives in terms of production, workflow and the final products themselves.

Fig. 5.23. Crowns and bridges manufactured using laser melting technology. (Courtesy of Concept Laser GmbH)

5.6.7.2 *Innovative Rowenta iron*

The stimulating partner in terms of know-how and competence to achieve these goals was found with the full-service-engineering provider Hofmann Werkzeugbau. The successful result of this co-operation led to the introduction of the Rowenta Perfect DX 9100 that takes a leading position among steam irons (see Figure 5.24). The mode of operation of the device is based on "Intra Steam," a new method that enables ironing with pulsed steam. The steam penetrates and moisturises the textile fibres evenly due to the short, intermittent pulses. For this to be achieved, highly complex geometric cores with cooling ducts have to be integrated within the injection mould.

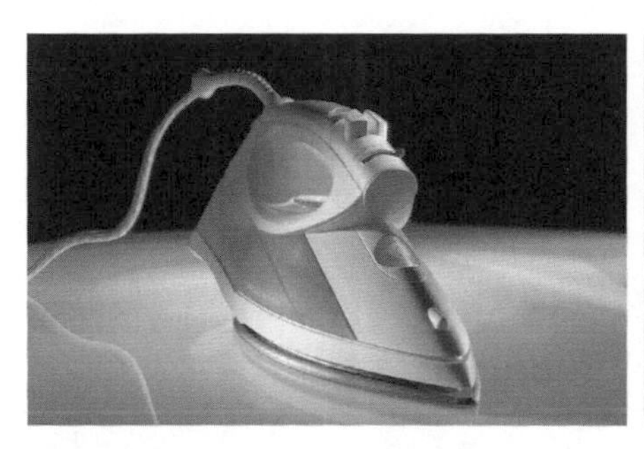

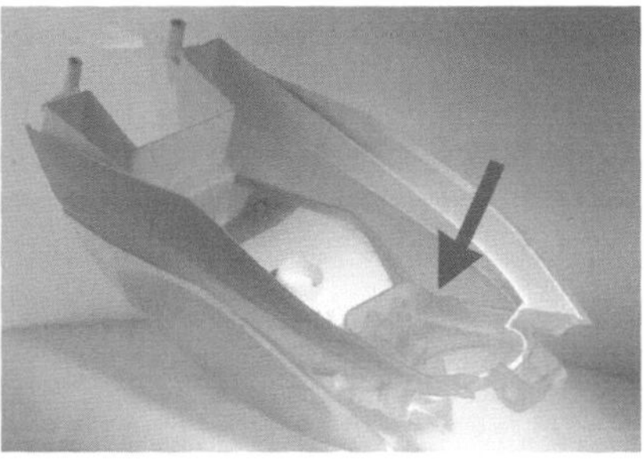

Fig. 5.24. Rowenta Perfect DX 9100 (left) and the internal geometry of the core insert with cooling ducts (right and arrowed). (Courtesy of Concept Laser GmbH and Hofmann Innovation Group)

5.6.7.3 *Intensive mould cooling with LaserCUSING®*

LaserCUSING by Concept Laser GmbH is able to minimise the deformations on the injection-moulded part due to the introduction of optimum mould cooling ducts in the mould (see Fig. 5.25). There is also

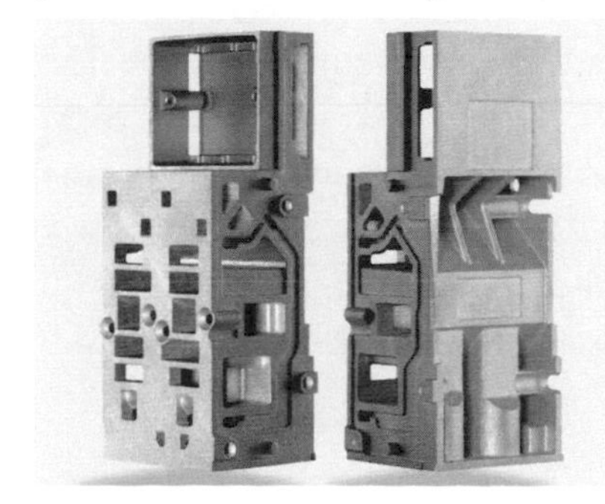

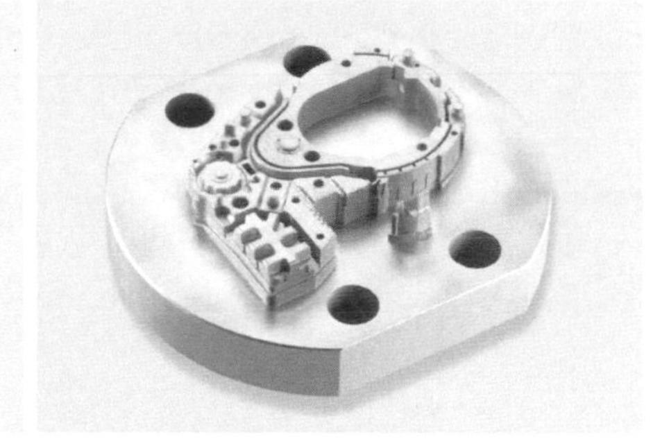

Fig. 5.25. Valve block (left) and mould insert (right) for series injection mould. (Courtesy of Concept Laser GmbH)

a significant reduction in the amount of reworking required on the mould contours. The accuracy achieved prior to secondary treatment was ±50 μm. The majority of surfaces only required rework with fine finish to obtain a higher degree of accuracy. The roughing-down process and pre-finishing process were omitted completely. As the distortion of the part was substantially reduced, the path from initial prototyping to creating a part ready for mass production was thus shorter. All these resulted in a considerable saving of time and cost.

5.7 SLM Solutions' Selective Laser Melting (SLM)

5.7.1 *Company*

SLM Solutions GmbH, formerly known as MTT Technologies GmbH, Lübeck, focusses on the development of their Selective Laser Melting (SLM) Systems. It started as one of the most popular processes - Vacuum Casting which the company introduced as the very first supplier in 1987. The Selective Laser Melting (SLM) Systems was first introduced to the market in 2000 and it built tools and single individual engineering and implant parts in pure, dense steel cobalt-chrome and titanium from CAD data files. The SLM Solutions GmbH head office address is Roggenhorster Strasse 9c, D-23556 Lübeck, Germany.

5.7.2 *Products*

Based on the Selective Laser Melting (SLM) technology, SLM Solutions has developed a series of 3D printers that can process most metals. They are SLM® 125HL (see Fig. 5.26), SLM® 280HL (see Fig. 5.27), SLM® 500HL and SLM® PSA500 (see Fig. 5.28). These machines make it possible to produce 100% dense metal parts from customary metal powder. The parts or specialty tools are built layer-by-layer (30 μm thickness). Metal powder (e.g. stainless steel 1.4404) is locally melted by

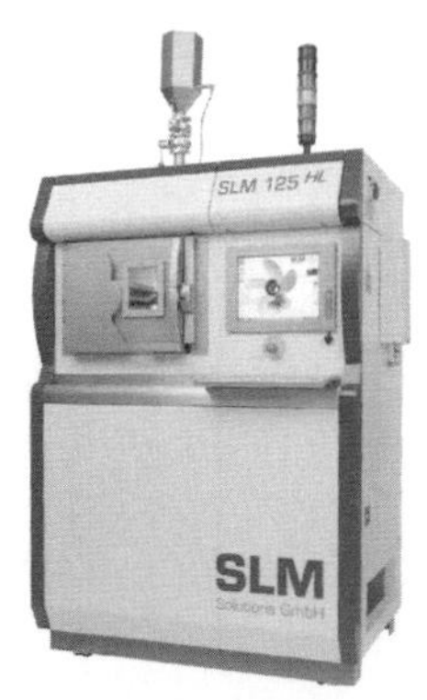

Fig. 5.26. SLM® 125HL printer. (Courtesy of SLM Solutions)

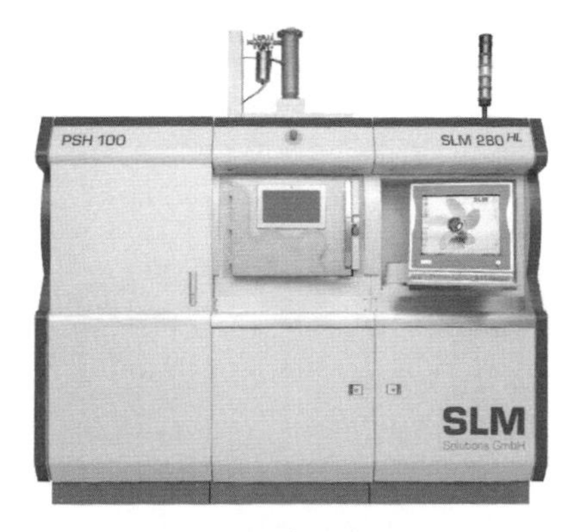

Fig. 5.27. SLM® 280HL printer. (Courtesy of SLM Solutions)

an intensive infrared laser beam that traces the layer geometry. These machines specifications are listed in Table 5.10.

Fig. 5.28. SLM® 500HL and SLM® PSA500 printers. (Courtesy of SLM Solutions)

Table 5.10. Specifications for SLM® 125HL, SLM® 280HL, SLM® 500HL and SLM® PSA500.

Models	SLM® 125HL	SLM® 280HL	SLM® 500HL	SLM® PSA500
Build Chamber (x/y/z) (mm)	125 x 125 x 75 (125)	280 x 280 x 350	500 x 280 x 325	Total Volume: ca. 60 - 100 L
Laser Power	100 / 200 W, YLR - Faser - Laser	400 / 1000 W, YLR - Faser - Laser	2 x 400 W, and optional 2 x 1000 W YLR - Faser - Laser	-
Build Speed	15 ccm / h	20 ccm/h or 35 ccm/h	70 ccm/h	-
Pract. Layer thickness	20 - 40 μm	20 - 75 μm Or 100 μm	20 - 200 μm	-

Table 5.10. *(Continued)* Specifications for SLM® 125HL, SLM® 280HL, SLM® 500HL and SLM® PSA500.

Models	SLM® 125HL	SLM® 280HL	SLM® 500HL	SLM® PSA500
Min. Scan Line / Wall Thickness	140 - 160 μm	150 / 1000 μm	160 - 180 μm	-
Operational Beam Focus variable	70 - 130 μm	70 - 120 μm/ 700 μm	80 - 150 μm/ 700 μm	-
Scan Speed	20 m/s	15 m/s	15 m/s	-
Inert Gas Consumption in Operation	Ar/ N_2, 0.5 L/min	Ar/ N_2, 2.5 L/min	Ar/ N_2, 5 L/min	-
Inert Gas Consumption Venting	Ar/ N_2, 1000L @100/min	Ar/ N_2, 1700L @100/min	Ar/ N_2, 2500L @100/min	-
Compressed Air Requirement	ISO 8573-1, 12.5 l/min, @ 1.5 bar	ISO 8573-1, 18 l/min, @ 1.5 bar	ISO 8573-1, 30 l/min, @ 1.5 bar	-
Dimensions in mm (B x H x T)	1350 x 1900 (2400) x 800	1800 x 1900 (2400) x 1000	3000 x 2000 (2500) x 1100	3500 x 2200 x 2700
Weight	~ 700 kg	~ 1000 kg	~ 2000 kg	~ 2500 kg
E - Connection / Consumption	400 Volt 3NPE, 32 A, 50/60 Hz, 4kW/h	400 Volt 3NPE, 32 A, 50/60 Hz, 8kW/h	400 Volt 3NPE, 32 A, 50/60 Hz, 8kW/h	400 Volt 3NPE, 32 A, 50/60 Hz, 8kW/h

5.7.3 *Process*

The Selective Laser Melting (SLM) process follows many of the conventions of the layer-based manufacturing processes. What differs is that the system can be used on a wide range of powdered metal materials. A component is split into layers and each of those layers is built on top of each other and fused together, layer after layer, until the part's form is built.

The SLM process uses a laser beam, controlled using optic lenses to pass a laser spot across the surface of a layer of powdered metal to build each layer. The metal powder is melted rather than just simply sintered together, thus giving parts which are 100% dense (or solid). The resulting part has much greater strength and dimensional accuracy than parts that are built by laser sintering [25, 26]. The schematic diagram in Figure 5.29 illustrates the SLM process.

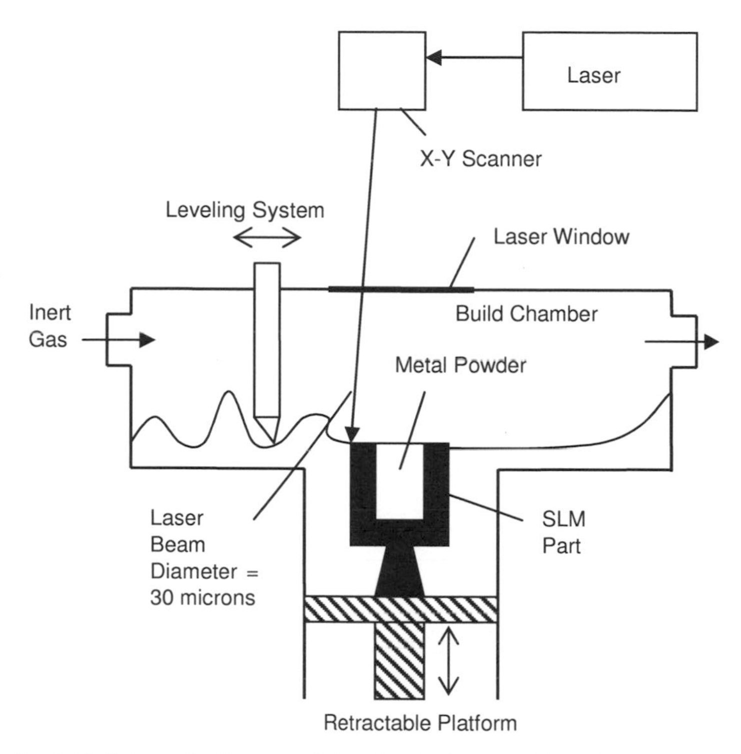

Fig. 5.29. Schematic diagram of the selective laser melting (SLM) process.

5.7.4 *Strengths and Weaknesses*

The main strengths of the SLM Solutions' Selective Laser Melting are:

(1) *High quality metal parts.* Builds high quality parts and tooling inserts from metal powders. Allows homogeneous build-up of components and tool cavities up to 100% density depending on requirements.
(2) *Large range of metal materials.* The system can be used to build almost any type of metal: stainless steel tool steel, titanium,

aluminium, cobalt-chrome, various nonferrous-metals. The system can also be used to shape gold.

(3) *Fast and low cost.* This is because no postprocessing such as heat treatment or infiltration required.

(4) *High accuracy.* High resolution process, dimensionally accurate, low heat generation no distortion of parts.

(5) *Complex geometries.* Produces tools and inserts with internal undercuts and channels for conformal cooling.

The major weaknesses of the SLM process are:

(1) *Large physical size of the unit.* The system requires a large space to house.

(2) *High power consumption.* The system consumes high power due to the high wattage of the laser required to perform direct metal powder sintering.

(3) *Relatively slow process.* Even with a build rate up to 70 cm^3/h, it is still a lengthy process compared to high speed machining [25].

5.7.5 *Applications*

Industrial applications include it being part of the process chain for building sheet metal forming and stamping tools. The SLM process can also be used for rapid tooling in building inserts, core and cavities for plastic injection moulding applications. Other applications include building functional prototypes and parts with specialised or dedicated metal, e.g. aluminium alloy, for the aerospace and biomedical industries.

5.7.6 *Examples*

5.7.6.1 *Planning osseointegrated implants*

Osseointegrated implants such as titanium screws [27] are used to drive into bones for dentures and facial prostheses. When planning where to place such implants, accurate AM models allow the depth of bone to be assessed, improving the selection of drilling sites before surgery. Though in many cases this process has improved the accuracy and reduced the theatre time, the drawback is that it incurs significant time and cost to produce the anatomical model, which is often damaged by test drilling.

To address this issue, the research team from PDR and Morriston Hospital decided to the plan the implant sites entirely on computer and only use AM to make templates to guide the surgeon in theatre. This allowed the clinicians to explore many different options without damaging the AM model of the skull. To be successful, the AM process has to produce strong, durable, dimensionally accurate parts that utilise materials that are safe to use in theatre while withstanding the demands of the surgical environment and sterilisation procedures.

Initial trials utilising the SLA and a medical standard resin proved that this approach can be successful. PDR then identified a new generative SLM process that can produce accurate parts directly from metal powders, in this case 316L stainless steel, a commonly used material for medical devices. As SLM produces fully dense stainless steel parts it can provide accurate, strong and durable surgical guides that can withstand contact with aggressive surgical instruments such as drills and oscillating saws. The material can also be sterilised by a number of commonly used processes like high temperature autoclaving. These advantages make SLM ideal for this approach and will enable PDR and its clinical partners to further explore and develop the application of computer generated surgical guides and templates (see Figure 5.30).

Fig. 5.30. A stainless steel drilling surgical template produced by Selective Laser Melting AM Technology. (Courtesy of SLM Solutions GmbH)

5.7.6.2 *Collaboration with NTU and DSO*

Nanyang Technological University (NTU) is in collaboration with Defence Science Organisation (DSO) Singapore on SLM of aluminium and copper alloys. Given the extensive research by the SLM community into laser power, scanning speed, as well as recent studies investigating the effect of varying hatch spacing in SLM [28, 29], these will serve as critical process parameters which will be varied as independent variables to achieve desired output parameters [30–32]. For this reason, this project focusses on the optimisation of SLM parameters on dimensional accuracies and properties [33–35], analysis of microstructure [36] and characterisation of the mechanical properties where parts were evaluated in accordance to ASTM standards. In addition, the effects of heat treatment on SLM parts were also studied to determine how certain properties such as ductility can be recovered through heat treatment [37].

Known for its low weight, high strength and good thermal properties, aluminium alloys, Al6061 and AlSi10Mg were studied in this project, targeted at aerospace applications. The copper alloys studied in this project were UNS C18400 and Hovadur K220, targeting at applications requiring high thermal and electrical properties. The results of this study showed that as built AlSi10Mg parts had an Ultimate Tensile Strength (UTS) of 459±2.7 MPa compared to a value of 300 MPa for bulk Al6061 [35, 38–40]. This demonstrated its potential in aerospace applications. As for copper, a thin wall feature of 200 μm could be produced with the optimised process parameter obtained in this study for Hovadur K220 [41]. This demonstrated the possibility of producing highly efficient heat exchangers with thin wall features. Figures 5.31 and 5.32 show some examples of SLM produced parts as a result of this project collaboration with DSO.

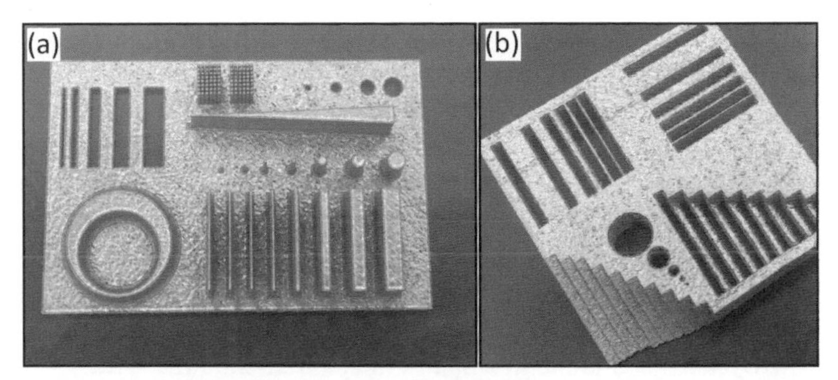

Fig. 5.31. (a) AlSi10Mg SLM part showing different features, (b) Al6061 SLM part showing thin varying groove dimensions.

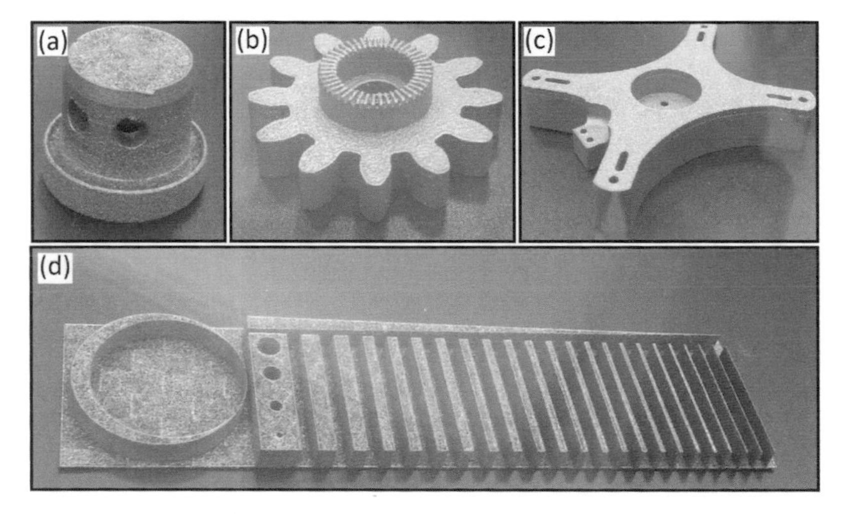

Fig. 5.32. (a) Mould insert with internal conforming cooling channels, (b) gear, (c) bracket with thermal dissipation properties and (d) varying thin wall features.

5.7.7 *Research and Development*

SLM Solutions always specifically addresses the latest developments in Selective Laser Melting (SLM) technology. It aims to keep improving the capabilities and coming up with new breakthroughs for SLM.

5.8 3D Systems' Phenix PX™

5.8.1 *Company*

Phenix Systems is a French company founded in 2000 that specialises in the design, production and sales of AM systems using the laser melting of metal and ceramic powders. The Phenix PX™ technology was developed in the early 1990s by a study group specialising in heterogeneous materials at Ecole Nationale Supérieure de Céramique Industrielle (ENSCI) in Limoges. This patented technology has gone through much research and development work by Phenix Systems concerning materials and software in order to reach a wide variety of industrial activities. In 2013, Phenix Systems was acquired by 3D Systems and the technology was renamed as Phenix PX™. 3D Systems' details are found in Section 3.1.1.

5.8.2 *Products*

3D Systems' direct metal production printers are PXS™, PXM™ and PXL™. 3D Systems' Phenix PX™ specialises in precise laser melting to produce metal pieces that meet industrial standards. Table 5.11 shows the specifications for 3D Systems' PXS™, PXM™ and PXL™ systems.

Table 5.11. Specifications for 3D Systems' PXS™, PXM™ and PXL™ systems.

Models	PXS™	PXM™	PXL™
Laser Power / type	50W / Fibre laser	300 W / Fibre laser	500 W / Fibre laser
Laser wavelength	1070 nm		
Layer thickness range	Adjustable, min 10 μm, max 50 μm		

Table 5.11. *(Continued)* Specifications for 3D Systems' PXS™, PXM™ and PXL™ systems.

Models	PXS™	PXM™	PXL™
Build envelope capacity (X x Y x Z)	100 x 100 x 80 mm (3.94 x 3.94 x 3.15 in)	140 x 140 x 100 mm (5.51 x 5.51 x 3.94 in)	250 x 250 x 300 mm (9.84 x 9.84 x 11.81 in)
Metal material choice	Stainless steels, tool steels, non-ferrous alloys, super alloys and others		
Ceramic material choice	Cermet (Al^2O3; TiO^2) and others		
Repeatability	x = 20 μm, y = 20 μm, z = 20μm		
Minimum detail resolution	x = 100 μm, y = 100 μm, z = 20μm		
Dimensions uncrated (W x D x H)	120 x 77 x 195 cm (48 x 31 x 77 in)	120 x 150 x 195 cm (48 x 59 x 77 in)	240 x 220 x 240 cm (95 x 87 x 95 in)
Weight uncrated	1000 kg (2200 lbs)	Approx. 1500 kg (3300 lbs)	Approx. 5000 kg (11000 lbs)
Electrical requirements	230 V / 2.7KVA / single phase	400 V / 8KVA / 3 phase	400 V / 15KVA / 3 phase
Compressed air requirements	6 - 8 bar CE		
Software tools	Phenix Processing - Phenix Manufacturing		
Control software	PX Control		
Operating system	Windows XP		
Input data file format	STL, IGES, STEP		
Network type and protocol	Ethernet 10 / 100, RJ - 45 Plug		
Recycling system	Optional external system (PX BOX)		Automatic
Loading system	Manual	Semiautomatic	Automatic
Certification	CE marked		

5.8.3 *Process*

This generative manufacturing process implements the combined effect of a fibre optic-based laser and a furnace on metallic or ceramic powders thereby making provision for nuances to conform to the materials openly available in the industrial market sectors. Figure 5.33 shows the schematic diagram of the 3D Systems' Phenix PX™ process. The main stages of the procedure are as follows:

(1) Processing from a 3D file then generating the manufacturing files of each scanned layer.
(2) Initialisation of the equipment depending on the materials being used.
(3) Production of a powder layer with a chosen thickness.
(4) Melting of the scanned section of the part.
(5) Automatic repetition of the two previous stages until the complete production of the part.

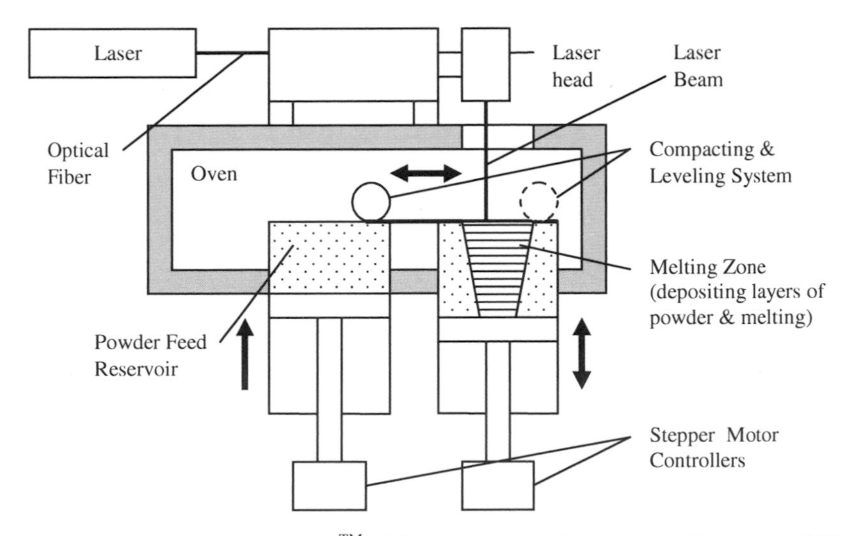

Fig. 5.33. 3D Systems' Phenix PX™ additive manufacturing process. (Courtesy of 3D Systems)

5.8.4 *Principles*

When certain metal and ceramic powders are subjected to local high temperatures caused by the laser beam, they tend to undergo melting and strengthening. On the Phenix PX™ machines, these are achieved as the laser beam follows the trajectory that matches the target trajectory. The melting operation is continuous, with the depositing of successive layers of powder. The process is accurate as it is possible to achieve details of 150 μm for metal parts and 300 μm for ceramic parts. The manufacturing procedure is fully automated. The operator works from the CAD file for the part and sets the manufacturing parameters, depending on the type of powder and the desired outcome.

The machines have a number of sub-assemblies. The most important is the laser head which provides the energy required for melting. The laser beam (50, 300W or 500W) is sent to the head by optical fibre. The head moves horizontally over the oven and has a galvanometric directional system for the beam to produce the desired geometry. The melting operation takes place inside the oven at 900°C, in a controlled atmosphere. The powder, held in a feed reservoir, is transferred by a mobile scrapper system and deposited in successive layers in the reservoir where the melting is performed. The powder is uniformly spread in the reservoir; only the area subjected to the laser beam is melted. The bottom of each of the reservoir is equipped with vertical movement pistons: by controlling these pistons it is possible to ensure the powder is flush on the surface. As there are no speed and torque requirements and only positioning is to be achieved, stepping motors are used, together with incremental encoders in order to guarantee the precision (with accuracy for the parts of ±20 μm).

5.8.5 *Strengths and Weaknesses*

The key strengths of the 3D Systems' Phenix PX™ are:

(1) *Small and accurate parts.* The Phenix PX™ machines focus on building small and accurate metal and ceramic parts.
(2) *Use of standard powders.* One of the main advantages is that the system is able to use standard powders rather than only limited to supplied powders as with that of the competition.
(3) *Low laser power.* The system uses relatively low laser power to melt metal and ceramic powders because of the application of high working temperatures (e.g. 900 °C for tool steels). Also, the use of fibre optics focusses the beam and minimises losses due to beam diffusion.

The main weaknesses of the Phenix PX™ are:

(1) *High temperature work envelopes required.* Care has to be taken as the general temperature of the work area will be extremely hot.
(2) *Postprocessing required.* For metals, further polishing is required while for ceramics, post-melting in the furnace is still required.

5.8.6 *Research and Development*

With the acquisition of Phenix Systems by 3D Systems, more resources are expected to be allocated to R&D to improve the print speeds, user-friendliness and the cost of the laser melting technology.

5.9 3D-Micromac AG's Laser Structuring Systems

5.9.1 *Company*

3D-Micromac AG Corporation, a supplier of customised laser micromachining facilities as well as coating and printing technologies, has gained an established position in the international market over the past years. The primary focus of 3D-Micromac's system is in microsintering. The address of the company's headquarters is Rosenbergstraße, 09126 Chemnitz, Germany.

5.9.2 *Product*

3D-Micromac's product is the microSTRUCT series that makes use of the Laser Structuring Systems. This is a processing technology for selective laser sintering under vacuum conditions specially developed for the production of prototypes as well as small and medium lots. This novel technology is capable of generating freeform micro parts made from metals and alloys with unequalled precision, shape diversity, and flexibility. It is done by overcoming the problems of oxidation and humidity by running the SLS process in a vacuum environment. The density of the micropart can be controlled and density gradients and compositional gradients can be bred by varying the sintering parameters. The microSTRUCT series consists of microSTRUCT vario, microSTRUCT compact, microSTRUCT OLED and microSTRUCT Excimer. Table 5.12 shows the specifications for the microSTRUCT series.

Table 5.12. Specifications for the microSTRUCT series. (Courtesy of 3D-Micromac AG Corporation)

Models	microSTRUCT vario	microSTRUCT compact	microSTRUCT OLED	microSTRUCT Excimer
System	Direct-driven positioning system		-	-
XY-traverse path	1150 mm x 450 mm	600 mm x 400 mm	600 mm x 400 mm	200 mm x 200 mm
Positioning accuracy	± 0.001 mm	± 0.01 mm	X-,Y- axis: ± 1 µm Z-axis: 10 µm / 50 mm	± 1 µm
Travel speed max.	500 mm/s	150 mm/s	-	2 m/s
Acceleration max.	400 mm/s^2	100 mm/s^2	-	-
Working areas	3 working areas for using scan systems, fixed optics or helical drilling optics	2 working areas for using scan systems, fixed optics or helical drilling optics	-	-
Working dimensions	450 mm x 450 mm	300 mm x 300 mm	-	-
Interfaces	GSD II, STL, DXF			
Laser source	Ps Laser 1064 nm with SHG and THG		Picosecond laser, wave length 355 nm, mean power > 3.5 W @ 100kHz. Repetition frequency: 0 - 2 MHz	193 nm (ArF) 248 nm (KrF) 308 nm (XeCl) Repetition rate: 10 Hz to 4kHz Beam profile TEM00 Energy 1 mJ = 1,200 mJ Average power: 0.5 - 20 W Pulse width: 10 - 500 ns

5.9.3 *Process*

The 3D-Micromac AG's Laser Structuring Systems are able to process powder materials with sub-micrometre grain sizes in layers as thin as one micrometre. The process is essentially a further development from the well-established laser sintering method. It uses a special recoating device to apply thin powder layers and a pulsed solid-state laser to locally melt the powder. In this way, micro parts can be produced in various metals giving excellent material properties, e.g. a detail resolution down to 30 micrometre is possible [42]. Fig. 5.34 shows an illustration of the Laser Structuring process.

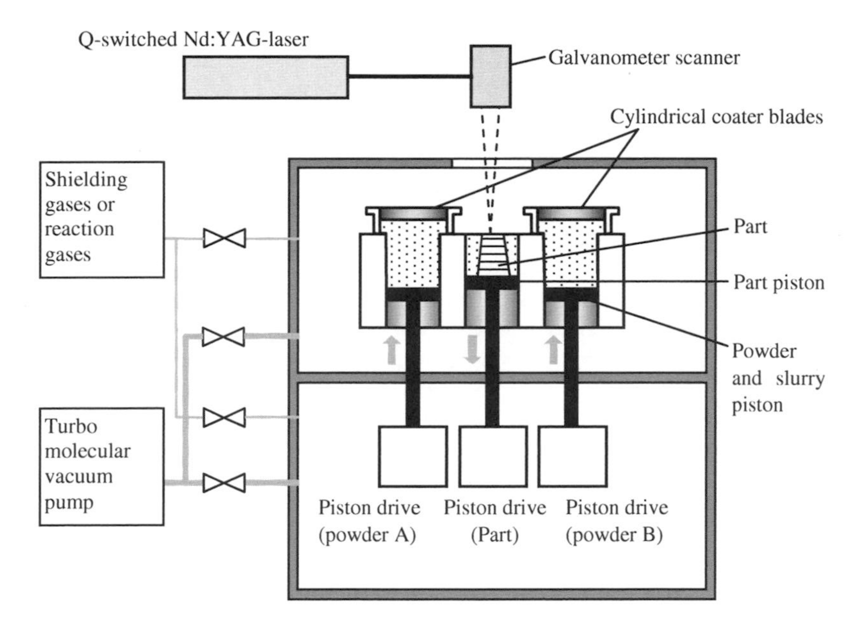

Fig. 5.34. Schematic illustration of the Laser Structuring process. (Courtesy of 3D-Micromac AG Corporation)

Materials used in the Laser Structuring process include tungsten, copper, aluminium, gold, titanium, silver, molybdenum composites and metal-ceramic-based composites. As an example, for tungsten material, the density of micro parts achievable is between 40% and 70%, or for tungsten/aluminium powder mixtures (composites), more than 95% density.

5.9.4 *Applications*

The micro-components produced are applied in areas such as micro-mechanics, telecommunications, medical, electronics and computer technologies. Other applications are found in industries such as the automotive, mould construction, hard-carbide tools, tool and mould making, photovoltaics and precision engineering industries.

5.9.5 *Strengths and Weaknesses*

The key strengths of the Laser Structuring process are:

(1) *Accurate and precise build parts.* As the Laser Structuring process is conducted in a vacuum chamber and with the use of a highly accurate Nd:YAG laser, accurate and precise build micro-structures can be achieved with the micron-size metal powder.
(2) *Fine surface finish.* The use of micro-size powder allows the process to attain surface finishes up to 1.5 μm R_a.
(3) *Wide range of submicron grained metal powder.* The system is able to process a wide range of submicron grained metal powder including: single component metals of tungsten, aluminium, copper, silver, titanium, molybdenum and steel; blend materials of copper/tungsten, aluminium/tungsten, copper/molybdenum and aluminium/molybdenum.

The main weaknesses of the Laser Structuring process are:

(1) *Need to maintain the vacuum chamber.* The process is carried out in a vacuum chamber that has to be maintained as it has direct impact on the quality of the part being built.
(2) *High power consumption.* The power consumed for using the Nd:YAG laser is relatively high.

5.9.6 *Research and Development*

3D-Micromac AG Corporation is working with EOS GmbH to integrate their respective technological know-how in the area of Micro Laser Sintering technology (MLS) into a new corporation. The establishment of 3D MicroPrint GmbH will further boost the development and commercialisation of the new MLS technology. This can also help to bring and invent more new solutions in the micro technology area [43].

5.10 ExOne's Digital Part Materialisation (DPM)

5.10.1 *Company*

ExOne has its origins with the Extrude Hone Corporation. The late Lawrence J. Rhoades, first founded Extrude Hone Corporation and when it expanded and grew, sold the technologies relating to abrasive flow machining, Surftran electrochemical deburring, and ThermoBurr deburring to Kennametal, Inc. in 2005. He then set up ExOne as a new business to serve as an incubator for inventive, new technologies that have the potential to improve manufacturing techniques. Currently, ExOne offers 3D printers that can produce sand and metal pieces in large scales. The address for ExOne's headquarters is 127 Industry Boulevard, N. Huntingdon, PA 15642 USA.

5.10.2 *Products*

ExOne focusses on 3D printing sand and metal parts with the use of its S-Max, S-Print and M-Print machines. The specifications of the S-Max, S-Print and M-Print machines are summarised in Table 5.13(a). S-Max machines emphasise on the flexibility and efficiency of the production process while S-Print machines aim to enhance productivity. M-Print machines are designed to achieve the industrial scale of production. ExOne Company's new product — M-Flex was also unveiled in 2012, which specialised in producing middle size models and prototypes. Meanwhile, ExOne also has X1-Lab and Orion Laser systems, which are customised products designed for researchers and educators. The specifications of the M-Flex, X1-Lab and Orion Laser systems are summarised in Table 5.13 (b).

Table 5.13 (a). Specifications of the S-Max, S-Print and M-Print machines.

Models	S-Max	S-Print Silicate	M-Print
Build volume (L x W x H)	1800 x 1000 x 700 mm (70.9 x 39.37 x 27.56 in.)	800 x 500 x 400 mm (29.5 x 15 x 15.75 in.)	
Build speed	60,000 to 85,000 cm^3/h (2.12 to 3.00 ft^3/h)	16,000 to 36,000 cm^3/h (0.56 to 1.2 ft^3/h)	-
Layer thickness	0.28 to 0.50 mm (0.011 to 0.0197 in.)	0.28 to 0.38 mm (0.011 to 0.015 in.)	Variable with minimum 0.15 mm
Print resolution	X/Y 0.1 mm / 0.1 mm (0.004 in. / 0.004 in.)	X/Y 0.1 mm / 0.1 mm (0.004 in. / 0.004 in.)	X/Y 0.07 mm / Z 0.15 mm (Set by layer thickness)
External dimension (L x W x H)	6900 x 3520 x 2860 mm (271.7 x 138.6 x 112.6 in.)	3270 x 2540 x 2860 mm (10.7 x 8.3 x 9.4 ft.)	
Weight	6500 kg (14,330 lbs)	3500 kg (7,717 lbs)	
Electrical requirements of printer	400 V 3-Phase/N/PE/ 50 - 60 Hz, max. 6.3kW	400 V 3-Phase/N/PE/ 50 - 60 Hz, max. 6.2kW	
Electrical requirements of heater	400 V 3-Phase/PE/ 50 - 60 Hz, max. 10.5kW	400 V 3-Phase /PE/ 50 - 60 Hz, max. 6.3kW	
Data interface	STL		
Consumable materials	FS001, FS003, FS005, FA001, FB001, FC005, MI001	FS001, FS003, FA901, FB901, FC901	PM-S4-60-MP, PM-S3-60-MP, PM-I-MP, PM-TSP-MP, PM-B-MPrint-01, PM-C-MPrint-01

Table 5.13 (b). Specifications of the M-Flex, X1-Lab and Orion Laser systems.

Models	M-Flex	X1-Lab	Orion Laser
Build volume (L x W x H)	400 x 250 x 250 mm (15.7 x 9.8 x 9.8 in.)	40 x 60 x 35 mm (1.5 x 2.3 x 1.3 in.)	Maximum workpiece size: 150 nm
Build speed	30 seconds/layer	1 minute/layer	X-Y-Z maximum travel speed: 200 mm/sec

Table 5.13 (b). *(Continued)* Specifications of the M-Flex, X1-Lab and Orion Laser systems.

Models	M-Flex	X1-Lab	Orion Laser
Layer thickness	Variable with minimum of 0.100 mm	Variable with minimum of 0.05 mm	-
Print resolution	X/Y 0.0635 mm, Z 0.100 mm (set by layer thickness)	X/Y 0.0635 mm, Z 0.100 mm (set by layer thickness)	Taper range: +/- 5% Roundness +/- 2% Surface roughness +/- 250 nm (PP)
External dimensions (L x W x H)	1674 x 1278 x 1552 mm (5.5 x 4.2 x 5.1 ft.)	965 x 711 x 1066 mm (3.2 x 2.3 x 3.5 ft.)	2300 x 1500 x 2000 mm (7.5 x 4.9 x 6.6 ft.)
Weight	TBD	-	Approximate 1200 to 1800 kg based on laser source
Electrical requirements	208 V - 240 V / 3 phases	120 VAC / 60Hz / 4.1 A	3 phases 300 - 480 VAC, 50/60 Hz, 40 A
Data interface	SLC, CLI	STL	-
Consumable materials	PM-S4-60-MP, PM-S3-60-MP, PM-I-MP, PM-B-MPrint-01, PM-C-MPrint-01	PM-R1-S4-30, PM-I-R1, PM-TSP-R116, PM-B-SR1-04, PM-C-R1-02	Stainless steel, Titanium, Nickel materials and alloys, Ceramic matrix composites, Molybdenum, Rhenium

5.10.3 *Process*

Utilising Digital Part Materialisation technology (DPM), formerly known as 3DP technology, the process has the ability to build metal components by selectively binding metal powder layer-by-layer with the chemical binder from the digital CAD file. The finished structural skeleton is then sintered and infiltrated to produce a finished part that is 60% dense. The process consists of the following steps. Figure 5.35 illustrates the DPM process schematically [44].

(1) A part is first designed on a computer using commercial CAD software.

(2) The CAD image is then transferred to the ExOne System's built-in software. Afterwards, the CAD file is automatically sliced into thin layers (0.1 - 0.15 mm).
(3) The CAD image is printed with an inkjet print head depositing millions of droplets of binder per second. These droplets dry quickly upon deposition.
(4) The process is repeated until the part is completely printed.
(5) The resulting "green" part, of about 60% density, is removed from the machine and excess powder is brushed away.
(6) The "green" part is next sintered in a furnace, while burning off the binder. It is then infiltrated with molten bronze via capillary action to obtain full density. This is carried out in an infiltration furnace.
(7) Postprocessing include machining, polishing and coating to enhance wear and chemical resistance e.g. nickel and chrome plating.

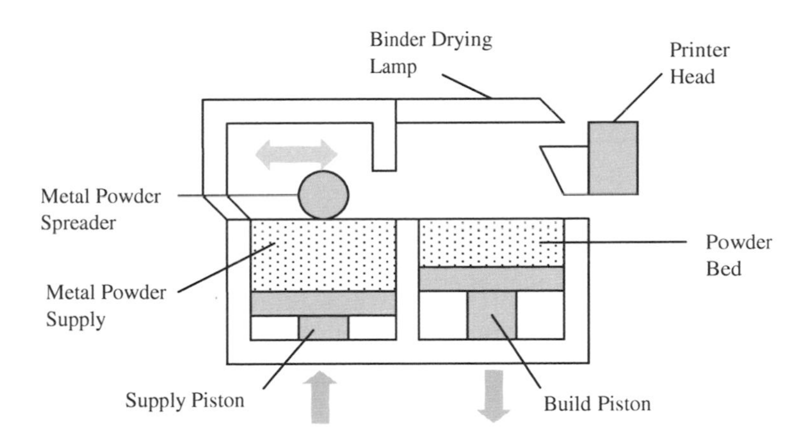

Fig. 5.35. Schematic illustration of the DPM process.

5.10.4 *Principle*

The working principle of ExOne's DPM uses an electrostatic inkjet printing head to deposit a liquid binder onto the powder metals. The part is built one layer at a time based on sliced cross-sectional data. The metal powder layer is spread on the build piston and a sliced layer is printed onto the powder layer by the inkjet print head depositing droplets of binder that are in turn dried by the binder drying lamp [45]. The process is repeated until the part build is completed.

5.10.5 *Strengths and Weaknesses*

The strengths of the DPM process are:

(1) *Fast.* The DPM machine creates multiple parts simultaneously and not sequentially like laser systems. Interchangeable build chambers allow quick turnaround between jobs. Build rates can be up to 85,000 cm^3 per hour.
(2) *Flexible.* Virtually no restriction of design flexibility, such as complex internal geometries and undercuts, can be created.
(3) *Reliable.* There is auto tuning and calibration for maximum performance and also built-in self-diagnostics and status reporting. The inkjet printing process is simple and reliable.
(4) *Large parts.* Large steel mould parts measuring 1800 x 1000 x 700 mm can be produced.

The weaknesses of the DPM process are:

(1) *Large space required.* The machine needs a very large area to house it.
(2) *Limited materials.* The system only prototypes parts with its own metal powder.

5.10.6 *Applications*

The ExOne's DPM is primarily used to rapidly fabricate complex stainless or tool steel tooling parts. Applications include injection moulds, extrusion dies, direct metal components and blow moulding [46]. The technology is also suitable for repairing worn out metal tools.

5.10.7 *Examples*

5.10.7.1 *ExOne helps US Navy to cut costs, improve quality and shorten lead time on production [47]*

US Naval Undersea Warfare Centre (NUWC) needed to replace the tail cones of their MK 30 anti-submarine mobile targets efficiently at a lower cost.

NUWC chose ExOne's DPM to produce the finished parts which was completed within 10 weeks including the design time. The entire project cost about $12,600 as compared to the traditional sand casting tooling method which would have required $20,000 and 25 weeks in total. Hence, ExOne's DPM effectively helped US NUWC to save cost and shorten the production period.

5.10.8 *Research and Development*

ExOne has opened its sixth Production Service Centre and at the time of writing had announced its plan for two new Sales Centre in Brazil and China [48].

ExOne is investing in non-traditional manufacturing and this promotes the research and development in additive manufacturing as well as in advancing the micromachining process.

5.11 Voxeljet AG's VX System

5.11.1 *Company*

Voxeljet AG was incorporated in 1999 by Ingo Ederer under the name Genesis GmbH with the aim of providing new generative processes for production. The special know-how of the company lies in the connection between high performance inkjet technology and rapid manufacturing.

The company developed a process which enabled tool-making to be revolutionised in the casting area. With the type designation GS 1500 now S15, the machine system was developed as one of the largest commercially available AM systems in the world with its unique productivity. Using the equipment, sand casting moulds can be generated without tools automatically. The system technology has been implemented successfully worldwide for customers such as BMW AG and Daimler-Chrysler AG.

In mid-2003, the company licensed the technology to Extrudehone Corporation. Voxeljet then focussed itself on process application, in order to be able to offer a fast generation of moulds to a broad client base. In 2004, the name was changed to Voxeljet Technology GmbH (with a large V) and just recently it became voxeljet AG. "AG" is the abbreviation for Aktiengesellschaft, which is German for a shareholders-owned corporation and can be public listed. Today, voxeljet AG operates in three fields: in the supply of services, moulds for casting, cast parts and plastic parts. In the field of printing technology, high performance inkjet systems are developed and produced for the most diverse applications. Industrial 3D printing systems for the manufacture of plastic components form the latest product line. The company address is Paul-Lenz-Straße 1, 86316 Friedberg, Germany.

5.11.2 *Products*

Voxeljet AG produces five machines, the VX200, VX500, VX800 (see Figure 5.36), VX1000 and VX4000 (see Figure 5.37), that produce thermoplastic models to order from 3D data, without tools automatically. A selectively bonded particle material is applied layer-by-layer to create models.

Once the building process is completed, the interchangeable container is removed from the system via a rail-guide system. As soon as a second interchangeable container (optional accessory) is inserted, the system is ready to start operation again.

With the aid of an unpacking station and an industrial vacuum cleaner, the components are removed from the surrounding powder before being dried in a convection oven (both optional accessories). A summary of the VX200, VX500, VX800, VX1000 and VX4000 machine specifications are summarised in Table 5.14.

Fig. 5.36. voxeljet's VX800.
(Courtesy of voxeljet AG)

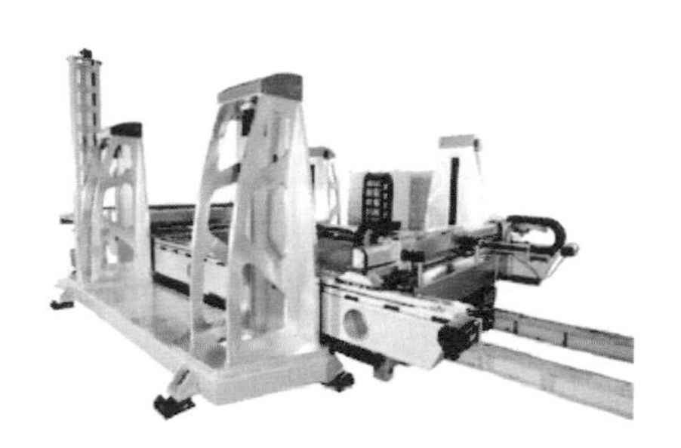

Fig. 5.37. voxeljet's VX4000.
(Courtesy of voxeljet AG)

Table 5.14. Specifications of voxeljet AG's machines.

Models	VX200	VX500	VX 800	VX1000	VX4000
Build volume, mm, WxDxH	300 x 200 x 150	500 x 400 x 300	850 x 450 x 1500 / 2000	1060 x 600 x 500	4000 x 2000 x 1000
Resolution x,y	300 dpi	42,3 x 42,3 μm	600 dpi	600 dpi	42,3 x 42,3 μm
Layer thickness, mm	150 μm	80 / 150 μm	300 μm	100 / 300 μm	120 / 300 μm
Build Speed, mm/ h	12	15	35	36	14.5
Parts build material	PolyPor-PMMA/ Sand				
Parts accuracy	0.3%; min ± 0.1mm				
Data interface	STL				
Network connection	Ethernet				
PC	Latest windows Operating software				
Outer dimensions, mm WxDxH	1700 x 900 x 1500	1800 x 1800 x1700	4000 x 280 x 2200	2400 x 2800 x 2000	19500 x 3800 x 7000
Weight, kg	450	1200	2500	3500	-

5.11.3 *Process*

The component is built layer-by-layer. This process requires 3D CAD data, which is transferred to process preparation software via an STL interface. That is where the model is virtually disassembled into individual layers. The moulds are placed in the virtual build space, and the building process is started. The loose basic material is evenly applied over the entire build width. A print head applies binder where the model is to be produced, whereby the binder infiltrates the most recently applied layer and connects it with the layer below. The building platform is lowered and the process starts again. Following the completion of the building process, the loose particle material is removed manually. Once the moulds have been cleaned, they can be mounted and prepared for casting. With sand casting, the moulds acts as the negative of the work piece. The detailed process is shown in Figure 5.38.

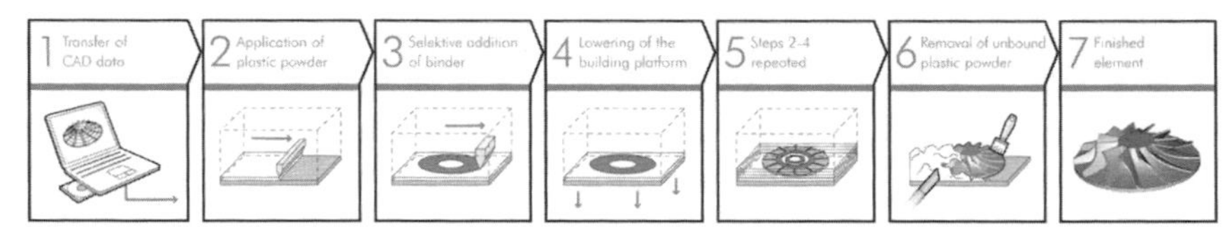

Fig. 5.38. Detailed working process of voxeljet's 3D printers. 1. Transfer of CAD data, 2. Application of plastic powder, 3. Selective addition of binder, 4. Lowering of the building platform, 5. Steps 2-4 repeated, 6. Removal of unbound plastic powder, 7. Finished element.

5.11.4 *Principle*

The principle of VX technology from voxeljet AG is based on the CJP technology from 3D Systems (see Section 5.2.).

5.11.5 *Strengths and Weaknesses*

The main strengths of the voxeljet AG VX technology are as follows:

(1) *Fast*: voxeljet's printers' incredible building speed places them among the fastest in the AM market.

(2) *Large production*: voxeljet's VX4000 has dimensions of 4000 x 2000 x 1000 mm offering the biggest build space in its class, thus allowing for the production of large-volume 3D models.

The weaknesses of the voxeljet AG VX technology are:

(1) *Large space required*: The machine needs a very large area to house it.
(2) *High power consumption*: The system consumes high power due to its increased printing capacity.

5.11.6 *Applications*

voxeljet AG produces plastic parts, sand moulds and cores for use in sand casting.

5.11.7 *Research and Development*

Powder binding technology typically has faster build speeds and lower material costs compared to other AM systems. The company is currently developing new material sets including ceramics, silicon carbide and tungsten carbide. The addition of ceramics would be significant to the medical implant/orthopaedic market for 3D printing.

Voxeljet AG recently launched the world's first continuous 3D printer called the VXC800 which replaced VX800. The system utilises patented technology consisting of a horizontal conveyor belt that controls the building layer-by-layer. Layers are built at the entrance of the conveyor while the unpacking is done at the exit. Hence, there are no restrictions on build length.

5.12 Other Notable Powdered-Based AM Systems

5.12.1 *Soligen's Direct Shell Production Casting (DSPC)*

5.12.1.1 *Company*

Soligen Technologies Inc. was founded by Yehoram Uziel, its President and CEO in 1991 and went public in 1993. It first installed its Direct Shell Production Casting (DSPC) System at three "alpha" sites in 1993. It bought the license to MIT's 3D printing patents for metal casting which was valid till 2006.

5.12.1.2 *Product*

Direct Shell Production Casting (DSPC) creates ceramic moulds for metal parts with integral coves directly and automatically from CAD files.

5.12.1.3 *Process*

The DSPC technology is derived from a process known as three-dimensional printing and was invented and developed at the MIT, USA. The process steps are illustrated in Fig. 5.39 and comprise of the following steps [49, 50]:

(1) A part is first designed on a computer, using commercial computer aided design (CAD) software.
(2) The CAD model is then loaded into the shell design unit—the central control unit of the equipment. The computer model for the casting mould is prepared by taking into consideration modifications such as scaling the dimensions to compensate for shrinkage, adding fillets, etc. The mould maker then decides on the number of mould cavities on each shell and the type of gating system. Once the CAD mould shells are modified, the shell design unit generates the necessary data files to be transferred to the shell production unit.

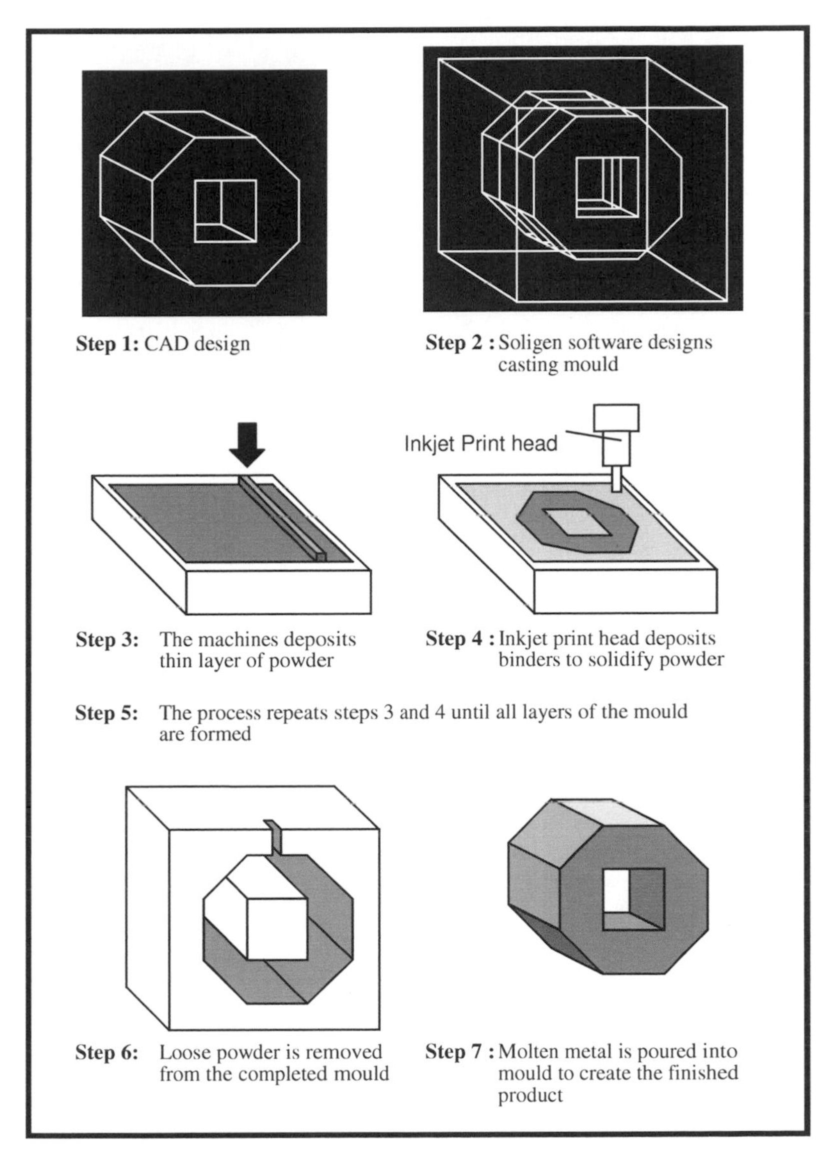

Fig. 5.39. Soligen Inc.'s DSPC process.

(3) The shell production unit begins depositing a thin layer of fine alumina powder over the shell working surface for the first slice of the casting mould. A roller follows the powder, levelling the surface.

(4) An inkjet print head moves over the layer, injecting tiny drops of colloidal silica binder onto the powder surface. The binder solidifies the powder into ceramic on contact and the unbounded alumina remains as support for the following layer. The work area lowers and another layer of powder is distributed.
(5) The process through steps 3 and 4 is repeated until the mould is complete.
(6) After the building process is completed, the casting shell remains buried in a block of loose alumina powder. The unbound excess powder is then separated from the finished shell. The shell can then be removed for postprocessing, which may include firing in a kiln to remove moisture or preheat to appropriate temperature for casting.
(7) Molten metal can then be poured in to fill the casting shell or mould. After cooling, the shell can be broken up to remove the cast which can then be processed to remove gatings, sprues, etc., thus completing the casting process.

The hardware of the DSPC system contains a PC computer, a powder holder, a powder distributor, rolls, a print head and a bin. The software includes a CAD system and a Soligen's slicing software.

5.12.1.4 *Principle*

The principle of Soligen's Direct Shell Production Casting (DSPC) is based on 3DP technology invented, developed and patented by MIT. The 3DP is licensed to Soligen on a worldwide basis for the field of metal casting. Using this technology, binder from the nozzle selectively binds the ceramic particles together to create the ceramic shell layer-by-layer until it is completed. The shell is removed from the DSPC machine and fired to ventrification temperatures to harden and remove all moisture. All excess ceramic particles are blown away.

In the process, parameters that influence performance and functionality are the layer thickness, powder's properties, the binders and the pressure of rollers.

5.12.1.5 *Strengths and weaknesses*

The key strengths of the DSPC process include the production of patternless casting and net-integral moulds. In patternless casing, direct tooling is possible — thus eliminating the need to produce any patterns. There are no parting lines, core prints or draft angles required. Integral gatings and chills can also be appropriately added.

The main limitation of the DSPC process is that the DSPC 300 only focusses on making ceramics moulds primarily for metal casting.

5.12.2 *Fraunhofer's Multiphase Jet Solidification (MJS)*

5.12.2.1 *Company*

Fraunhofer-Gessellschaft is Germany's leading organisation of applied research, maintaining 46 research establishments at 31 locations [51]. Two of these establishments, the Fraunhofer Institute for Applied Materials Research (IFAM) and the Fraunhofer Institute for Manufacturing Engineering and Automation (IPA) cooperated in developing a RP process named multiphase jet solidification (MJS).

5.12.2.2 *Process*

The MJS process is one that is able to produce metallic or ceramic parts. It uses low melting point alloys or powder-binder mixture which is squeezed out through a computer-controlled nozzle to build the part layer-by-layer [52].

The MJS process comprises two main steps: data preparation and model building.

Data Preparation. In the first step, the part is designed on a 3D CAD system. The data is then imported to the MJS system, and together with process parameters like machining speed and materials flow rate added, a controller file for the machine is generated. This file is subsequently downloaded onto the controller of the machine for the build process.

Model Building. The material used is usually a powder-binder-mixture but it can also be a liquefied alloy. At the beginning of the build process, the material is heated to beyond its melting point in a heated chamber. It is then squeezed out through a computer-controlled nozzle by a pumping system, and deposited layer-by-layer onto a platform [52, 53]. The melted material solidifies when it comes into contact with the platform or the previous layer as both temperature and pressure decrease and heat is dissipated to the part and the surrounding. The contact of the liquefied material leads to partial remelting of the previous layer and thus bonding between layers results. After one layer is finished, the extrusion jet moves in the Z-direction and the next layer is built. The part is built layer-by-layer until it is complete.

The main components of the apparatus used for the MJS process comprises a PC, a computer-controlled 3-axes positioning system and a heated chamber with a jet and a hauling system. The machine precision of the positioning system is ±0.01 mm and has a work volume of 500 × 540 × 175 mm. The chamber is temperature stabilized within ±1°C. The material is supplied as powder, pellets or bars. Extrusion temperature can reach up to 200°C. Extrusion orifices vary from 0.5 mm to 2.0 mm.

5.12.2.3 *Principle*

The working principle of the MJS process [54] is shown in Figure 5.40. The basic concept applies the extrusion of low viscosity materials through a jet layer-by-layer, similar to the fused deposition modelling process. The main differences between the two processes are in the raw material used to build the model and the feeding system. For the MJS process, the material is supplied in different phases using power-binder-mixture or liquefied alloys instead of using material in the wire-form. As the form of the material is different, the feed and nozzle systems are also different. The material used is a wax loaded with up to approximately 50% volume fraction metal powder.

In the MJS process, parameters that influence its performance and functionality are the layer thickness, the feed material, i.e. whether it is liquefied alloys (usually low melting point metals) or powder-binder-mixture (usually materials with high melting point), the chamber pressure, the machining speed (build speed), the jet specification, the material flow, and the operating temperature.

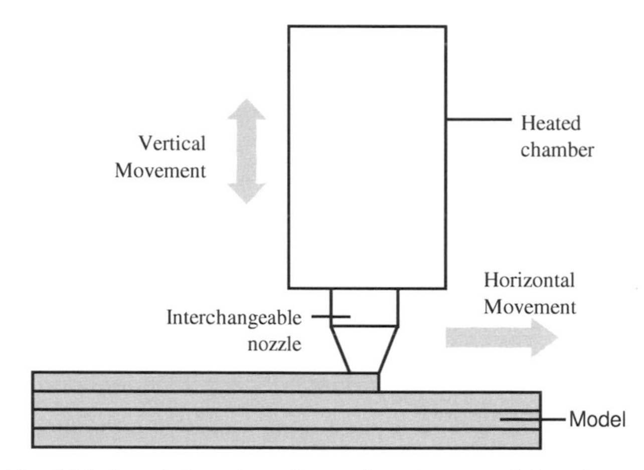

Material is heated above its melting point and squeezed through a nozzle and deposited layer-by-layer.

Fig. 5.40. Working principle of the MJS process.

5.12.3 *AeroMet Corporation's LASFORM Technology*

5.12.3.1 *Company*

In 1997, AeroMet™ was formed as a subsidiary of MTS Systems Corporation (MTS). The initial Lasform™ system installed at its Eden Prairie, Minnesota facility is operated in collaboration with U.S. Army Research Laboratory (ARL) of Aberdeen Proving Ground, Maryland. The AeroMet™ Lasform™ technology is based on research performed jointly by the Applied Physics Laboratory of John Hopkins University, the Applied Research Laboratory of Penn State University and MTS Systems Corporation. AeroMet Corporation ceased operation in 2005.

5.12.3.2 *Products*

The Lasform™ process uses commercially available precursor materials and creates parts that require minimal post-machining or heat treatment prior to use. As the precursor material is in the form of metal powders, it is also possible to produce "graded alloys" across the geometry of a component via real-time mixing of elemental constituents. This is a very unique feature of great interest to AM users and designers [55]. Since the AeroMet™ process takes place in an inert environment, it is possible to use the Lasform process in niobium, rhenium and other reactive materials which require protective processing atmospheres.

5.12.3.3 *Process*

The Lasform™ process is described as follows [56]:

(1) The AeroMet™ laser forming process starts with a 3D CAD representation of the part. This is then translated via proprietary software to generate trajectory paths for the laser forming system. These paths are transmitted as machine instructions to the laser forming system.
(2) The focussed laser beam traces out the structural shape pattern of the desired part by moving the titanium target plate beneath the beam in the approximate x-y trajectories.
(3) Titanium pre-alloyed powder is introduced into the molten metal head and provides for the buildup of the desired shape as the molten spot is traversed over a target plate in the desired pattern.
(4) The 3D structure is fabricated by repeating the pattern, layer-by-layer over the desired geometry and indexing the focal point up one layer for the repeat pattern. This layer-by-layer registry with metallurgical integrity between layers generates the desired integral ribbed structure called a machining pre-form. Post-processes include heat treatment, machining and inspection.

5.12.3.4 *Principle*

The Lasform™ process uses gas atomised and hydride-dehydride titanium alloy powders introduced into the focus region of the CO_2 laser beam [57]. The focus region is shifted in the X-Y plane as determined by the CAD slice. This is achieved by driving a numerical controlled manipulator to reproduce the desired shape. A solid titanium deposit layer remains and the process is repeated for the next layer in the Z-direction. The new layer is also fused with the previous one, building layer upon layer until the part is complete. The process is carried out in an argon-filled environment. The production of high quality titanium shapes via laser direct metal deposition requires the integration of several technologies. These include high power laser beam generation & delivery, metal powder handling, robotics, process sensing and control and environmental management.

5.12.3.5 *Strengths and weaknesses*

The strengths of the Lasform™ includes the following. It allows very high quality titanium parts to be rapidly produced. Parts mechanical tested showed that AMS standards for commercially pure Ti, Ti-6Al-4V, and Ti-5Al-2.5Sn are met. As it has a very large work volume, large parts can be seamlessly accommodated, while maintaining an oxygen-free, inert atmosphere. The laser-formed shapes also require minimal post-machining and heat treating. This provides substantial cost and time savings by eliminating high materials waste, costly manufacturing tooling and long machining times. The process is flexible for its ability to vary the composition of the material throughout the part that could result in the formation of a functionally gradient material where the microstructure and mechanical properties can vary depending on the composition. Thus the composition of different regions of the part can be customised according to functional and economic requirements.

The Lasform™ however requires a very large space to be dedicated to house the system. It also specialises only in the fabrication of titanium and other metals parts. It therefore cannot be used to produce polymer parts or parts to be used as non-functional models.

References

[1] 3D Systems. (2007). *Products: SLS® Systems.* Retrieved from http://www.3dsystems.com/products/sls/index.asp

[2] 3D Systems. *Product Brochures and Datasheets (Material Specifications).*

[3] Johnson, J. L. (1994). *Principles of Computer Automated Fabrication.* Palatino Press, Chapter 3, 75-84.

[4] Sun, M. S. M., Nelson, J. C., Beaman, J. J. & Barlow, J. J. (1991). A model for partial viscous sintering. In *Proceedings of the Solid Freeform Fabrication Symposium*, University of Texas.

[5] Hug, W. F., & Jacobs, P. F. (1991). Laser technology assessment for stereolithographic systems. In *Proceedings of the Second International Conference on Rapid Prototyping,* June 23-26, 29-38.

[6] Barlow, J. J., Sun, M. S. M. & Beaman, J. J. (1991). Analysis of selective laser sintering. In *Proceedings of the Second International Conference on Rapid Prototyping*, June 23-26, 29-38.

[7] 3D Systems (DTM Corp). (1999). *Horizons Q4*, 6–7.

[8] 3D Systems. (2005). *SLS® Technology Featured at the "Extreme Textiles" Exhibition at the Smithsonian's Cooper-Hewitt, National Design Museum.* Retrieved from http://www.3dsystems.com/newsevents/newsreleases/ pdfs/3D_Systems_SLS_technology_featured_at_Extreme.pdf

[9] 3D Systems. (2007). *Bertrandt.* Retrieved from http://www.3dsystems.com/appsolutions/casestudies/pdf/CS_Bertrandt.pdf

[10] 3D Systems. (2006). *4D Concepts for Hekatron.* Retrieved from http://www.3dsystems.com/appsolutions/casestudies/pdf/CS_4D_Concepts.pdf

[11] 3D Systems. (2014). *Customer Success Story – Hankook Tire.* Retrieved from http://www.3dsystems.com/sites/www.3dsystems.com/files/cs-hankook-01-2014.pdf

[12] Z Corporations. (2008). *Solutions.* Retrieved from https://www.zcorp.com/Solutions/Rapid-Prototypes—CAD/spage.aspx

[13] Goldsberry, C. (2007). *Rapid processes gaining ground.* Modern Plastics Worldwide, **84**(9): 35-38.

[14] Grochowski, A. (2000). *Rapid prototyping – rapid tooling*. CADCAM Forum, pp. 39-41.

[15] Serbin, J. C. W., Pretsch, C. & Shellabear, M. (1995). *STEREOS and EOSINT 1995–New Developments and State of the Art. EOS* GmbH.

[16] EOS GmbH. (Jul 2013). *The Challenge of Customised Products*. Retrieved from http://www.eos.info/additive_manufacturing/for_your_individual_design/challenge_customised_products

[17] Festo AG & Co. KG. (Jul 2013). *Industry: EOS Technology Enables Automation Specialist Festo to Design its Bionic Assistance System*. Retrieved from http://www.eos.info/press/customer_case_studies/festo

[18] Goebner, J. *A Peek into the EOS Lab: Micro Laser Sintering*. Retrieved from http://ip-saas-eos-cms.s3.amazonaws.com/public/edd2302bd9ae390e/c51e3234355cfc4834e0ecca4f70ed9a/eos_microlasersintering.pdf

[19] Xiong, Y., Smugeresky, J. E., Ajdelsztajn, L., & Schoenung, J. M. (2008). *Fabrication of WC-Co cermets by laser engineered net shaping*. Materials Science and Engineering, **A 496**(1-2): 261-266.

[20] Hwang, J. Y., Neira, A., Scharf, T. W., Tiley, J., & Banerjee, R. (2008). *Laser-deposited carbon nanotube reinforced nickel matrix composites*. Scripta Materialia, **59**(5): 487-490.

[21] Arcam AB. (2007). *Electron Beam Melting*. Retrieved from http://www.arcam.com/technology/tech_ebm.asp

[22] Gibbons, G. J., & Hansell, R. G. (2005). *Direct tool steel injection mold inserts through the Arcam EBM free-form fabrication process*. Assembly Automation, **25**(4): 300-305.

[23] Arcam AB. *New orthopaedic implants improve people's quality of life*. Rapid News.

[24] *Arcam AB Nears Completion of Fast EBM 3D Printing Research*. Engineering.com. Retrieved from http://www.engineering.com/3DPrinting/3DPrintingArticles/ArticleID/6062/Arcam-AB-Nears-Completion-of-Fast-EBM-3D-Printing-Research.aspx

[25] Prototype Magazine. (2007). *MCP SLM Machines*. Retrieved from http://www.prototypemagazine.com/index.php?option=com_content&task=view&id=102&Itemid=2

[26] Van Elsen, M., Al-Bender, F., & Krith, J. (2008). *Application of dimensional analysis to selective laser melting*. Rapid Prototyping Journal, **14**(1): 15-22.

[27] Yang, Y., Huang, Y., & Wu, W. (2008). One-step shaping of NiTi biomaterial by selective laser melting. In *Proceedings of SPIE – The International Society for Optical Engineering,* art. No. 68250C.

[28] Li, R., Liu, J., Shi, Y., Wang, L., & Jiang, W. (2012). *Balling behavior of stainless steel and nickel powder during selective laser melting process.* International Journal of Advanced Manufacturing Technology, vol. 59, pp. 1025-1035.

[29] Guan, K., Wang, Z., Gao, M., Li, X., & Zeng, X. (2013). *Effects of processing parameters on tensile properties of selective laser melted 304 stainless steel.* Materials & Design, vol. 50, pp. 581-586.

[30] Chua, C. K., Liu, A., & Leong, K. F. (2008). State of the Art in Rapid Metallic Manufacturing. In *3rd International Conference on Rapid Prototyping and Manufacturing and the 2nd International Conference for Bio-manufacturing*, ICRPM-BM. Beijing, China.

[31] Averyanova, M., Bertrand, P., & Verquin, B. (2011). *Studying the influence of initial powder characteristics on the properties of final parts manufactured by the selective laser melting technology.* Virtual and Physical Prototyping 6, 215-223.

[32] Delgado, J., Ciurana, J., & Serenó, L. (2011). *Comparison of forming manufacturing processes and selective laser melting technology based on the mechanical properties of products.* Virtual and Physical Prototyping 6, 167-178.

[33] Liu, Z. H., Chua, C. K., & Leong, K. F. (2010). *Physical and Dimensional Quantification of SLM Parts.* 4th International Conference PMI. Ghent, Belgium.

[34] Liu, Z. H., Zhang, D. Q., Chua, C. K., & Leong, K. F. (2013). Melt Characterisation of M2 High Speed Steel in Selective Laser Melting. In *6th International Conference on Advanced Research in Virtual and Rapid Prototyping*, Leiria, Portugal.

[35] Liu, A., Chua, C. K., & Leong, K. F. (2010). *Properties of Test Coupons Fabricated by Selective Laser Melting*. Advanced Precision Engineering. Zhao, J., Kunieda, M., Yang, G., & Yuan, X. M. Stafa-Zurich, Trans Tech Publications Ltd. 447-448: 780-784.

[36] Liu, Z. H., Zhang, D. Q., Chua, C. K., & Leong, K. F. (2013). *Crystal structure analysis of M2 high speed steel parts produced by selective laser melting.* Materials Characterization 84(0): 72-80.

[37] Liu, A., Chua, C. K., & Leong, K. F. (2012). Heat Treatment of SLM M2 High Speed Steel Parts. In *5th International Conference PMI.* Ghent, Belgium.

[38] Zhang, D. Q., Liu, Z. H., Loh, L. E., & Chua, C. K. (2012). Microstructure Characterisation of AlSi10Mg Parts Produced by Selective Laser Melting. In *2012 International Additive Manufacturing Forum and 6th China National Conference of Rapid Prototyping & Manufacturing*. Wuhan, China.

[39] Loh, L. E., Chua, C. K., Liu, Z. H., & Zhang, D. Q. (Oct 2013). Effect of Laser Beam Profile on Melt Track in Selective Laser Melting. In *6th International Conference on Advanced Research in Virtual and Rapid Prototyping*, Leiria, Portugal.

[40] Loh, L. E., Liu, Z. H., Zhang, D. Q., Mapar, M., Sing, S. L., Chua, C. K., & Yeong, W. Y. (2014). *Selective Laser Melting of aluminium alloy using a uniform beam profile.* Virtual and Physical Prototyping 9(1): 11-16.

[41] Zhang, D. Q., Liu, Z. H., & Chua, C. K. (Oct 2013). Investigation on Forming Process of Copper Alloys via Selective Laser Melting. In *6th International Conference on Advanced Research in Virtual and Rapid Prototyping*, Leiria, Portugal.

[42] Sintermask Technologies AB. (2007). *Sintermask Technologies*. Retrieved from http://www.sintermask.se/page.php?p=7

[43] 3D MicroPrint GmbH. (2014). *EOS and 3D-Micromac are bringing Micro Laser Sintering technology into a newly-established business enterprise*. Retrieved from http://3d-micromac.com/newsdetails/items/id-3d-microprint-gmbh-eos-and-3d-micromac-are-bringing-micro-laser-sintering-technology-into-a-newly-established-business-enterp.html

[44] The Editors. (2001). *Added options for producing 'impossible' shapes, rapid traverse–Technology and trends spotted by the Editors of Modern Machine Shop*. MMS Online.

[45] Vasilash, G. S. (2001). *A quick look at rapid prototyping*. Automotive Design and Production.

[46] Waterman, P. J. (2000). *RP3: Rapid prototyping, pattern making and production.* DE Online.

[47] Exone. *ExOne Cut Costs, Improve Quality & Shorten Lead Time on Production.* Retrieved from http://www.exone.com/sites/default/files/Case_Studies/sand_US Navy2.pdf

[48] ExOne. *ExOne Opens Sixth Production Service Centre*. Retrieved from http://www.exone.com/en/news/exone-opens-sixth-production-service-center

[49] Uziel, Y. (1995). *Art to part in 10 days*. Machine Design, pp. 56-60.

[50] Gregor, A. (1994). *From PC to factory*. Los Angeles Times.

[51] Fraunhofer-Gesellschaft, Profile of the Fraunhofer-Gesellschaft: its purpose, capabilities and prospects, Fraunhofer-Gesellschaft, 1995.

[52] Geiger, M., Steger, W., Greul, M., & Sindel, M. (1994). *Multiphase jet solidification, European action on rapid prototyping*. EARP Newsletter, **3**, Aarhus.

[53] Rapid Prototyping Report. (1994). *Multiphase Jet Solidification (MJS)*. CAD/CAM Publishing Inc. **4**(6): 5.

[54] Greulick, M., Greul, M., & Pitat, T. (1995). *Fast, functional prototypes via Multiphase Jet Solidification.* Rapid Prototyping Journal, **1**(1): 20-25.

[55] Arcella, F. G., Whitney, E. J., & Krantz, D. (1995). *Laser forming near shapes in titanium.* ICALEO '95: Laser Materials Processing, **80**, 178-183.

[56] Abott, D. H., & Arcella, F. G. (1998). *AeroMet implementing novel Ti process.* Metal Powder Report, 53(2).

[57] Arcella, F. G., Abbot, D. H., & House, M. A. (1998). Rapid laser forming of titanium structures. In *Metallurgy World Conference and Exposition*, Grenada, Spain.

Problems

1. Using a sketch to illustrate your answer, describe the Selective Laser Sintering process.

2. Discuss the types of materials available for the Sinterstation HiQ Series.

3. Describe the differences between the Selective Laser Sintering process and the colorjet printing process.

4. List the advantages and disadvantages of the ColorJet Printing process.

5. Present the types of EOS systems and their capabilities in terms of materials processing.

6. Compare and contrast the laser-based EOS process with the non-laser-based Direct Shell Production Casting for the production of foundry casting moulds.

7. What are the critical factors that influence the performance and applications of the following AM processes?

   (a) 3D Systems' SLS,

   (b) 3D Systems CJP,

   (c) Optomec's LENS,

   (d) Arcam's EBM,

8. Name three laser powder-based AM systems and three non-laser powder-based AM systems.

9. Compare and contrast EOS's EOSINT M system with Optomec's LENS system. What are the advantages and disadvantages for each of these systems?

10. Describe the system and process for Optomec's Aerosol jet system.

11. Describe Arcam's Electron Beam Melting (EBM) technology.

12. List the advantages and disadvantages for the Concept Laser's LaserCUSING process.

13. Discuss the advantages and disadvantages of powder-based AM systems compared with:

    (a) liquid-based AM systems,

    (b) solid-based AM systems

14. List the significances of MCP-HEK's selective laser melting process.

15. Discuss possible applications for the 3D Systems' PX^{TM} Series.

16. Describe the process and principles of the Selective Mask Sintering process by Speed Part.

17. What are the key features of the 3D Micromac AG's MicroSINTERING process?

18. Discuss the differences between ExOne's DPM process and that of the Selective Laser Sintering process for the production of metal parts.

19. List voxeljet's field of operation for their AM system.

20. Discuss the processes, strengths and limitations of the following AM processes:

    (a) Direct Shell Production Casting

    (b) Multiphase Jet Solidification

    (c) Lasform Technology

Chapter 6

ADDITIVE MANUFACTURING DATA FORMATS

6.1 STL Format

Representation methods used to describe CAD geometry vary from one system to another. A standard interface is needed to convey geometric descriptions from various CAD packages to Additive Manufacturing (AM) systems. For the last three decades, the STL (**ST**ereo**L**ithography) file format, as the de facto standard, has been used in many, if not all, AM systems to exchange information between design programs and AM systems [1].

The STL file [1-3], conceived by 3D Systems, USA, is created from the CAD database via an interface on the CAD system. This file consists of an unordered list of triangular facets representing the outside skin of an object. There are two STL file formats. One is the ASCII format and the other is the binary format. The size of ASCII STL file is larger than that of the binary format, but is human-readable. In a STL file, triangular facets are described by a set of X, Y and Z coordinates for each of the three vertices and a unit normal vector with X, Y and Z to indicate the side of the facet, which is inside or outside the object. An example is shown in Figure 6.1

Because the STL file is a facet model derived from precise CAD drawings, it is an approximate model of the part. Moreover, many commercial CAD models are not robust enough to generate the facet model (STL file) and frequently have problems as a result.

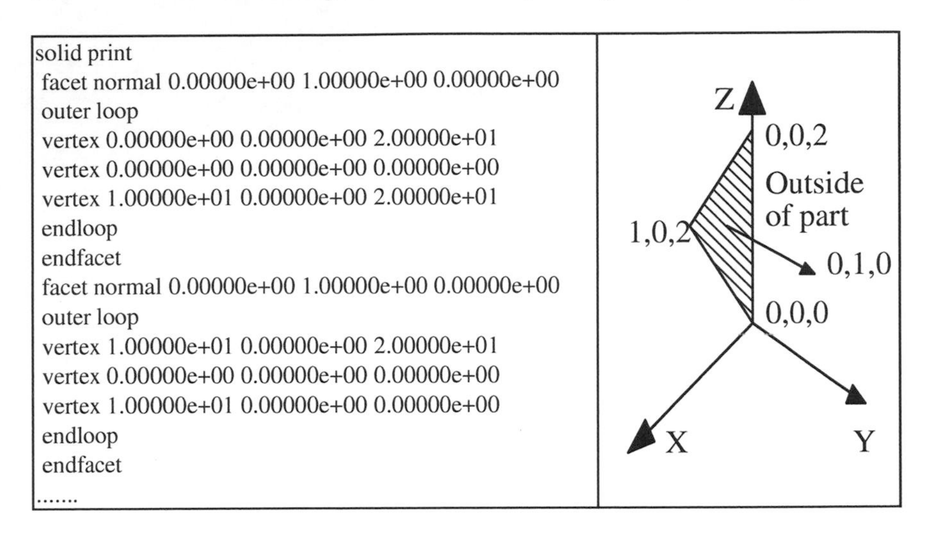

```
solid print
 facet normal 0.00000e+00 1.00000e+00 0.00000e+00
 outer loop
 vertex 0.00000e+00 0.00000e+00 2.00000e+01
 vertex 0.00000e+00 0.00000e+00 0.00000e+00
 vertex 1.00000e+01 0.00000e+00 2.00000e+01
 endloop
 endfacet
 facet normal 0.00000e+00 1.00000e+00 0.00000e+00
 outer loop
 vertex 1.00000e+01 0.00000e+00 2.00000e+01
 vertex 0.00000e+00 0.00000e+00 0.00000e+00
 vertex 1.00000e+01 0.00000e+00 0.00000e+00
 endloop
 endfacet
.......
```


Fig. 6.1. A sample STL File.

Nevertheless, there are several advantages of the STL file. First, it provides a simple method of representing 3D CAD data. Second, it is already a de facto standard and has been used by most CAD systems and AM systems. Finally, it can provide small and accurate files for data transfer for certain shapes.

On the other hand, several disadvantages of the STL file exist. First, the STL file is many times larger than the original CAD data file for a given accuracy parameter. The STL file carries much redundant information such as duplicate vertices and edges, as shown in Figure 6.2. Second, geometry flaws exist in STL files because many commercial tessellation algorithms used by CAD vendors today are not sufficiently robust. This gives rise to the need for a 'repair software', which slows the production cycle time. Third, the STL file carries limited information to represent colour, texture material, substructure, and other properties of the manufactured end object. Finally, the subsequent slicing of large STL files can take many hours. However, some AM processes can slice while they are building the previous layer, and this will alleviate this disadvantage.

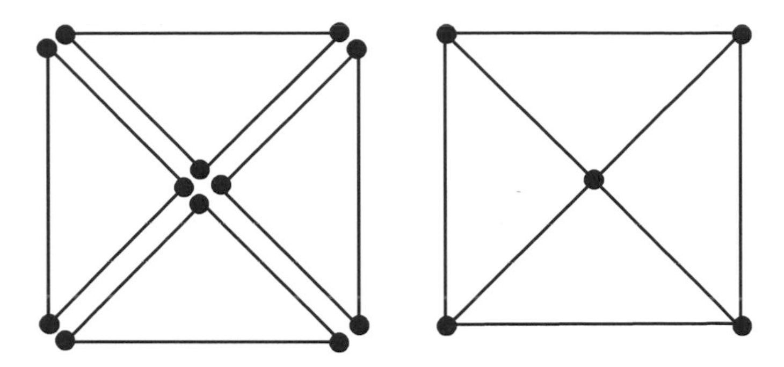

Fig. 6.2. Edge and vertex redundancy in STL format.

6.2 STL File Problems

Several problems plague STL files and they are due to the very nature of STL files as they contain no topological data. Many commercial tessellation algorithms used by CAD vendors today are also not robust [4-6], and as a result they tend to create polygonal approximation models which exhibit the following types of errors:

(1) Gaps (cracks, holes, punctures), that is, missing facets
(2) Degenerate facets (where all its edges are collinear)
(3) Overlapping facets
(4) Non-manifold topology conditions

The underlying problem is due, in part, to the difficulties encountered in tessellating trimmed surfaces, surface intersections and controlling numerical errors. This inability of the commercial tessellation algorithm to generate valid facet model tessellations makes it necessary to perform model validity checks before the tessellated model is sent to the AM equipment for manufacturing. If the tessellated model is invalid, procedures become necessary to determine what the specific problems are, whether they are due to gaps, degenerate facets or overlapping facets, etc.

Early research has shown that repairing invalid models is difficult and not at all obvious [7]. However, before proceeding any further into discussing the procedures that can be generated to resolve these difficulties, the following sections shall clarify what the problems, as mentioned earlier, are. In addition, an illustration will be presented to show the consequences brought about by a model having a missing facet, that is, a gap in the tessellated model.

6.2.1 *Missing Facets or Gaps*

Tessellation of surfaces with large curvature can result in errors at the intersections between such surfaces, leaving gaps or holes along the edges of the part model [8]. A surface intersection anomaly which results in a gap is shown in Figure 6.3.

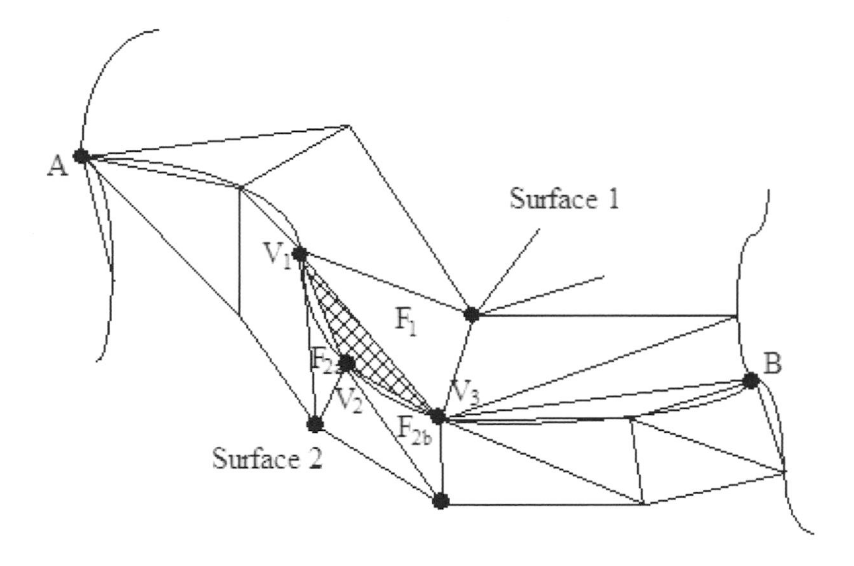

Fig. 6.3. Gaps due to missing facets [4].

6.2.2 *Degenerate Facets*

A geometrical degeneracy of a facet occurs when all of the facets' edges are collinear even though all its vertices are distinct. This might be caused by stitching algorithms that attempt to avoid shell punctures as shown in Figure 6.4(a) below [9].

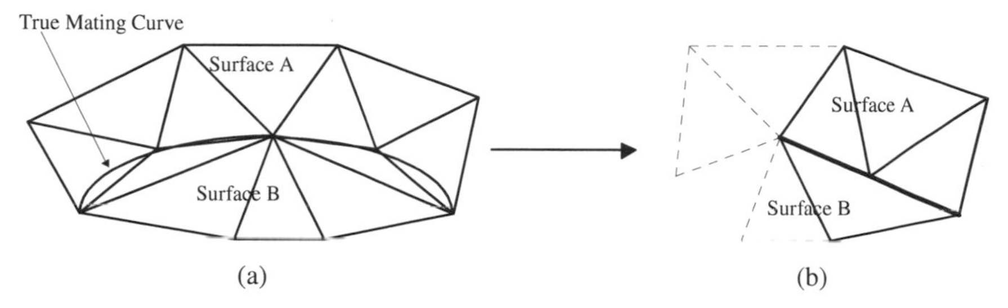

Fig. 6.4. Shell punctures (a) created by unequal tessellation of two adjacent surface patches along their common mating curve and (b) eliminated at the expense of adding a degenerate facet.

The resulting facets generated, shown in Figure 6.4(b), eliminate the shell punctures. However, this is done at the expense of adding a degenerate facet. While degenerate facets do not contain valid surface normals, they do represent implicit topological information on how two surfaces are mated. This important information is consequently stored prior to discarding the degenerate facet.

6.2.3 *Overlapping Facets*

Overlapping facets may be generated due to numerical round-off errors that occur during tessellation. The vertices are represented in 3D space as floating point numbers instead of integers. Thus the numerical round-off can cause facets to overlap if tolerances are set too liberally. An example of an overlapping facet is illustrated in Figure 6.5.

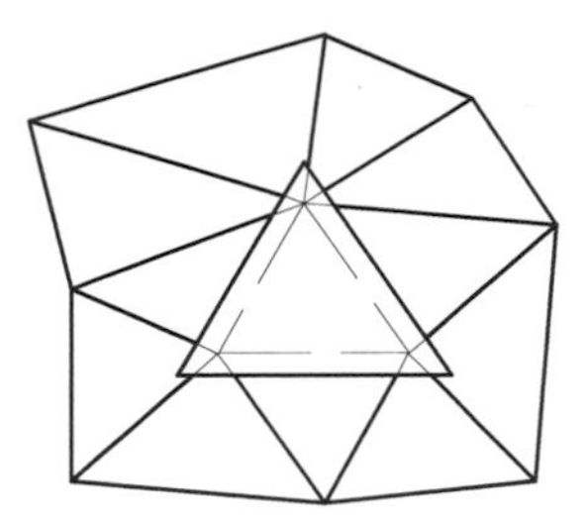

Fig. 6.5. Overlapping facets.

6.2.4 *Non-Manifold Conditions*

There are 3 types of non-manifold conditions, namely:

(1) A non-manifold edge
(2) A non-manifold point
(5) A non-manifold face

These may be generated because tessellations of fine features are susceptible to round-off errors. An illustration of a non-manifold edge is shown in Figure 6.6 (a). Here, the non-manifold edge is actually shared by four different facets, as shown in Figure 6.6 (b). A valid model would be one whose facets have only an adjacent facet each, that is, one edge shares two facets only. Hence the non-manifold edges must be resolved such that each facet has only one neighbouring facet along each edge, that is, by reconstructing a topologically manifold surface [4]. Shown in Figures 6.6 (c) and 6.6 (d) are two other types of non-manifold conditions.

All problems that have been mentioned previously are difficult for most slicing algorithms to handle, and cause fabrication problems for AM processes which essentially require valid tessellated solids as input. Moreover, these problems arise because tessellation is a first-order approximation of more complex geometric entities. Thus, such problems have become almost inevitable as long as solid models are represented using the STL format, which inherently has these limitations.

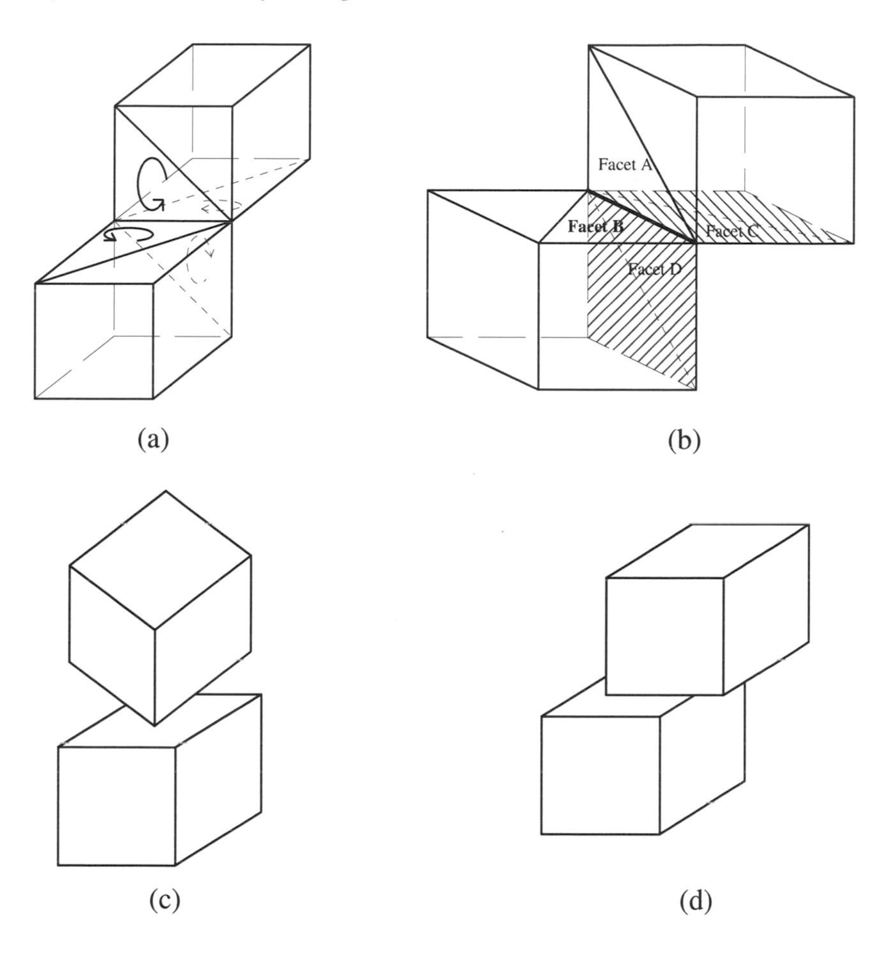

Fig. 6.6. A non-manifold edge (a) whereby two imaginary minute cubes share a common edge and (b) whereby four facets share a common edge after tessellation. (c) Non-manifold point and (d) non-manifold face.

6.3 Consequences of Building a Valid and Invalid Tessellated Model

Each of the following sections presents an example of the outcome of a model built using a valid and invalid tessellated model, when used as an input to AM systems:

6.3.1 *A Valid Model*

A tessellated model is said to be valid if there are no missing facets, degenerate facets, overlapping facets or any other abnormalities. When a valid tessellated model (see Figure 6.7 (a)) is used as an input, it will first be sliced into 2D layers, as shown in Figure 6.7 (b). Each layer would then be converted into unidirectional (or 1D) scan lines for the laser or other AM techniques to commence building the model, as shown in Figure 6.7 (c). The scan lines act as on/off points for the laser beam controller so that the part model can be built accordingly without any problems.

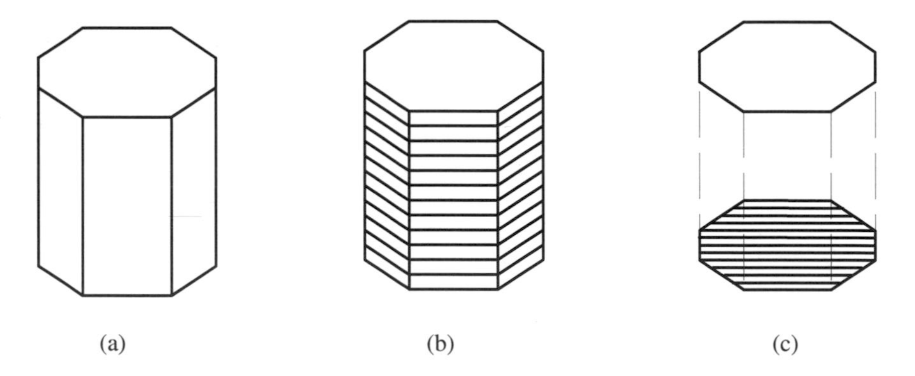

Fig. 6.7. (a) A valid 3D model, (b) a 3D model sliced into 2D planar layers and (c) conversion of 2D layers into 1D scan lines.

6.3.2 *An Invalid Model*

However, if the tessellated model is invalid, a situation may develop as shown in Figure 6.8.

A solid model is tessellated non-robustly and results in a gap as shown in Figure 6.8(a). If this error is not corrected and the model is subsequently sliced, as shown in Figure 6.8(b), in preparation for it to be built layer by layer, the missing facet in the geometrical model would cause the system to have no pre-defined stopping boundary on the particular slice. Thus the building process would continue right to the physical limit of the AM machine, creating a stray physical solid line and ruining the part being produced, as illustrated in Figure 6.8(c). Therefore, it is of utmost importance that the model is 'repaired' before it is sent for building. Thus, the model validation and repair problem are stated as follows:

Given a facet model (a set of triangles defined by their vertices), in which there are gaps, i.e., missing one or more sets of polygons, generate 'suitable' triangular surfaces which 'fill' the gaps [4].

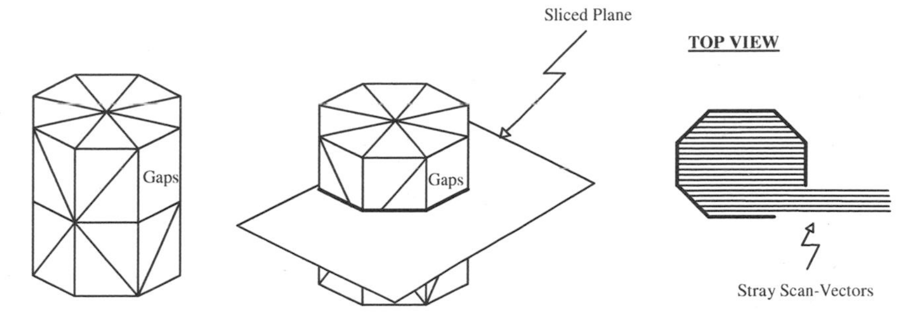

Fig. 6.8. (a) An invalid tessellated model, (b) an invalid model being sliced and (c) a layer of an invalid model being scanned.

6.4 STL File Repair

The STL file repair can be implemented using a generic solution and dedicated solutions for special cases.

6.4.1 *Generic Solution*

In order to ensure that the model is valid and can be robustly tessellated, one solution is to check the validity of all the tessellated triangles in the model. This section presents the basic problem of missing facets and a proposed generic solution to solve the problem.

In existing AM systems, when a punctured shell is encountered, the course of action taken usually requires a skilled technician to manually repair the shell. This manual shell repair is frequently done without any knowledge of the designer's intent. The work can be very time-consuming and tedious, thus negating the advantages of AM as the cost would increase and the time taken might be longer than that taken if traditional prototyping processes were used.

The main problem of repairing the invalid tessellated model would be that of matching the solution to the designer's intent when it may have been lost in the overall process. Without knowledge of the designer's intent, it would indeed be difficult to determine what the 'right' solution should be. Hence, an 'educated' guess is usually made when faced with ambiguities of the invalid model.

The algorithm for a generic solution to solve the 'missing facets' problem aims to match, if not exceed, the quality of repair done manually by a skilled technician when information of the designer's intent is not available. The basic approach of the algorithm would be to detect and identify the boundaries of all the gaps in the model. Once the boundaries of the gap are identified, suitable facets would then be generated to repair and 'patch up' these gaps. The size of the generated facets would be restricted by the gap's boundaries while the orientation of its normal would be controlled by comparing it with the rest of the shell. This is to ensure that the generated facet orientation is correct and consistent throughout the gap closure process.

The orientation of the shell's facets can be obtained from the STL file which lists its vertices in an ordered manner following Mobius' rule. The algorithm exploits this feature so that the repair carried out on the invalid model, using suitably created facets, would have the correct orientation. Thus, this generic algorithm can be said to have the ability to make an inference from the information contained in the STL file so that the following two conditions can be ensured:

(1) The orientation of the generated facet is correct and compatible with the rest of the model.
(2) Any contoured surface of the model would be followed closely by the generated facets due to the smaller facet generated. This is in

contrast to manual repair whereby, in order to save time, fewer facets generated to close the gaps are desired, resulting in large generated facets that do not follow closely to the contoured surfaces.

Finally, the basis for the working of the algorithm is due to the fact that in a valid tessellated model, there must only be two facets sharing every edge. If this condition is not fulfilled, then this indicates that there are some missing facets. With the detection and subsequent repair of these missing facets, the problems associated with an invalid model can then be eliminated.

6.4.1.1 *Solving the 'Missing Facets' problem*

The following procedure illustrates the detection of gaps in the tessellated model and its subsequent repair. It is carried out in four steps:

Step 1: Checking for approved edges with adjacent facets

The checking routine executes as follows for Facet A, as seen in Figure 6.9:

(a)
   i. Read in first edge {vertex 1-2} from the STL file.
   ii. Search file for a similar edge in the opposite direction {vertex 2-1}.
   iii. If edge exists, store this under a temporary file (eg. file B) for approved edges.

(b)
   i. Read in second edge {vertex 2-3} from the STL file.
   ii. Search file for a similar edge in the opposite direction {vertex 3-2}.
   iii. Perform as in *a(iii)* above.

(c)
   i. Read in third {vertex 3-1} from the STL file.
   ii. Search file for a similar edge in the opposite direction {vertex 1-3}.
   iii. Perform as in *a(iii)* above.

This process is repeated for the next facet until all the facets have been searched.

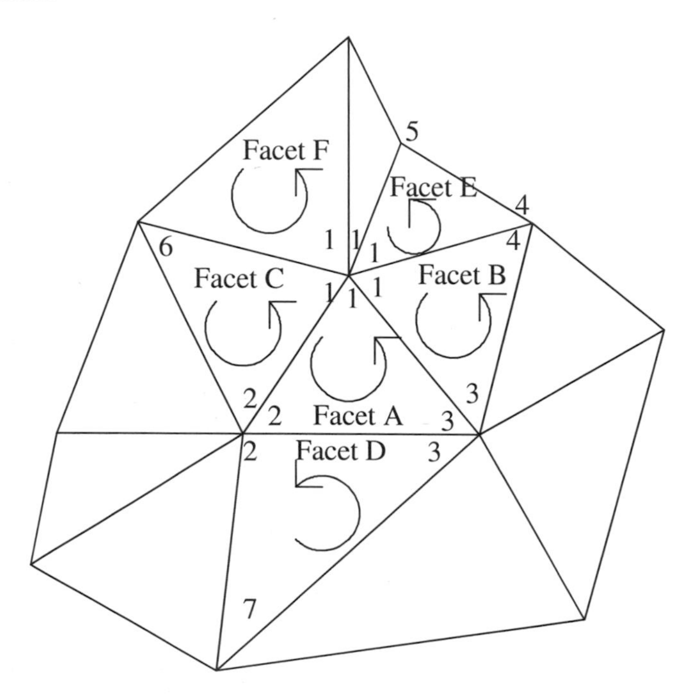

Fig. 6.9. A representation of a portion of a tessellated surface without any gaps.

Step 2: Detection of gaps in the tessellated model

With reference to Fig. 6.10, the detection routine executes as follows:

(a) i. For Facet A, read in edge {vertex 2-3} from the STL file.
 ii. Search file for a similar edge in the opposite direction {vertex 3-2}.
 iii. If edge does not exist, store edge {vertex 3-2} in another temporary file (eg. file C) for suspected gap's bounding edges and store vertex 2-3 in file B1 for existing edges without adjacent facets (this would be used later for checking the generated facet orientation).

(b) i. For Facet B, read in edge {vertex 5-2} from the STL file.
ii. Search file for a similar edge in the opposite direction {vertex 2-5}.
iii. If it does not exist, perform as in a(iii) above.

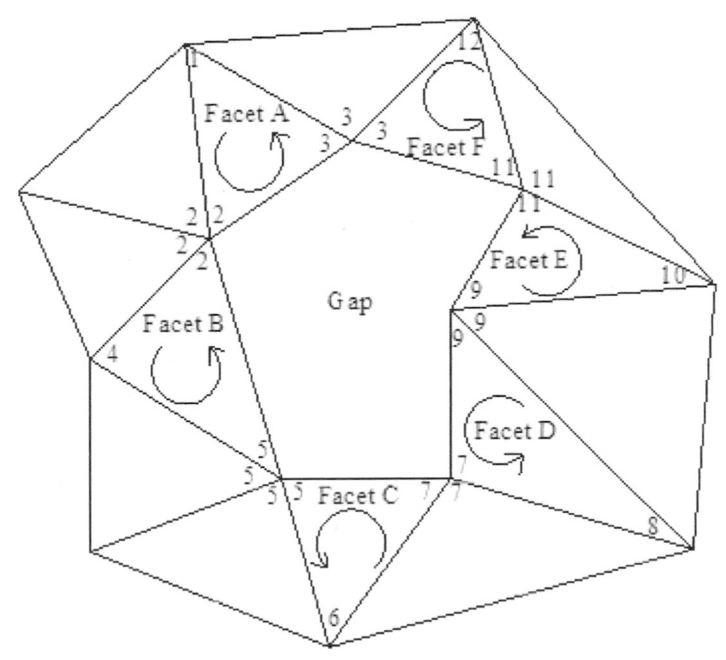

Fig. 6.10. A representation of a portion of a tessellated surface with a gap present.

(c) i. Repeat for edges: 5-2; 7-5; 9-7; 11-9; 3-11.
ii. Search for edges: 2-5; 5-7; 7-9; 9-11; 11-3.
iii. Store all the edges in the temporary file B1 for edges without any adjacent facet and store all the suspected bounding edges of the gap in temporary file C. File B1 can appear as in Table 6.1.

Table 6.1. File B1 contains existing edges without adjacent facets.

Edge						
Vertex	First	Second	Third	Fourth	Fifth	Sixth
First	2	7	3	5	9	11
Second	3	5	11	2	7	9

Step 3: Sorting of erroneous edges into a closed loop

When the checking and storing of edges (both with and without adjacent facets) are completed, a sort would be carried out to group all the edges without adjacent facets to form a closed loop. This closed loop would represent the gap detected and be stored in another temporary file (e.g., file D) for further processing. The following is a simple illustration of what could be stored in file C for edges that do not have an adjacent edge:

Assuming all the 'erroneous' edges are stored according to the detection routine (see Figure 6.10 for all the erroneous edges), then file C can appear as in Table 6.2.

Table 6.2. File C containing all the 'erroneous' edges that would form the boundary of each gap.

	Edge								
Vertex	First	Second	Third	Fourth	Fifth	Sixth	Seventh	Eighth	Ninth
First	3	5	*	11	2	7	*	9	*
Second	2	7	*	3	5	9	*	11	*

* represents all the other edges that would form the boundaries of other gaps

As can be seen in Table 6.2, all the edges are all unordered. Hence, a sort would have to be carried out to group all the edges into a closed loop. When the edges have been sorted, it would then be stored in a temporary file, say file D. Table 6.3 is an illustration of what could be stored in file D.

Table 6.3. File D containing sorted edges.

	Edge					
Vertex	First	Second	Third	Fourth	Fifth	Sixth
First	3	2	5	7	9	11
Second	2	5	7	9	11	3

Figure 6.11 is a representation of the gap, with all the edges forming a sorted closed loop.

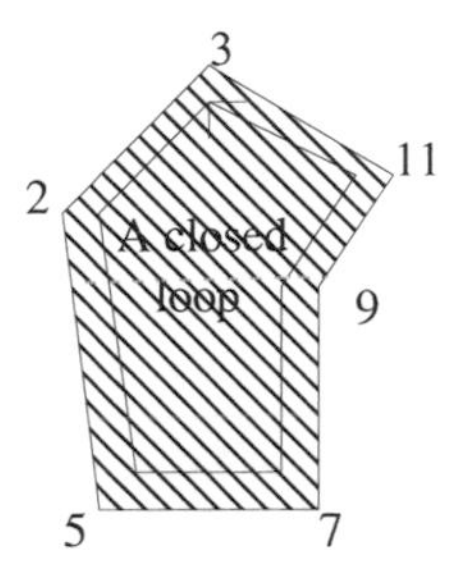

Fig. 6.11. A representation of a gap bounded by all the sorted edges.

Step 4: Generation of facets for the repair of the gaps

When the closed loop of the gap is established with its vertices known, facets are generated one at a time to fill up the gap. This process is summarised in Table 6.4 and illustrated in Figure 6.12.

Table 6.4. Process of facet generation.

		V3	V2	V5	V7	V9	V11
Generation of facets	F1	1	2	-	-	-	3
	F2	E	1	-	-	2	3
	F3	E	1	2	-	3	E
	F4	E	E	1	2	3	E

V = vertex, F = facet, E = eliminated from the process of facet generation

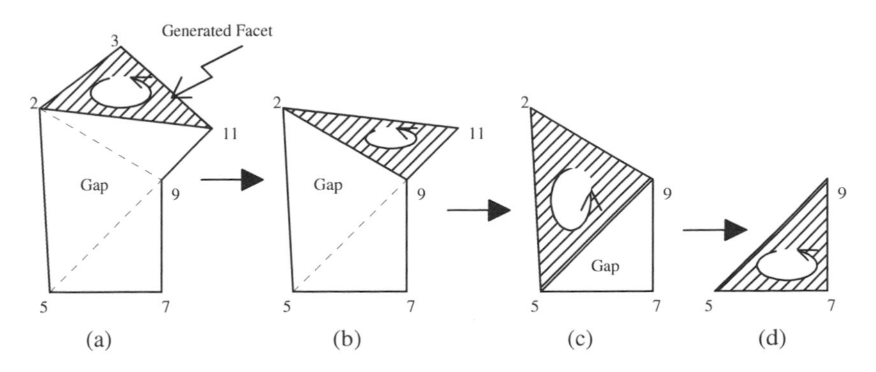

Fig. 6.12. (a) First facet generated, (b) second facet generated, (c) third facet generated and (d) fourth facet generated.

With reference to File D,

(1) Generating the first facet: First two vertices (V3 and V2) in the first two edges of file D will be connected to the first vertex in the last edge (V11) in file D, and the facet will be stored in a temporary file E (see Table 6.5 on how the first generated facet would be stored in file E). The facet is then checked for its orientation using the information stored in file B1. Once its orientation is determined to be correct, the first vertex (V3) from file D will be temporarily removed.
(2) Generating the second facet: Of the remaining vertices in file D, the previous second vertex (V2) will become the first edge of file D. The second facet is formed by connecting the first vertex (V2) of the first edge with that of the last two vertices in file D (V9, V11), and the facet is stored in temporary file E. It is then checked to confirm if its orientation is correct. Once it is determined to be correct, the vertex (V11) of the last edge in file D is then removed temporarily.
(3) Generating the third facet: The whole process is repeated as it was done in the generation of facets 1 and 2. The first vertex of the first two edges (V2, V5) is connected to the first vertex of the last edge (V9) and the facet is stored in temporary file E. Once its orientation is confirmed, the first vertex of the first edge (V2) will be removed from file D temporarily.
(4) Generating the fourth facet: The first vertex in the first edge will then be connected to the first vertices of the last two edges to form the fourth facet and it will again be stored in the temporary file E. Once the number of edges in file D is less than 3, the process of facet generation will be terminated. After the last facet is generated, the data in file E will be written to file A and its content (file E's) will be subsequently deleted. Table 6.5 shows how file E may appear.

The above procedures work for both types of gaps whose boundaries consist either of an odd or even number of edges. Figure 6.13 and Table 6.6 illustrate how the algorithm works for an *even* number of edges or vertices in file D.

Table 6.5. Illustration of how data could be stored in File E.

Generated	First Edge		Second Edge		Third Edge	
Facet	First vertex	Second vertex	First vertex	Second vertex	First vertex	Second vertex
First	V3	V2	V2	V5	V5	V3
Second	V2	V9	V9	V11	V11	V2
Third	V2	V5	V5	V9	V9	V2
Fourth	V5	V7	V7	V9	V9	V5

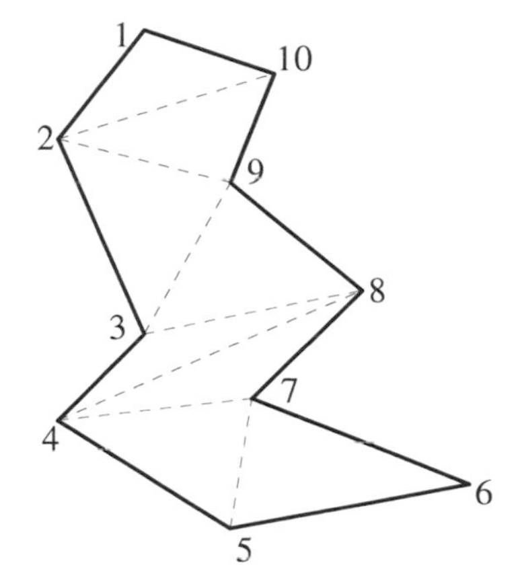

Fig. 6.13. Gaps with an even number of edges.

Table 6.6. Process of facet generation for gaps with even number of edges.

Facets	**Vertices**									
	V1	V2	V3	V4	V5	V6	V7	V8	V9	V10
F1	1	2								3
F2	E	1							2	3
F3	E	1	2						3	E
F4	E	E	1					2	3	E
F5	E	E	1	2				3	E	E
F6	E	E	E	1			2	3	E	E
F7	E	E	E	1	2		3	E	E	E
F8	E	E	E	E	1	2	3	E	E	E

With reference to Table 6.6,

First facet generated:
Edge 1 → V1, V2
Edge 2 → V2, V10
Edge 3 → V10, V1

Second facet generated:
Edge 1 → V2, V9
Edge 2 → V9, V10
Edge 3 → V10, V1

and so on until the whole gap is covered. Similarly, Figure 6.14 and Table 6.7 illustrate how the algorithm works for an *odd* number of edges or vertices in file D.

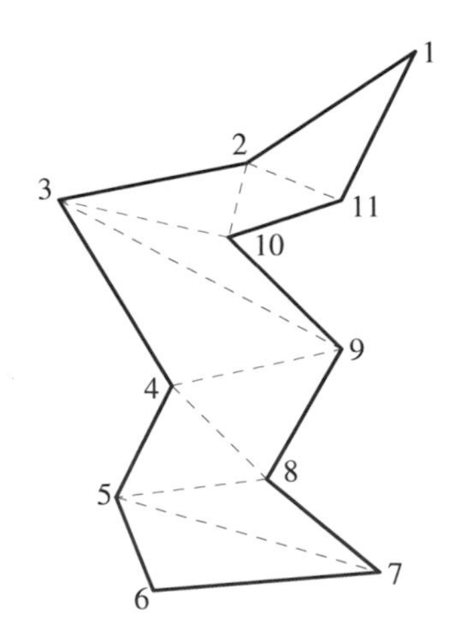

Fig. 6.14. Gaps with odd number of edges.

Table 6.7. Process of Facet Generation for Gaps with Odd Number of Edges.

Facets	**Vertices**										
	V1	V2	V3	V4	V5	V6	V7	V8	V9	V10	V11
F1	1	2									3
F2	E	1								2	3
F3	E	1	2							3	E
F4	E	E	1						2	3	E
F5	E	E	1	2					3	E	E
F6	E	E	E	1				2	3	E	E
F7	E	E	E	1	2			3	E	E	E
F8	E	E	E	E	1		2	3	E	E	E
F9	E	E	E	E	1	2	3	E	E	E	E

The process of facet generation for *odd* vertices is also done in the same way as *even* vertices. The process of facet generation has the following pattern:

F1 → First and second vertices are combined with the last vertex. Once completed, eliminate first vertex. The remainder is 10 vertices.

F2 → First vertex is combined with last 2 vertices. Once completed, eliminate the last vertex. The remainder is 9 vertices.

F3 → First and second vertices are combined with the last vertex. Once completed, eliminate first vertex. The remainder is 8 vertices.

F4 → First vertex is combined with last 2 vertices. Once completed, eliminate the last vertex. The remainder is 7 vertices.

This process is continued until all the gaps are patched.

6.4.1.2 *Solving the 'wrong orientation of facets' problem*

In the case when the generated facet's orientation is wrong, the algorithm should be able to detect it and corrective action be taken to rectify this error. Figure 6.15 shows how a generated facet with a wrong orientation can be corrected.

It can be seen that facet Z (vertices 1, 2, 11) is oriented in a clockwise direction and this contradicts the right-hand rule adopted by the STL format. Thus, this is not acceptable and needs corrections.

This can be done by shifting the last record in file D of Table 6.8 to the position of the first edge in file D of Table 6.9. All the edges, including the initial first one will be shifted one position to the right (assuming that the records are stored in the left to right structure). Once this is done, step 4 of facet generation can be implemented.

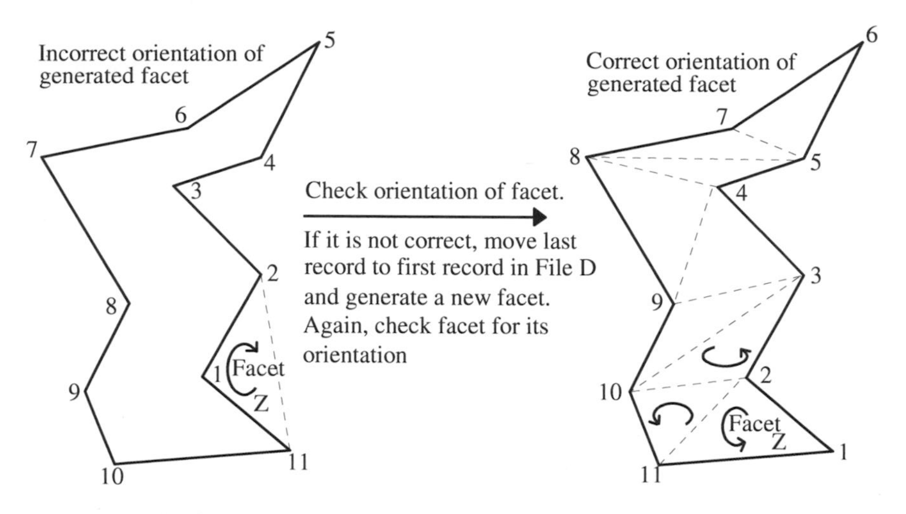

Fig. 6.15. Incorrectly generated facet's orientation and its repair.

Before the shift:

Table 6.8. Illustration showing how File D is manipulated to solve orientation problems.

	Edge										
Vertex	First	Second	Third	Fourth	Fifth	Sixth	Seventh	Eighth	Ninth	Tenth	Eleventh
First	1	2	3	4	5	6	7	8	9	10	11
Second	2	3	4	5	6	7	8	9	10	11	1

After the shift:

Table 6.9. Illustration showing the result of the shift to correct the facet orientation.

	Edge										
Vertex	First	Second	Third	Fourth	Fifth	Sixth	Seventh	Eighth	Ninth	Tenth	Eleventh
First	11	1	2	3	4	5	6	7	8	9	10
Second	1	2	3	4	5	6	7	8	9	10	11

As can be seen from the above example, vertices 1 and 2 are used initially as the first edge to form a facet. However, this resulted in a facet having a clockwise direction. After the shift, vertices 11 and 1 are used as the first edge to form a facet.

Facet Z, as shown on the right-hand-side of Figure 6.15, is again generated (vertices 1, 2, 11) and checked for its orientation. When its orientation is correct (i.e. in the anti-clockwise direction), it is saved and stored in temporary file E.

All subsequent facets are then generated and checked for their respective orientations. If any of its subsequent generated facets has incorrect orientation, the whole process will be restarted using the initial temporary file D. If all the facets are in the right orientation, it will then be written to the original file A.

6.4.1.3 *Comparison with an existing algorithm for facet generation*

An illustration of an existing algorithm that might cause a very narrow facet (shaded) to be generated is shown in Figure 6.16.

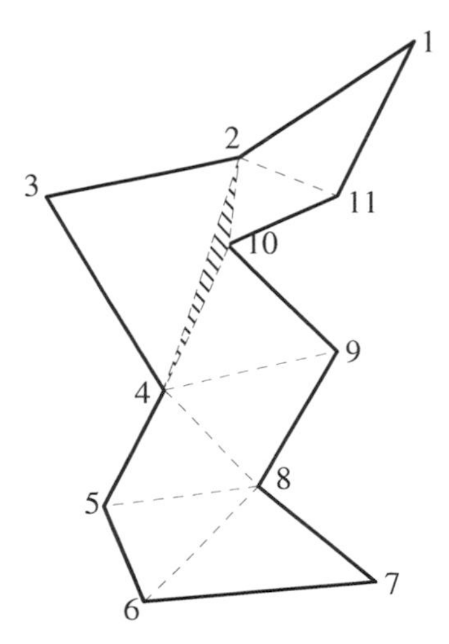

Fig. 6.16. Generation of facets using an algorithm that uses the smallest angle between edges.

This results from using an algorithm that uses the smallest angle to generate a facet. In essence, the problem is caused by the algorithm's search for a time-local rather than a global-optimum solution [10]. Also, calculation of the smallest angle in 3-D space is very difficult.

Figure 6.16 is similar to Figure 6.14. However, in this case, the facet generated (shaded) can be very narrow. In comparing the algorithms, the result obtained would match, if not, exceed the algorithm that uses the smallest angle to generate a facet.

6.4.2 *Special Algorithms*

The generic solution presented could only cater to gaps (whether simple or complex) that are isolated from one another. However, should any of the gaps meet at a common vertex, the algorithm may not be able to work properly. In this section, the algorithm is expanded to include solving some of these special cases. The special cases include:

(1) Two or more gaps are formed from a coincidental vertex
(2) Degenerate facets
(3) Overlapping facets

The special cases are classified as such because these errors are not commonly encountered in the tessellated model. Hence it is not advisable to include this expanded algorithm in the generic solution as it can be very time-consuming to apply during a normal search. However, if there are still problems in the tessellated model after the generic solution's repair, the expanded algorithm can then be used to detect and solve the special case problems.

6.4.2.1 *Two or more gaps are formed from a coincidental vertex*

The first special case deals with problems where two or more gaps are formed from a coincidental vertex. Appropriate modifications to the general solution may be made according to the solutions discussed as follows:

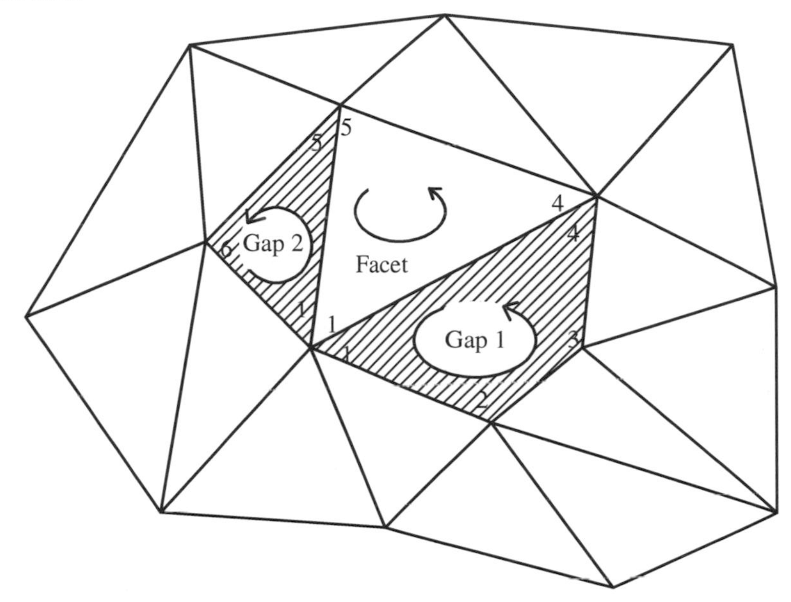

Fig. 6.17. Two gaps sharing one coincidental vertex.

As can be seen from Figure 6.17, there exists two gaps that are connected to vertex 1. The algorithm given in Section 6.4.1 would have a problem identifying which vertex to go when the search reaches vertex 1 (either vertex 2 in gap 1 or vertex 5 in gap 2). Table 6.10 illustrates what file C would look like, given the two gaps.

Table 6.10. File C containing 'erroneous' edges that would form the boundaries of gaps.

	Edge						
Vertex	First	Second	Third	Fourth	Fifth	Sixth	Seventh
First	V3	V5	V4	***V1***	V6	V2	***V1***
Second	V4	V6	***V1***	***V5***	V1	V3	***V2***

When the search starts to find all the edges that would form a closed loop, the previous algorithm might mistakenly connect edges: 3-4; 4-1; and *1-5*; instead of *1-2*. This is clearly an error as the edge that is supposed to be included in file D should be *1-2, and not 1-5*. It is therefore pertinent that for every edge searched, the second vertex (e.g. V1 of the third edge, shaded, in Table 6.10) of that edge should be searched against the first vertex of subsequent edges (e.g. V1 of edge 1-5 in the fourth edge) and this should not be halted the first time the first vertex of subsequent edges are found. The search should continue to check if there are other edges with V1 (e.g. edge 1-2, in the seventh edge). Every time, say for example, vertex 1 is found, there should be a count. When the count is more than two, it indicates that there is more than one gap sharing the same vertex; this may be called the coincidental vertex. Once this happens, the following procedure would then be used:

Step 1: Conducting a normal search (for the boundary of Gap 1)

At the start of the normal search, the first edge of file C, vertices 3 and 4, as seen as in Table 6.11 (a) is saved into a temporary file C1.

Table 6.11 (a). Representation of how Files C and C1 would look like.

File C							
Edge							
Vertex	First	Second	Third	Fourth	Fifth	Sixth	Seventh
First	V3	V5	V4	V1	V6	V2	V1
Second	V4	V6	V1	V5	V1	V3	V2

File C1			
Edge			
Vertex	First	Second	Third
First	V3	?	?
Second	V4	?	?

The second vertex in the first edge (V4) of file C1 is searched against the first vertex of subsequent edge in file C (refer to Table 6.11a, shaded box). Once it is found to be the same vertex (V4) and that there are no other edges sharing the same vertex, the edge (i.e. vertex 4-1) is stored as the second edge in file C1 (refer to Table 6.11 (b)).

Table 6.11 (b). Representation of how files C and C1 would look like during the normal search (special case).

Count 1 ⇓ (Fourth edge) Count 2 ⇓ (Seventh edge)

File C							
Edge							
Vertex	First	Second	Third	Fourth	Fifth	Sixth	Seventh
First	V3	V5	V4	V1	V6	V2	V1
Second	V4	V6	V1	V5	V1	V3	V2

File C1			
Edge			
Vertex	First	Second	Third
First	V3	V4	?
Second	V4	V1	?

Step 2: Detection of more than one gap

The second vertex of the second edge (V1) in file C1 is searched for an equivalent first vertex of subsequent edges in file C (refer to Table 6.11 (b), shaded box containing vertex). Once it is found (V1, first vertex of fourth edge), a count of one is registered and at the same time, that edge is noted.

The search for the same vertex is continued to determine if there are other edges sharing the same vertex 1. If there is an additional edge sharing the same vertex 1 (Table 6.11 (b), seventh edge), another count is registered, making a total of two counts. Similarly, the particular record in which it happens again is noted. The search is continued until there are no further edges sharing the same vertex. When completed, the following is carried out for the third edge in file C1.

For the first count, reading from file C, all the edges that would form the first alternative closed loop are sorted. These first alternatives in file C2 (refer to Table 6.11 (c) and Figure 6.18 (a) for a graphical representation of how the first alternatives may look like) are then saved. A closed loop is established once the second vertex of the last edge (V3) is the same as the first vertex of the first edge (V3).

Table 6.11 (c). First alternative closed loop that may represent a gap boundary.

File C2							
	Edge						
Vertex	First	Second	Third	Fourth	Fifth	Sixth	Seventh
First	V3	V4	V1	V5	V6	V1	V2
Second	V4	V1	V5	V6	V1	V2	V3

For the second count, the edges are sorted to form a second alternative of a closed loop that will represent the boundary of the gap. These edges are then saved in another temporary file C3 (refer to Table 6.11 (d) and Figure 6.18 (b) for a graphical representation of how the second alternatives may look like). Once a closed loop is established, the search is stopped.

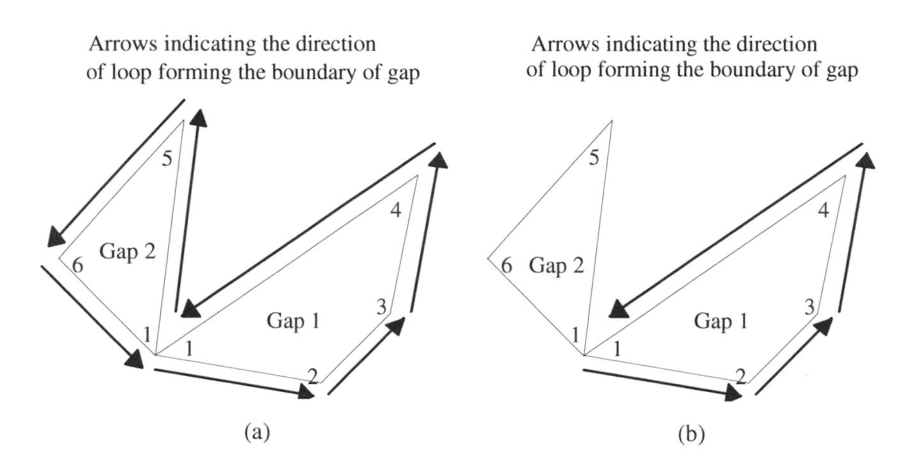

Fig. 6.18. Graphical representation of the two gaps sharing (a) a coincidental vertex (first alternative) and (b) the same vertex (second alternative).

Table 6.11 (d). Second alternative closed loop that may represent a gap boundary.

File C3				
	Edge			
Vertex	First	Second	Third	Fourth
First	V3	V4	V1	V2
Second	V4	V1	V2	V3

Step 3: Comparison of alternatives for least record to form gap boundary

From File C2 (first alternative) and File C3 (second alternative), it can bc seen that the second alternative has the least record to form the boundary of gap 1. Hence the second alternative data would be written to file D for the next stage of facet generation and the two temporary files C2 and C3 would be discarded. Once gap 1 is repaired, gap 2 can be repaired by using the generic solution.

For further illustration, the following algorithm can cater to more than two gaps sharing the same vertex, as shown in Figure 6.19 (a), or even three gaps arranged differently, as shown in Figure 6.19 (b).

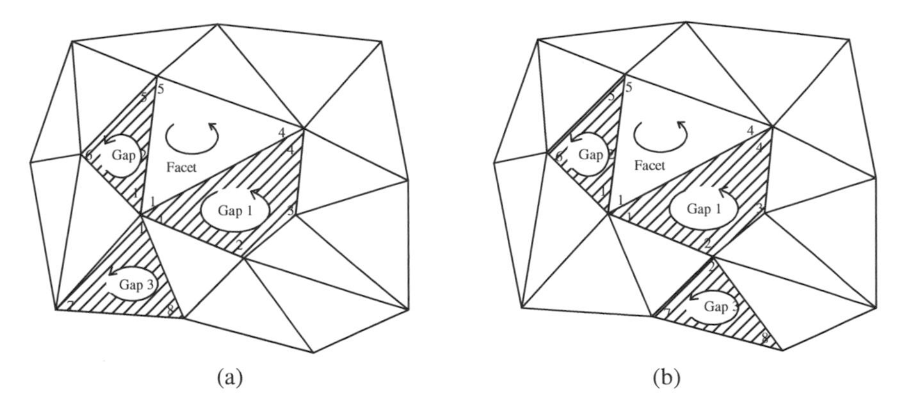

Fig. 6.19. Three facets sharing (a) one coincidental vertex and (b) two coincidental vertices.

Step 1: Conducting of normal search

For the case shown in Figure 6.19 (b), Tables 6.12 (a) and 6.12 (b) show how file C and file C1 may appear respectively.

Table 6.12 (a). Illustration of how File C may appear, given three gaps sharing same two vertices.

				Count 1 for V1 ⇓		Count 1 for V2 ⇓	Count 2 for V1 ⇓	Count2 for V2 ⇓		
File C										
Edge										
Vertex	First	Second	Third	Fourth	Fifth	Sixth	Seventh	Eighth	Ninth	Tenth
First	V3	V5	V4	V1	V6	V2	V1	V2	V8	V7
Second	V4	V6	V1	V5	V1	V3	V2	V7	V2	V8

Table 6.12 (b). File C1 during normal search.

File C1				
Edge				
Vertex	First	Second	Third	Fourth
First	V3	V4	?	?
Second	V4	V1	?	?

Step 2: Detection of multiple gaps

Referring to Table 6.12 (a), it can be seen that:
For V1 → there are 2 counts
For V2 → there are 2 counts

For Count 1 for V1 and Count 1 for V2, from File C (Table 6.12 (a)), the first alternative closed loop can be generated and the file is shown in Table 6.12 (c). Figure 6.20 (a) illustrates a graphical representation of the gap's boundary.

Table 6.12 (c). First alternative of closed loop that may represent the gap boundary.

	Edge									
Vertex	First	Second	Third	Fourth	Fifth	Sixth	Seventh	Eighth	Ninth	Tenth
First	V3	V4	V1	V5	V6	V1	V2	-	-	-
Second	V4	V1	V5	V6	V1	V2	V3	-	-	-

For Count 1 for V1 and Count 2 for V2, the second alternative of a closed loop can be sorted and is shown in Table 6.12 (d). Figure 6.20 (b) illustrates a graphical representation of the gap's boundary.

Table 6.12 (d). Second alternative of closed loop that may represent a gap boundary.

Vertex	**Edge**									
	First	Second	Third	Fourth	Fifth	Sixth	Seventh	Eighth	Ninth	Tenth
First	V3	V4	V1	V5	V6	V1	V2	V7	V8	V2
Second	V4	V1	V5	V6	V1	V2	V7	V8	V2	V3

For Count 2 for V1 and Count 1 for V2, the third alternative of a closed loop can again be sorted and is shown in Table 6.12 (e). Figure 6.20 (c) illustrates a graphical representation of the gap's boundary.

Table 6.12 (e). Third alternative of closed loop that may represent a gap boundary.

Vertex	**Edge**									
	First	Second	Third	Fourth	Fifth	Sixth	Seventh	Eighth	Ninth	Tenth
First	V3	V4	V1	V2	-	-	-	-	-	-
Second	V4	V1	V2	V3	-	-	-	-	-	-

For Count 2 for V1 and Count B for V2, the fourth alternative of a closed loop can also be sorted and is shown in Table 6.12 (f). Figure 6.20 (d) illustrates a graphical representation of the gap's boundary.

Table 6.12 (f). Fourth alternative of closed loop that may represent a gap boundary.

Vertex	Edge									
	First	Second	Third	Fourth	Fifth	Sixth	Seventh	Eighth	Ninth	Tenth
First	V3	V4	V1	V2	V7	V8	V2	-	-	-
Second	V4	V1	V2	V7	V8	V2	V3	-	-	-

Step 3: Comparison of the four alternatives

As can be seen from the four alternatives (see Figure 6.20), the third alternative, as shown in Figure 6.20c, is considered the best solution and the correct solution to fill gap 1. This correct solution can be found by comparing which alternative uses the least edges to fill gap 1 up. Once the solution is found, the edges would be saved to file D for the next stage, which is facet generation.

After gap 1 is filled, gaps 2 and 3 can then be repaired using the basic generic solution.

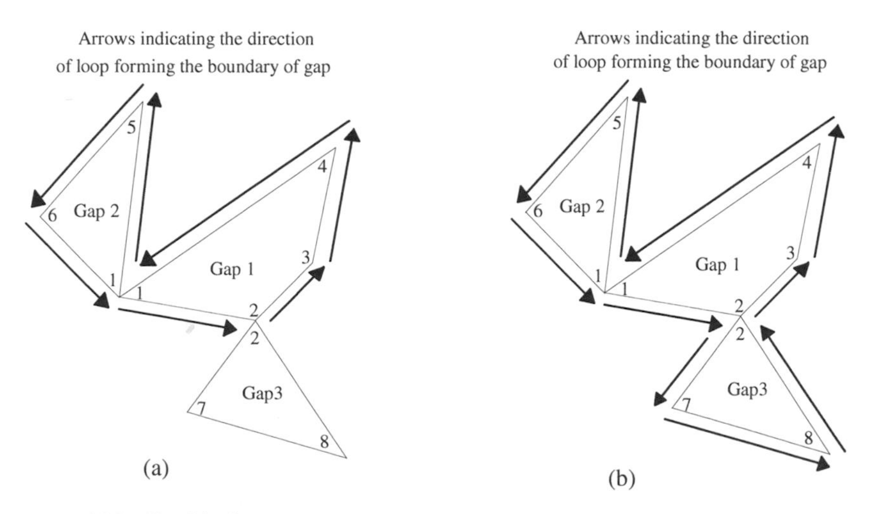

Fig. 6.20. Graphical representation of the three gaps sharing two coincidental vertices.

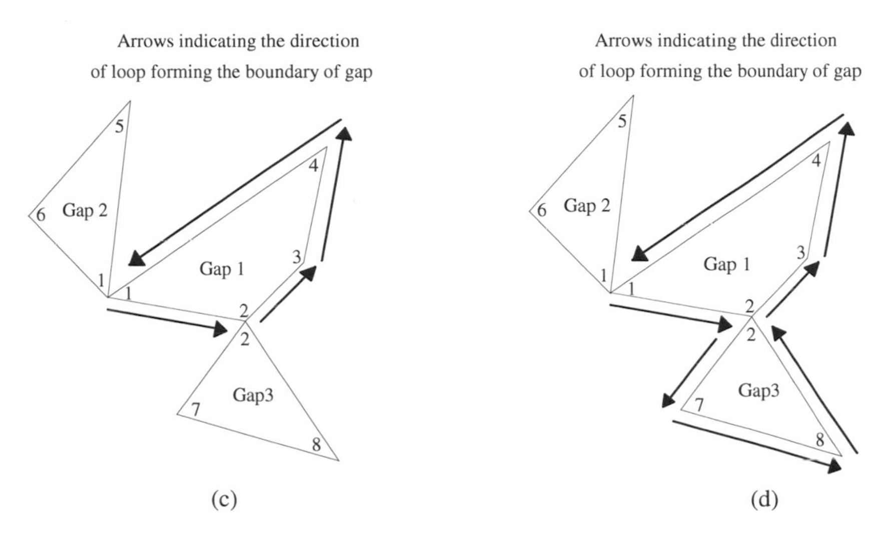

Fig. 6.20. *(Continued)* (a) First alternative, (b) second alternative, (c) third alternative and (d) fourth alternative — graphical representation of the three gaps sharing two coincidental vertices.

6.4.2.2 *Degenerate facets*

When dealing with a degenerate facet such as Facet A, shown in Figure 6.21, which shares the same common edge (a-c) with two different facets B and C, vector algebra is applied to solve it. The following steps are taken:

Step 1: Edge a-c is converted into vectors.

$$(c_1\boldsymbol{i} + c_2\boldsymbol{j} + c_3\boldsymbol{k}) - (a_1\boldsymbol{i} + a_2\boldsymbol{j} + a_3\boldsymbol{k}) \tag{6.1}$$

Step 2: Collinear vectors are checked.

$$\text{Let} \quad \boldsymbol{x} = \text{ac} \Rightarrow x_1\boldsymbol{i} + x_2\boldsymbol{j} + x_3\boldsymbol{k}; \tag{6.2}$$

$$\boldsymbol{y} = \text{ce} \Rightarrow y_1\boldsymbol{i} + y_2\boldsymbol{j} + y_3\boldsymbol{k}; \tag{6.3}$$

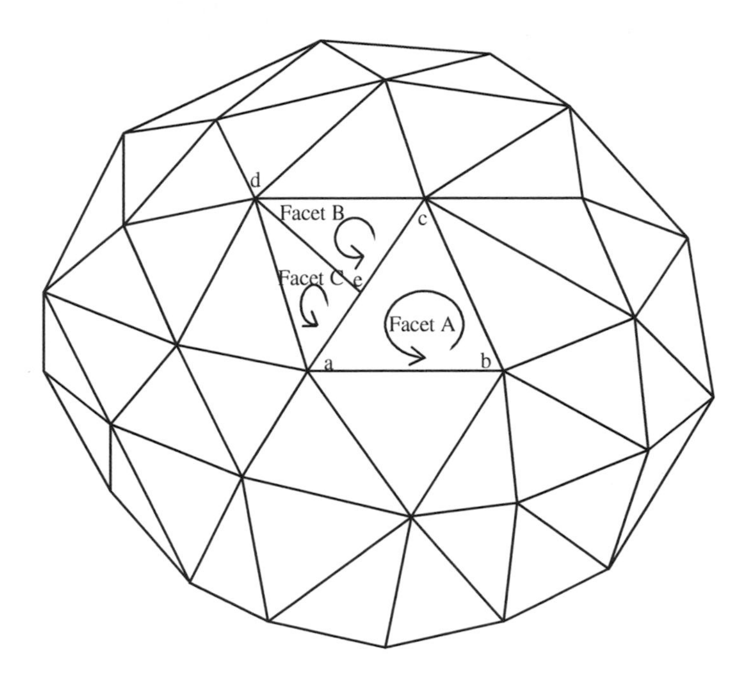

Fig. 6.21. An illustration showing a degenerate facet.

By mathematical definition, the two vectors $\boldsymbol{x}$ and $\boldsymbol{y}$ are said to be collinear if there exists scalars s and t, both non-zero, such that:

$$s\boldsymbol{x} + t\boldsymbol{y} = 0 \tag{6.4}$$

However, when applied in computer, it is only necessary to have:

$$s\boldsymbol{x} + t\boldsymbol{y} \leq \varepsilon \tag{6.5}$$

where ε is a definable tolerance

Step 3: If the two vectors are found to be collinear vectors, the position of vertex *e* is generated.

Step 4: Facet A is split into two facets (see Figure 6.22). The two facetsare generated using the three vertices in facet A and vertex *e*.

First facet vertices are: a, b, e
Second facet vertices are: b, c, e

The orientation is checked and the new facets are stored in a temporary file.

Step 5: Search and delete facet A from original file.

Step 6: The two new facets are stored in the original file and the data are deleted from the temporary file.

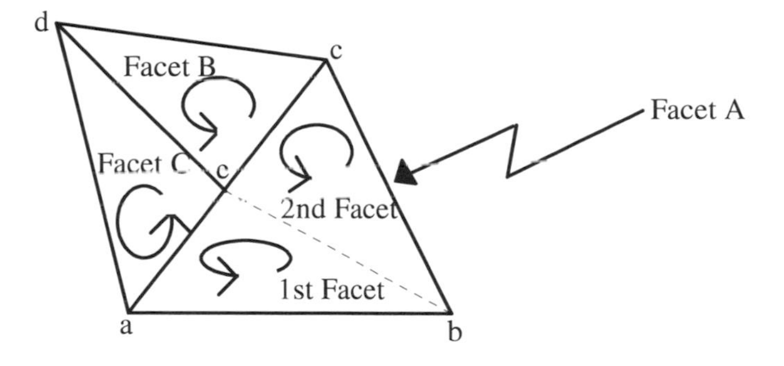

Fig. 6.22. Illustration on how a degenerate facet is solved.

6.4.2.3 *Overlapping facets*

The condition of overlapping facets can be caused by errors introduced by inconsistent numerical round-off. This problem can be resolved through vertex merging where vertices within a pre-determined numerical round-off tolerance of one another can be merged into just one vertex. Figure 6.23 illustrates one example of how this solution can be applied. Figure 6.24 illustrates another example of an overlapping facet.

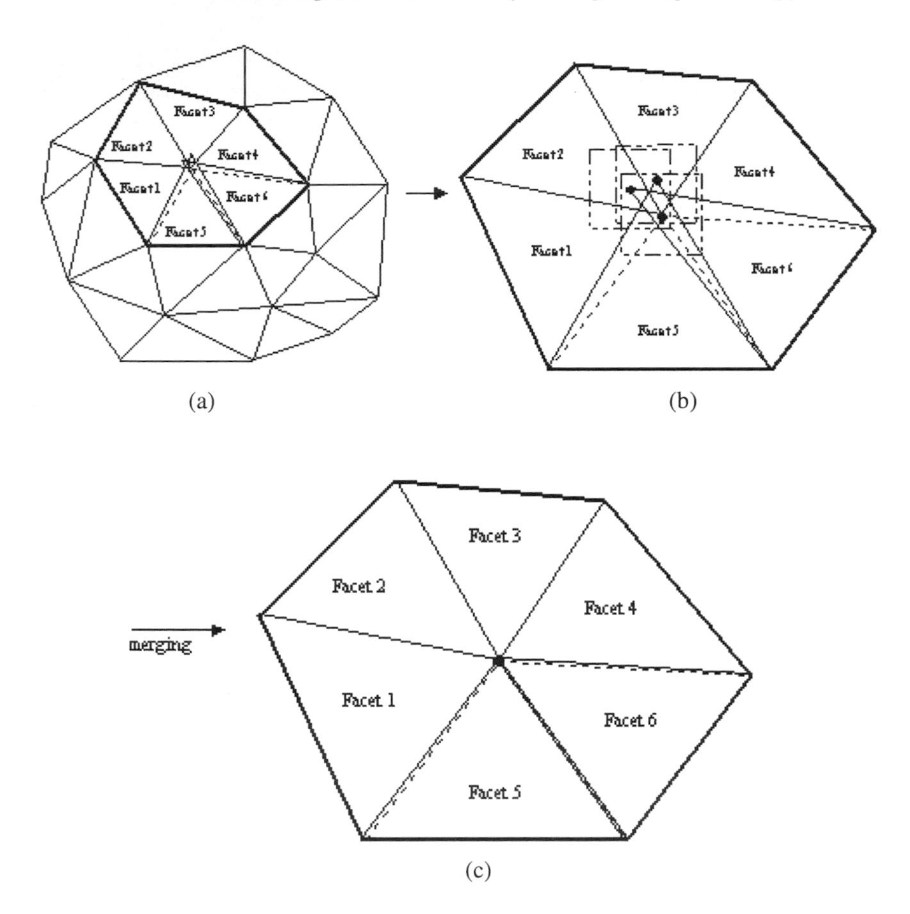

Fig. 6.23. (a) Overlapping facets, (b) numerical round-off equivalence region and (c) vertices merged.

It is recommended that this merging of vertices be done before the searching of the model for gaps. This will eliminate unnecessary detection of erroneous edges and save substantial computational time expended in checking whether the edges can be used to generate another facet.

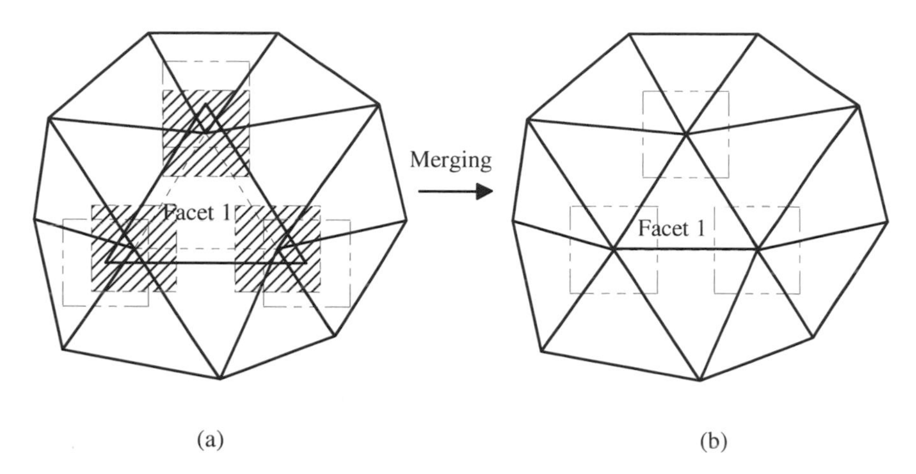

Fig. 6.24. (a) An overlapping facet and (b) facet's vertices merged with vertices of neighbouring facets.

6.4.3 *Performance Evaluation*

Computational efficiency is an issue whenever CAD model repair of solids that have been finely tessellated is considered. This is due to the fact that for every unit increase in the number of facets (finer tessellation), the additional increase in the number of edges is 3. Thus, the computational time required for the checking of erroneous edges would correspondingly increase.

6.4.3.1 *Efficiency of the detection routine*

Assuming that there are 12 triangles in the cube (Figure 6.25), the number of edges = $12 \times 3 = 36$. The number of searches is computed as follows:

(1) Read first edge, search 35 edges and remove 2 edges.

(2) Read second edge, search 33 edges and remove 2 edges.

(3) Read third edge, search 31 edges and remove 2 edges and so on.

$$\text{Number of searches} = 1 + 3 + \ldots + 35$$

In general,

$$\begin{aligned}\text{Number of searches} &= 1 + \ldots + (n-1) \\ &= n^2 / 4 \end{aligned} \tag{6.6}$$

where n = number of edges

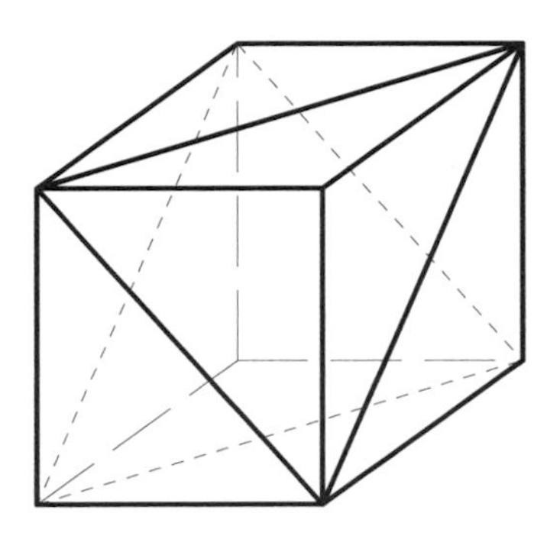

Fig. 6.25. A cube tessellated into 12 triangles.

Although this result does not seem satisfactory, it is both an optimum and a robust solution that can be obtained given the inherent nature of the STL format (such as its lack of topological information).

However, if topological information is available, the efficiency of the routine used to detect the erroneous edges can definitely be increased significantly. Some additional points worth noting are: first, binary files are far more efficient than ASCII files because they are only 20-25% as large and thus reduce the amount of physical data that needs to be transferred because they do not require the subsequent translation into a binary representation. Consequently, one easily saves several minutes per file by using binary instead of ASCII file formats [10].

As for vertex merging, the use of one dimensional AVL-tree can significantly reduce the search-time for sufficiently identical vertices [10]. The AVL-trees, which are usually twelve to sixteen levels deep, reduces each search from O(n) to O(log n) complexity, and the total search-time from close to an hour to less than a second.

6.4.3.2 *Estimated computational time for shell closure*

The computational time required for shell closure is relatively fast. The estimated time can range from a few seconds to less than a minute and is arrived at based on the processing time obtained by Jan Helge Bohn [10] that uses a similar shell closure algorithm.

6.4.3.3 *Limitations of current shell closure process*

The shell closure process developed thus far does not have the ability to detect or solve the problems posed by any of the non-manifold conditions. However, the detection of non-manifold conditions and their subsequent solutions would be the next focus in ongoing research.

A limitation of the algorithm involves the solving of coplanar (see Figure 6.26 (a)) and non-coplanar facets (see Figure 6.26 (b)) whose intersections result in another facet. The reasons for such errors are related to the application that generated the faceted model, the application that generated the original 3D CAD model and the user.

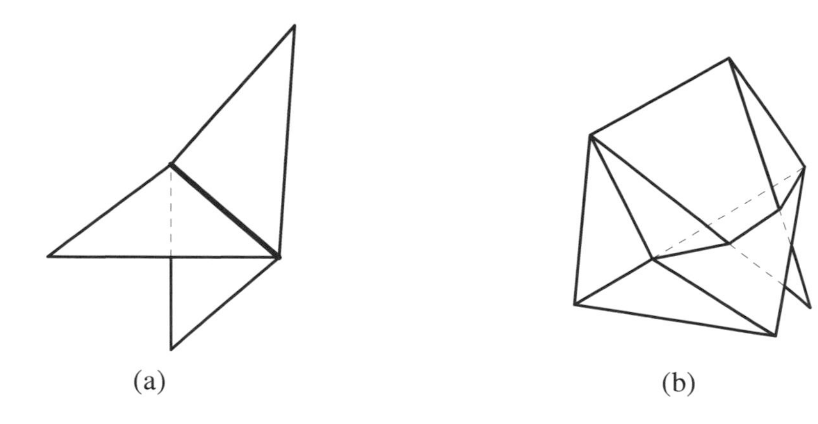

Fig. 6.26. (a) Incorrect triangulation (coplanar facet) and (b) non-coplanar whereby facets are split after being intersected.

Another limitation involves the incorrect triangulation of parametric surface (see Figure 6.27). One of the overlapping triangles, T_b = BCD should not be present and should thus be removed while the other triangle, T_a = ABC should be split into two triangles so as to maintain the correct contoured surface. The proposed algorithm is presently unable to solve this problem [11].

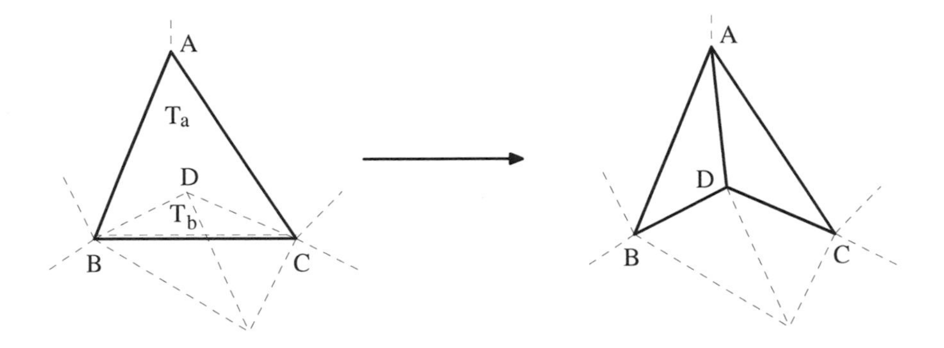

Fig. 6.27. Incorrect triangulation of parametric surface.

Finally, as mentioned earlier, the efficiency of $n^2/4$ (when n is the number of edges) is a major limitation especially when the number of facets in the tessellated model becomes very large (e.g. greater than 40,000). There are differences in the optimum use of computational resources. However, further work is being carried out to ease this problem by using topological information which is available in the original CAD model.

6.5 Other Translators

6.5.1 *IGES File*

IGES (Initial Graphics Exchange Specification) is a standard used to exchange graphics information between commercial CAD systems. It was set up as American National Standard in 1981 [12, 13]. The IGES file can precisely represent CAD models. It includes not only the geometry information (Parameter Data Section) but also topological information (Directory Entry Section). In the IGES, surface modelling, constructive solid geometry (CSG) and boundary representation (B-rep) are introduced. Especially, the ways of representing the regularised operations for union, intersection, and difference have been also defined.

The advantages of the IGES standard are its wide adoption and comprehensive coverage. Since IGES was set up as American National Standard, virtually every commercial CAD/CAM system has adopted IGES implementations. Furthermore, it provides the entities of points, lines, arcs, splines, NURBS surfaces and solid elements. Therefore, it can precisely represent CAD models.

However, several disadvantages of the IGES standard in relation to its use as an AM format include the following objections:

(1) Because IGES is the standard format to exchange data between CAD systems, it also includes much redundant information that is not needed for AM systems.
(2) The algorithms for slicing an IGES file are more complex than the algorithms slicing a STL file.
(3) The support structures needed in AM systems such as SLA cannot be created according to the IGES format.

IGES is a generally used data transfer medium which interfaces with various CAD systems. It can precisely represent CAD models. Advantages of using IGES over current approximate method include precise geometry representations, few data conversions, smaller data files and simpler control strategies. However, the problems are lack of transfer standards for variety CAD systems and system complexities.

6.5.2 *HP/GL File*

HP/GL (Hewlett-Packard Graphics Language) is a standard data format for graphic plotters [1, 2]. Data types are all two-dimensional, including lines, circles, splines, texts, etc. The approach, as seen from a designer's point of view, would be to automate a slicing routine which generates a section slice, invoke the plotter routine to produce a plotter output file and then loop back to repeat the process.

The advantages of the HP/GL format are that a lot of commercial CAD systems have the interface to output HP/GL format and it is a 2D geometry data format which does not need to be sliced.

However, there are two distinct disadvantages of the HP/GL format. First, because HP/GL is a 2D data format, the files would not be appended, potentially leaving hundreds of small files needing to be given logical names and then transferred. Second, all the support structures required must be generated in the CAD system and sliced in the same way.

6.5.3 *CT Data*

Computerised Tomography (CT) scan data is a particular approach for medical imaging [1, 14]. This is not standardised data. Formats are proprietary and somewhat unique from one CT scan machine to another. The scan generates data as a grid of three-dimensional points, where each point has a varying shade of grey indicating the density of the body tissue found at that particular point. Data from CT scans have been used to build skull, femur, knee, and other bone models on Stereolithography systems. Some of the reproductions were used to generate implants, which have been successfully installed in patients. The CT data consist essentially of raster images of the physical objects being imaged. It is used to produce models of human temporal bones.

There are three approaches to make models out of CT scan information:

(1) Via CAD Systems,
(2) STL-interfacing and
(3) Direct Interfacing.

The main advantage of using CT data as an interface of AM is that it is possible to produce structures of human body by the AM systems. However, some disadvantages of CT data include: first, the increased difficulty in dealing with image data as compared with STL data and second, the need for a special interpreter to process CT data.

6.6 Standards for Representing Additive Manufactured Objects

6.6.1 *Additive Manufacturing File format (AMF)*

For decades, research and development has been going on for the Additive Manufacturing industry to specify an international standard. As AM technology experienced fast and furious changes from producing primarily homogeneous products of simple geometry to producing heterogeneous objects of complex shapes, textures and colours, the industry standard for representing AM objects has evolved [15]. In 2011, the two most prominent standard developers worldwide, American Society for Testing and Materials (ASTM) International and the International Standards Organisation (ISO) signed a Partner Standards Developing Organisation (PSDO) cooperative agreement to establish common standards based on the work of the ASTM International Committee F42 and ISO Technical Committee 261 on Additive Manufacturing [16]. The ASTM International Committee F42 on AM Technologies was formed in 2009 to develop standards which would play a critical role in all aspects of the AM industry [17] while the ISO/TC 261 was created in 2011. The F42 committee has published several standards regarding terminology, Additive Manufacturing File (AMF) and part production of several metals using AM technologies. In the current manufacturing environment, the standards adopted by the industry are categorised under two groups: informal standards and formal standards.

STL format is grouped under informal standards commonly called industry or *de facto* standards. For the several past decades in the AM industry, STL format is the most common interface between computer design programs and AM systems due to its simplicity and ease of use (see Section 6.1). However, as STL format only contains information on the surface mesh and has no provisions for representing colour, texture, material, substructure, and even more complex 3D shapes using multiple materials are created by an expanding AM industry, new standard interchange file formats have to be developed to deal with these problems and address the needs.

The ISO 10303 series, a group of standards widely known as STEP (STandard for the Exchange of Product model data), is considered under the formal category. They are approved internationally and supported by a public standards-making authority for the electronic exchange of product data between computer-based product life cycle systems under the ISO Technical Committee ISO/TC 184/SC 4 on industrial data [18]. STEP covers a broad range of different product types and life cycle stages. STEP internally consists of several Application Protocols (APs), which are used to describe a particular life cycle stage of a particular product type. The exchange of the actual information indicating description and the method used is purely based on APs. To implement the description, a set of Integrated Resources (IRs) are constructed to form the APs and the construction of the IRs is significantly applied for specific purposes. Developers have long noticed the importance of tackling the emerging problems due to the unprecedented scale of the ISO 10303 standard. To track the new evolvements in technology while they are occurring, STEP is currently subjected to revision. One example would be the strong emphasis on the use of XML as a means of exchanging STEP information.

Additive Manufacturing File format (AMF) Version 1.1 is the new standard issued under the collaboration between ASTM and ISO in 2013 which answers the growing need of a standard interchange file format that can provide detailed properties of the target product. The new standard, ISO/ASTM 52915:2013, will greatly benefit the whole AM industry, allowing CAD designers, scanners and 3D graphical editors to conveniently work with AM equipment [15]. Similar to the STL format, the AMF format is based on triangles and may be sorted, expressed and transferred via paper or electronic means, as long as the data required by this specification is collected. When prepared in a systematic electronic format, an extensible markup language (XML), a standard which provides for tagging of information content within documents, must be closely followed in order to support standards-compatible interoperability.

AMF format aims to address the following aspects [15]:

Technology independence – AMF allows any AM machines to build the part to the best of their capability without using exclusive technologies.

Simplicity – AMF allows easy implementation and understanding. The format can be displayed and rectified in common ASCII text viewers.

Scalability – AMF could tackle increasing in part complexity and size and with the improving resolution and precision of manufacturing machines.

Performance – AMF allows appropriate duration for read-and-write process and moderate file size.

Backwards compatibility – Interchangeable between existing STL file and new AMF file.

Future compatibility – AMF caters to future changes, allowing new features to be amounted to the parts using the old machines.

AMF has its significant advantage in that it is more flexible. As the required information is stored in XML format which is a widely accepted ASCII text file, amendments could be made to the file as long as the add-on adheres to the XML standard [15, 19]. Moreover, XML is both machine and human readable which offers convenience in coding and debugging.

The general structure of an AMF file begins with the XML declaration line specifying the XML version and encoding. Blank lines and standard XML comments can be interspersed in the file and these will be ignored by the typical interpreter. The main file is enclosed between an opening </amf> element and a closing </amf> element. These elements are necessary to denote the file type, as well as to fulfil the requirement that all XML files have a single-root element. The unit system can also be specified (mm, inch, ft, meters, or micrometers). In the absence of a unit specification, millimetres are assumed.

Within the AMF brackets, there are five top level elements and these include:

<object>

The object element specifies the geometry and it defines a volume or volumes of material, each of which is associated with a material identification (ID) for printing.

The top level <object> element specifies a unique id and contains two child elements: <vertices> and<volume>. The required <vertices> element lists all vertices that are used in this object. Each vertex is implicitly assigned a number in the order in which it was declared starting at zero. The required child element <coordinates> gives the position of the point in three-dimensional (3D) space using the <x>, <y>, and <z> elements.

After the vertex information, at least one <volume> element shall be included. Each volume encapsulates a closed volume of the object. Multiple volumes can be specified in a single object. While volumes may share vertices at interfaces they may not overlap. In each volume, the child element <triangle> shall be used to define triangles that tessellate the surface of the volume. Each <triangle> element will comprise of three vertices from the set of indices of the previously defined vertices. The indices of the three vertices of the triangles are specified using the <v1>, <v2>, and <v3> elements according to the right-hand rule such that vertices are listed in counter-clockwise order as viewed from outside the volume. Each triangle is implicitly assigned a number in the order in which it is declared starting from zero.

Note that the file does not describe support structure, only the intended object to be manufactured.

Only a single object element is required for a fully functional AMF file; the other top level elements are optional.

<material>

The optional material element defines one or more materials for printing with an associated material ID. If no material element is included, a single default material is assumed.

Each material is assigned a unique ID. Geometric volumes are associated with materials by specifying a material ID within the <volume> element. The material ID "0" is reserved for no material (void).

Material attributes are contained within each <material>. The element <colour> is used to specify the red/green/blue/alpha (RGBA) values in a specified colour space for the appearance of the material. It can be inserted at the material level to associate a colour with a material, the object level to colour an entire object, the volume level to colour an entire volume, a triangle level to colour a triangle, or a vertex level to associate a colour with a particular vertex. Object colour overrides material colour specification, a volume colour overrides an object colour, vertex colours override volume colours, and triangle colouring overrides a vertex colour.

Additional material properties can be specified using the <metadata> element, such as the material name for operational purposes or elastic properties for equipment that can control such properties.

<texture>

The optional texture element defines one or more images or textures for colour or texture mapping each with an associated texture ID.

<constellation>

The optional constellation element hierarchically combines objects and other constellations into a relative pattern for printing. If no constellation elements are specified, each object element will be imported with no relative position data.

A constellation can specify the position and orientation of objects to increase packing efficiency and describe large arrays of identical objects. The <instance> element specifies the displacement and rotation an existing object in order for it to be specified in its position in the constellation. The displacement and rotation are always defined relatively to the original position and orientation in which the object was originally defined. Rotation angles are specified in degrees and are applied first to rotation about the x axis, then the y axis, and then the z axis.

A constellation can refer to another constellation, with multiple levels of hierarchy. Recursive or cyclic definitions of constellations, though, are not permissible.

<metadata>

The optional metadata element specifies additional information about the object(s) geometries, and materials being defined in the file. Information can include a name, textual description, authorship, copyright information, and special instructions.

The AMF file can be stored either in plain text or be compressed. If compressed, the compression shall be in ZIP archive format (3) and can be done manually or at write time using any one of several open compression libraries. Both the compressed and uncompressed version of this file will have the AMF extension, and it is the responsibility of the parsing program to determine whether or not the file is compressed and if so, to perform decompression during read.

6.6.2 *Other Standards on Additive Manufacturing*

There are several international standards that are being developed and published for Additive Manufacturing.

The ISO/ASTM 52921:2013: Standard Terminology for Additive Manufacturing – Coordinate Systems and Test Methodologies is a standard for terminology that describes terms, definitions of terms, descriptions of terms, nomenclature, and acronyms associated with coordinate systems and testing methodologies for AM technologies. This is for AM users, producers, researchers, educators, press/media, and others to have a common understanding of these terminology, particularly when reporting results from the testing of parts made on AM systems.

The ASTM F2792-12a: Standard Terminology for Additive Manufacturing Technologies is an ASTM international standard that formalised the terms, definitions of terms, descriptions of terms, nomenclature, and acronyms associated with AM technologies to provide standardised terminology for usage by AM users, producers, researchers, educators, press/media and others.

The ASTM F2971 – 13: Standard Practice for Reporting Data for Test Specimens Prepared by Additive Manufacturing describes a standard procedure for reporting results by testing or evaluation of specimens produced by AM. This practice provides a common format for presenting data for AM specimens, for: (a) to establish further data reporting requirements, and (b) to provide information for the design of material property databases.

The ASTM F2924 – 14: Standard Specification for Additive Manufacturing Titanium-6 Aluminium-4 Vanadium with Powder Bed Fusion covers additively manufactured titanium-6aluminum-4vanadium

(Ti-6Al-4V) components using full-melt powder bed fusion such as electron beam melting and laser melting. The components produced by these processes are used typically in applications that require mechanical properties similar to machined forgings and wrought products. Components manufactured to this specification are often, but not necessarily, post processed via machining, grinding, electrical discharge machining (EDM), polishing, and so forth to achieve desired surface finish and critical dimensions.

The ASTM F3001 – 14: Standard Specification for Additive Manufacturing Titanium-6 Aluminium-4 Vanadium ELI (Extra Low Interstitial) with Powder Bed Fusion establishes the requirements for additively manufactured Ti-6Al-4V with extra low interstitials (ELI) components using full-melt powder bed fusion such as electron beam melting and laser melting. The standard covers the classification of materials, ordering information, manufacturing plan, feedstock, process, chemical composition, microstructures, mechanical properties, thermal processing, hot isostatic pressing, dimensions and mass, permissible variations, retests, inspection, rejection, certification, product marking and packaging, and quality program requirements.

The ASTM F3056 – 14: Standard Specification for Additive Manufacturing Nickel Alloy (UNS N06625) with Powder Bed Fusion covers additively manufactured nickel alloy components using full-melt powder bed fusion such as electron beam melting and laser melting. The components produced by these processes are used typically in applications that require mechanical properties similar to machined forgings and wrought products. Components manufactured to this specification are often, but not necessarily, post processed via machining, grinding, electrical discharge machining (EDM), polishing, and so forth to achieve desired surface finish and critical dimensions. This specification is intended for the use of purchasers or producers, or both, of additively manufactured nickel alloy components for defining the requirements and ensuring component properties.

There are other standards that are under discussion at ISO/ASTM technical committees and there will be more of these standards published as AM becomes more popular and before widely used in the industry.

References

[1] Jacobs, P. F. (1992). *Rapid Prototyping and Manufacturing*. Society of Manufacturing Engineers.

[2] Famieson, R., & Hacker, H. (1995). *Direct Slicing of CAD Models for Rapid Prototyping*. Rapid Prototyping Journal, ISATA94, Aachen, Germany.

[3] Donahue, R. J. (1991). CAD Model and Alternative Methods of Information Transfer for Rapid Prototyping Systems. In *Proceedings of the Second International Conference on Rapid Prototyping*, pp, 217-235.

[4] Wozny, M. J. (1992). Systems Issues in Solid Freeform Fabrication. In *Proceedings, Solid Freeform Fabrication Symposium 1992*, Texas, USA, pp. 1-5.

[5] Leong, K. F., Chua, C. K., & Ng, Y. M. (1995). A Study of Stereolithography File Errors and Repair Part 1 - Generic Solutions. In *International Journal of Advanced Manufacturing Technologies.*

[6] Leong, K. F., Chua, C. K., & Ng, Y. M. (1995). A Study of Stereolithography File Errors and Repair Part 2 - Special Cases. In *International Journal of Advanced Manufacturing Technologies.*

[7] Rock, S. J., & Wozny, M. J. (1991). A Flexible Format for Solid Freeform Fabrication. In *Proceedings, Solid Freeform Fabrication Symposium 1991*, Texas, USA, pp. 1-12.

[8] Crawford, R. H. (1993). Computer Aspects of Solid Freeform Fabrication: Geometry, Process Control and Design. In *Proceedings, Solid Freeform Fabrication Symposium 1993*, Texas, USA, pp. 102-111.

[9] Bohn, J. H., & Wozny, M. J. (1992). Automatic CAD-Model Repair: Shell-Closure. In *Proceedings, Solid Freeform Fabrication Symposium 1992*, Texas, USA, pp 86-94.

[10] Bohn, J. H. (1993). *Automatic CAD-Model Repair.* Ann Arbor, Mich., USA, UMI.

[11] Dolenc, A., & Malela, I. (1992). A Data Exchange Format for LMT Processes. In *Proceedings of the Third International Conference on Rapid Prototyping*, pp. 4-12.

[12] Reed, K., Harrvd, D., & Conroy, W. (1990). *Initial Graphics Exchange Specification (IGES) version 5.0.* CAD-CAM Data Exchange Technical Centre.

[13] Jinghon, L. (1992). Improving Stereolithography Parts Quality - Practical Solutions. In *Proceedings of the Third International Conference on Rapid Prototyping*, pp. 171-179.

[14] Swaelens, B., & Kruth, J. P. (1993). Medical Applications of Rapid Prototyping Techniques. In *Proceedings of the Fourth International Conference on Rapid Prototyping*, pp. 107-120.

[15] *ISO ASTM International, Standard Specification for Additive Manufacturing File Format (AMF) Version 1.1*, ISO/ASTM 52915:2013(E), ISO, Case postate 56, CH-1211, Geneva 20, Switzerland, and ASTM International, 100 Barr Harbor Drive, PO Box C700, West Conshohocken, PA 19428-2959, US.

[16] ASTM International, Standardisation News. (2013). *Additive Manufacturing to Benefit from Standards Agreement: ASTM and ISO Sign Additive Manufacturing PSDO Agreement.* Retrieved from http://www.astm.org/standardisation-news/outreach/astm-and-iso-sign-additive-manufacturing-psdo-agreement-nd11.html

[17] ASTM International, Technical Committees. (2013). *Committee F42 on Additive Manufacturing Technologies.* Retrieved from http://www.astm.org/COMMITTEE/F42.htm

[18] ISO Standards Development, Technical Committee. (2013). *ISO/TC 184/SC 4 Industrial data.* Retrieved from http://www.iso.org/iso/home/standards_development/list_of_iso_technical_committees/iso_technical_committee.htm?commid=54158

[19] The World Wide Web Consortium. (2014). *Extensible Markup Language (XML) 1.0 (Fifth Edition).* Retrieved from http://www.w3.org/TR/REC-xml/

Problems

1. What is the *de facto* industry format adopted by Additive Manufacturing systems for the past three decades? Describe the format and illustrate with an example. What are the pros and cons of using this format?

2. Referring to Figure 6.28, write a sample STL file for the shaded triangle.

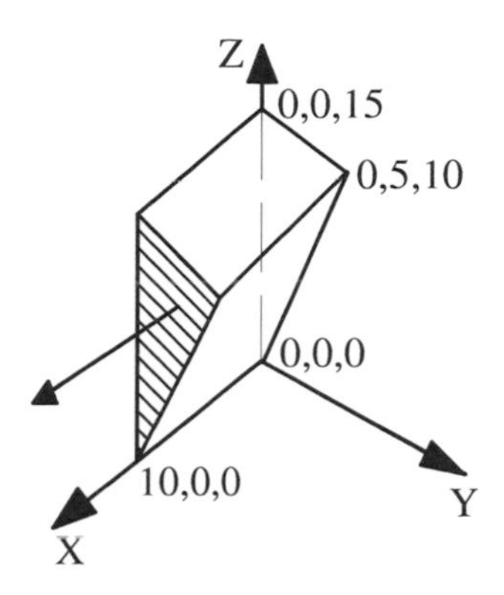

Fig. 6.28. Sample STL File.

3. Based on the STL format, how many triangles and coordinates would a cube contain?

4. What causes *missing facets or gaps* to occur?

5. Illustrate, with diagrams, the meaning of *degenerate facets*.

6. Explain *overlapping facets*.

7. What are the three types of non-manifold conditions?

8. What are the consequences of building a valid and invalid tessellated model?

9. What problems can the generic solution solve?

10. Describe the algorithm used to solve the *missing facets* problem.

11. In Figure 6.29, facet X is incorrectly oriented. Describe how the problem can be resolved. Draw the newly generated facet X with the corrected orientation.

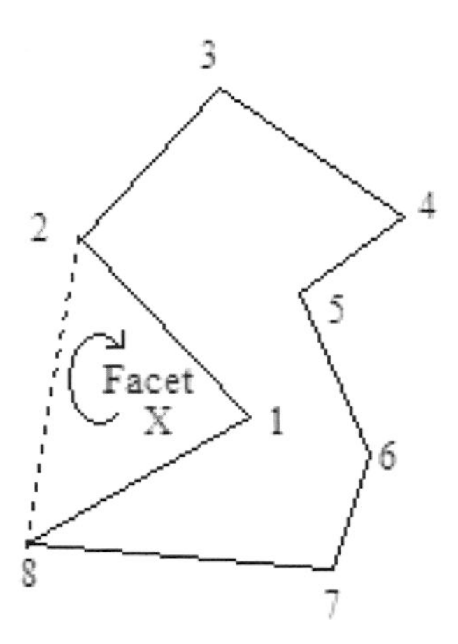

Fig. 6.29. Incorrectly generated facet's orientation.

12. Describe the algorithm used to solve special case 1 where two or more gaps are formed from a coincidental vertex.

13. Prove and illustrate how a degenerate facet can be repaired by the use of vector algebra.

14. How can the problem of overlapping facets be solved?

15. What is the efficiency of the detection routine? Illustrate using the example of a cube.

16. What are some of the limitations of the solutions, both generic and special cases, used to solve STL-related problems?

17. Name some other translators used in place of STL.

18. What is the new ASTM International standard file format for Additive Manufacturing systems? Discuss its advantages and disadvantages compared to STL file format.

19. What are the five top elements within the AMF file?

20. Discuss some of the future features that might be developed based on the AMF at the current state.

Chapter 7

APPLICATIONS AND EXAMPLES

7.1 Application–Material Relationship

Areas of applications are closely related to the purposes of prototyping or manufacturing, and consequently, the materials used. As such, the closer the AM materials are to traditional prototyping or manufacturing materials in physical and behavioural characteristics, the wider will be the range of applications. Unfortunately, there are marked differences in these areas between current AM materials and traditional materials in manufacturing. The key to increasing the applicability of AM technologies therefore lies in widening the range of materials.

In the early developments of AM systems, the emphasis of the tasks at hand was orientated towards the creation of "visualisation" and "touch-and-feel" models to support design, i.e. creating 3D objects with little or no regard to their function and performance. These are broadly classified as "Applications in Design." It is a result that is influenced and in many cases limited by the materials available in these AM systems. However as the initial costs of the machines are high, vendors are constantly on the look-out for more areas of applications, with the logical search for functional evaluation and testing applications, and eventually tooling. This not only calls for improvements in AM technologies in terms of the process to create stronger and more accurate parts, but also in terms of developing an even wider range of materials, including metals and ceramic composites. Applications of AM prototypes were first extended to "Applications in Engineering, Analysis and Planning" and later extended further to "Applications in Manufacturing and Tooling." These typical applications areas are summarised in Figure 7.1 and discussed in the following sections.

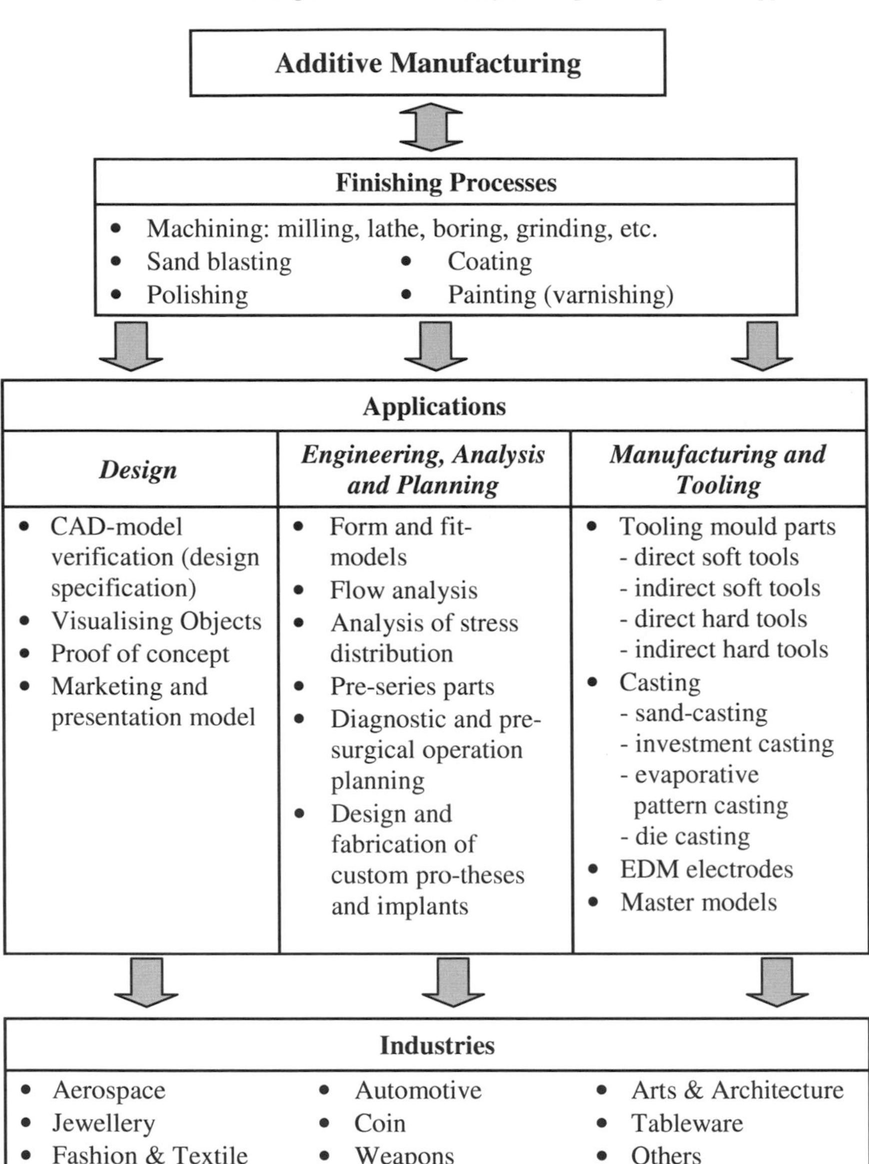

Fig. 7.1. Typical application areas of AM.

The major breakthrough in AM technologies in manufacturing has been their abilities in enhancing and improving product development while at the same time reducing costs and time required to take the product from conception to market. Sections 7.6 to 7.18 contain examples of applications in the aerospace, automotive, jewellery, coin, tableware, arts and architecture, construction, fashion and textile, weapons, music, food and movie industries. These examples are by no means exhaustive, but they do represent their applications in a wide cross-section of the industry.

7.2 Finishing Processes

As there are various influencing factors such as shrinkage, distortion, curling and accessible surface smoothness, it is necessary to apply some post-AM finishing processes to the parts just after they have been produced. These processes can be carried out before the AM parts are used in their desired applications. Furthermore, additional processes may be necessary in specific cases, e.g., when creating screw threads.

7.2.1 *Cutting Processes*

In most cases, the resins or other materials used in the AM systems can be subjected to conventional cutting processes, such as milling, boring, turning, and grinding.

These processes are particularly useful in the following cases:

(1) Deviations in geometrical measurements or tolerances due to unpredictable shrinkage during the curing or bonding stages of the AM process.
(2) Incomplete generation of selected form features. This could be due to fine or complex-shaped features that are difficult to achieve.
(3) Clean removal of necessary support structures or other remainder materials attached to the AM parts.

In all these cases, it is possible to achieve economic surface finishing of the objects generated with a combination of NC machining and computer-aided NC programming.

7.2.2 *Sandblasting and Polishing*

Sandblasting or abrasive jet deburring can be used as an additional cleaning operation or process to achieve better surface quality. However, there is a trade-off in terms of accuracy. Should better finishing be required, additional polishing by mechanical means with super-fine abrasives or simple rotary sanding can also be used after sandblasting.

7.2.3 *Coating*

Appropriate surface coatings can be used to further improve the physical properties of the surface of plastic AM parts. One example is galvano-coating, a coating which provides very thin metallic layers to plastic AM parts.

7.2.4 *Painting*

Painting is applied fairly easily on AM parts made of plastics or paper. It is carried out mainly to improve the aesthetic appeal or for presentation purposes, e.g. for marketing or advertising presentations.

Once the AM parts are appropriately finished, they can then be used for the various areas of application as shown in Figure 7.1.

7.3 Applications in Design

7.3.1 *CAD Model Verification*

This is the initial objective and strength of AM systems, in that designers often need the physical part to confirm the design that they have created in the CAD system. This is especially important for parts or products designed to fulfil aesthetic functions or are intricately designed to fulfil functional requirements.

7.3.2 *Visualising Objects*

Designs created on CAD systems need to be communicated not only amongst designers within the same team, but also to other departments, like manufacturing, and marketing. Thus, there is a need to create objects from the CAD designs for visualisation so that all these people will be referring to the same object in any communication. Tom Mueller in his paper entitled, "Application of Stereolithography in injection moulding" [1] characterises this necessity by saying:

> *"Many people cannot visualise a part by looking at print. Even engineers and toolmakers who deal with print everyday require several minutes or even hours of studying a print. Unfortunately, many of the people who approve a design (typically senior management, marketing analysts, and customers) have much less ability to understand a design by looking at a drawing."*

7.3.3 *Proof of Concept*

Proof of concept relates to the adaptation of specific details to an object environment or aesthetic aspects, e.g. verifying that a car telephone design is suitable and blends in well within a specific car. It also relates to the specific details of the design on the functional performance of a desired task or purpose, e.g. how a lever arm of a CD player opens the door of the CD tray.

7.3.4 *Marketing and Commercial Applications*

Frequently, the marketing or commercial departments require a physical model for presentation and evaluation purposes, especially for assessment of the project as a whole. The mock-up or presentation model can even be used to produce promotional brochures and related materials for marketing and advertising even before the actual product becomes available.

7.4 Applications in Engineering, Analysis and Planning

Other than creating a physical model for visualisation or proofing purposes, designers are also interested in the engineering aspects of their designs. This invariably relates to the functions of the design. AM technologies become important as they are able to provide the information necessary to ensure sound engineering and function of the product. What makes it more attractive is that it also helps to save development time and reduce costs. Based on the improved performance of processes and materials available in current AM technologies, some applications for functional models are presented in the following sections.

7.4.1 *Scaling*

AM technology allows easy scaling down (or up) of the size of a model by scaling the original CAD model. In a case of designing bottles for perfumes with different holding capacities, the designer can simply scale the CAD model appropriately for the desired capacities and view the renderings on the CAD software. With the selected or preferred capacities determined, the CAD data can be modified accordingly to create the corresponding AM model for visualisation and verification purposes (see Figure 7.2).

Fig. 7.2. Perfume bottles with different capacities.

7.4.2 *Form and Fit*

Other than dealing with sizes and volumes, forms have to be considered from the aesthetics and functional standpoints as well. How a part fits into a design and its environment are important aspects which have to be addressed. For example, the wing mirror housing for a new car design has to be of a form that augments well with the general appearance of the exterior design. This will also include how it fits to the car door. The model will be used to evaluate how it satisfies both aesthetic and functional requirements.

Form and fit models are used not just in the automotive industries. They can also be used for industries involved in aerospace and others like consumer electronic products and appliances.

7.4.3 *Flow Analysis*

Designs of components that affect or are affected by air or fluid flow cannot be easily modified if produced by the traditional manufacturing routes. However, if the original 3-D design data can be stored in a computer model, then any change of object data based on some specific tests can be realised with computer support. The flow dynamics of these products can be computer simulated with software. Experiments with 3-D physical models are frequently required to study product performance in air and liquid flow. Such models can be easily built using AM technology. Modifications in design can be done on computer and rebuilt for re-testing very much faster than using traditional prototyping methods. Flow analyses are also useful for studying the inner sections of inlet manifolds, exhaust pipes, replacement heart valves [2], or similar products that at times can have rather complex internal geometries. Should it be required, transparent parts can also be produced using rapid tooling methods to aid visualisation of internal flow dynamics. Typically, flow analyses are necessary for products manufactured in the aerospace, automotive, biomedical and shipbuilding industries.

7.4.4 *Stress Analysis*

In stress analysis using mechanical or photo-optical methods or otherwise, physical replicas of the part being analysed is necessary. If the material properties or features of the AM technologies generated objects are similar to those of the actual functional parts, they can be used in these analytical methods to determine the stress distribution of the product.

7.4.5 *Mock-Up Parts*

"Mock-up" parts, a term first introduced in the aircraft industry, are used for final testing of parts from different aspects. Generally, mock-up parts are assembled into the complete product and functionally tested at pre-determined conditions e.g. for fatigue. Some AM techniques are able to generate "mock-ups" very quickly to fulfil these functional tests before the design is finalised.

7.4.6 *Pre-Production Parts*

In cases where mass production will be introduced once the prototype design has been tested and confirmed, pilot-production runs of 10 or more parts are usual. The pilot-production parts are used to confirm tooling design and specifications. The necessary accessory equipment, such as fixtures, chucks, special tools and measurement devices required for the mass production process are prepared and checked. Many of the AM methods are able to quickly produce pilot-production parts, thus helping to shorten the process development time, thereby accelerating the overall time-to-market process.

7.4.7 *Diagnostic and Surgical Operation Planning*

In combining engineering prototyping methodologies with surgical procedures, AM models can complement various imaging systems, such as magnetic resonance imaging (MRI) and computed tomography (CT) scanning, to produce anatomical models for diagnostic purposes. These

AM models can also be used for surgical and reconstruction operation planning. This is especially useful in surgical procedures that have to be carried out by different teams of medical specialists and inter-departmental communication is of essence. Several related examples and case studies can be found in Chapter 8.

7.4.8 *Design and Fabrication of Custom Prostheses and Implants*

AM can be applied to design and fabricate customised prostheses and implants. A prosthesis or implant can be made from anatomical data inputs from imaging systems, e.g. laser scanning and CT. In cases such as producing ear prostheses, a scan profile can be taken of the good ear to create a computer-mirrored exact replica replacement using AM technology. These models can be further refined and processed to create the actual prostheses or implants to be used directly on a patient. The ability to efficiently customise and produce such prostheses and implants is important, as standard sizes are not always an ideal fit for the patient. Also, a less than ideal fit, especially for artificial joints and weight bearing implants, can often result in accumulative problems and damage to the surrounding tissue structures. More examples and case studies in these areas can be found in Chapter 8.

7.5 Applications in Manufacturing and Tooling

Central to the theme of rapid tooling is the ability to produce multiple copies of a prototype with functional material properties in short lead times. Apart from mechanical properties, materials can also include functionalities such as colour dyes, transparency, flexibility and the like. Two issues are to be addressed here: tooling proofs and process planning. Tooling proofs refer to getting the tooling right so that there will not be a need to do a tool change during production because of process problems. Process planning is meant for laying down the process plans for the manufacture as well as assembly of the product based on the prototypes produced.

Rapid tooling can be classified as soft and hard tooling, direct and indirect tooling [3], as schematically shown in Figure 7.3. Soft tooling, typically made of silicon rubber, epoxy resins, low melting point alloys and foundry sands, generally allows for only single casts or for small batch production runs. Hard tooling, on the other hand, usually made from tool steels, generally allows for longer production runs.

Direct tooling is referred to when the tool or die is created directly by the AM process. As an example in the case of injection moulding, the main cavity and cores, runner, gating and ejection systems, can be produced directly using the AM process. In indirect tooling, on the other hand, only the master pattern is created using the AM process. A mould, made of silicone rubber, epoxy resin, low melting point metal, or ceramic, is then created from the master pattern.

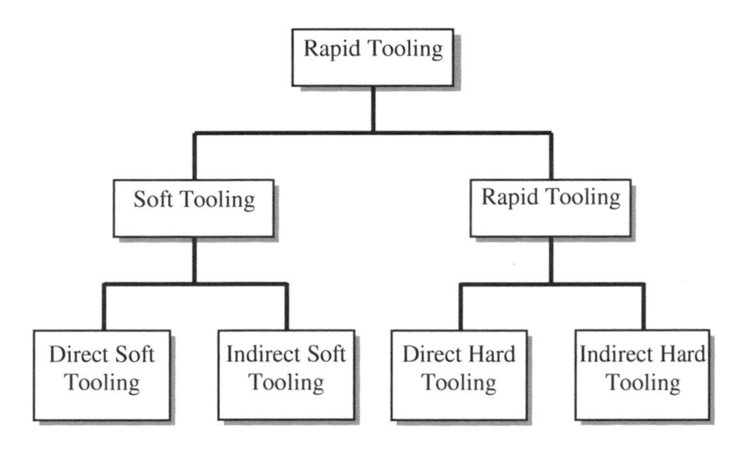

Fig. 7.3. Classification of rapid tooling.

7.5.1 *Direct Soft Tooling*

This is where the moulding tool is produced directly by the AM systems. Such tooling can be used for liquid metal sand casting, in which the mould is destroyed after a single cast. Other examples, such as composite moulds, can be made directly using stereolithography. These are generally used in injection moulding of plastic components and can withstand between 100 and 1000 shots. As these moulding tools can typically only support a single cast or small batch production before breaking down, they are classified as soft tooling. The following section lists several examples of direct soft tooling methods.

7.5.1.1 *Selective laser sintering® of sand casting moulds*

Sand casting moulds can be produced directly using the selective laser sintering (SLS) process. Individual sand grains are coated with a polymeric binder. Laser energy is applied to melt this binder, thereby coating the individual sand grains and bonding the grains of sand together in the shape of a mould [4]. Accuracy and surface finish of the metal castings produced from such moulds are similar to those produced by conventional sand casting methods. Functional prototypes can be produced this way, and if modifications are necessary, a new prototype can be produced within a few days.

7.5.1.2 *Direct AIM (ACES injection moulding)*

A rapid tooling method developed by 3D CAD/CAM systems uses the SLA to produce resin moulds that allow direct injection of thermoplastic materials. Known as the Direct AIM (ACES injection moulding) [5], this method is able to produce high levels of accuracy. However, build times using this method is relatively slow on the standard stereolithography (SLA) machine. Also, because the mechanical properties of these moulds are generally poor, tool damage can occur during the ejection of the part. This is more evident when producing geometrically complex parts using these moulds.

7.5.1.3 *SLA composite tooling*

This method builds moulds with thin shells of resin with the required surface geometry, which is then backed-up with aluminium powder-filled epoxy resin to form the rest of the mould tooling [6]. This method is advantageous in that higher mould strengths can be achieved when compared to those produced by the Direct AIM method which builds a solid SLA resin mould. To further improve the thermal conductivity of the mould, aluminium shot can be added to back the thin shell, thus promoting faster build times for the mould tooling. Other advantages of this method include higher thermal conductivity of the mould and lower tool development costs when compared to moulds produced by the Direct AIM method.

7.5.2 *Indirect Soft Tooling*

In this rapid tooling method, a master pattern is first produced using AM. From the master pattern, a mould tooling can be built out of an array of materials such as silicone rubber, epoxy resin, low melting point metals, and ceramics.

7.5.2.1 *Arc spray metal tooling*

Using metal spraying on the AM model, it is possible to create very quickly an injection mould that can be used to mould a limited number of prototype parts. A typical metal spray process for creating injection mould is shown in Figure 7.4. The metal spraying process is operated manually with a hand-held gun. An electric arc is introduced between two wires, which melt the wires into tiny droplets [7]. Compressed air blows out the droplets in small layers of approximately 0.5 mm of metal.

Using the master pattern produced by any AM process, this is mounted onto a base and bolster, which are then layered with a release agent. A coating of metal particles using the arc spray is then applied to the master pattern to produce the female form cavity of the desired tool. Depending on the type of tooling application, a reinforcement backing is selected and applied to the shell. Types of backing materials include filled epoxy resins, low melting point metal alloys and ceramics. This method of producing soft tooling is cost and lead time-saving.

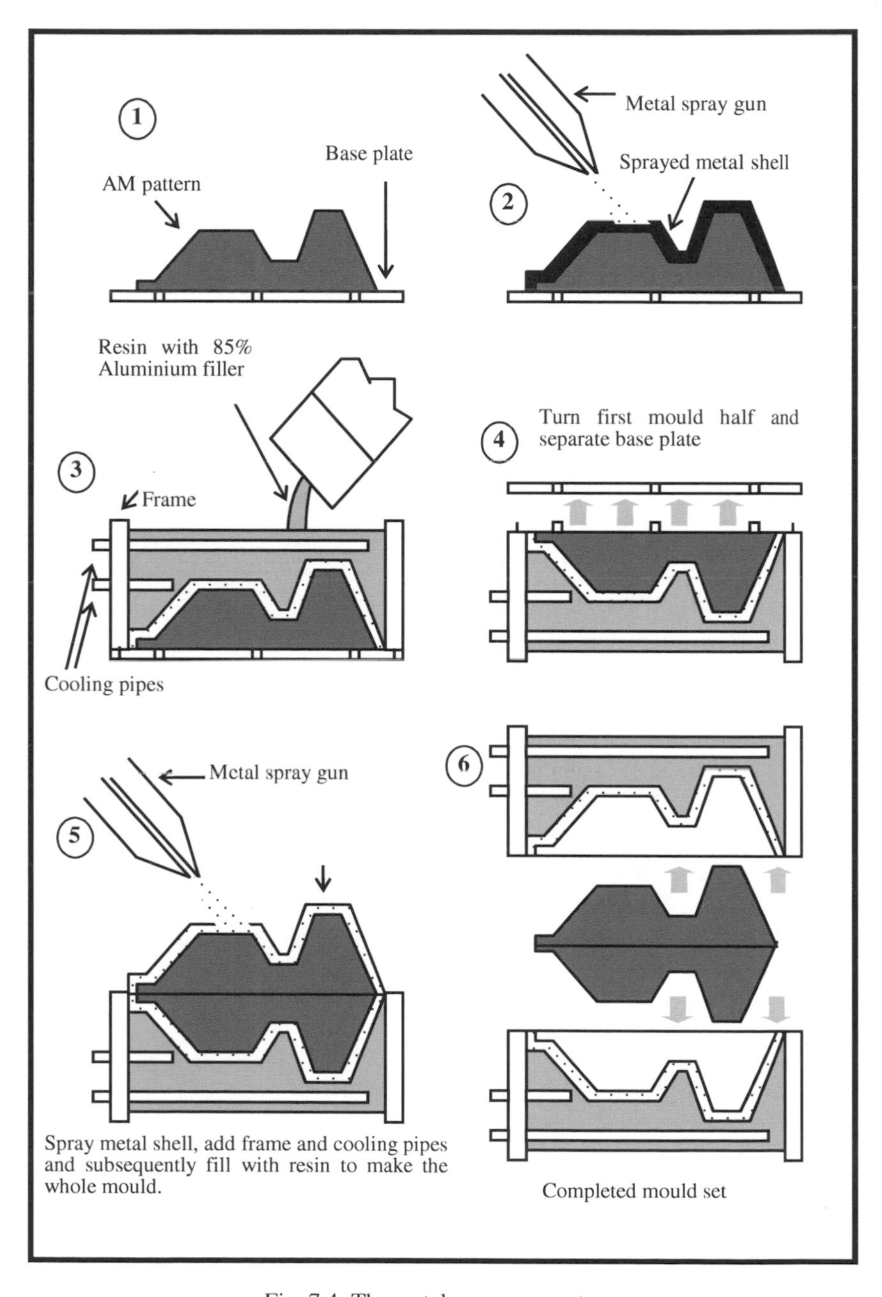

Fig. 7.4. The metal arc spray system.

7.5.2.2 *Silicone rubber moulds*

In manufacturing functional plastic, metal and ceramic components, vacuum casting with the silicone rubber mould has been the most flexible rapid tooling process and the most used to date. They have the following advantages:

(1) Extremely high resolution of master model details can be easily copied to the silicone cavity mould.

(2) Gross reduction of backdraft problems (i.e. die lock, or the inability to release the part from the mould cavity because some of the geometry is not within the same draw direction as the rest of the part).

The master pattern, attached with a system of sprue, runner, gating and air vents, is suspended in a container. Silicone rubber slurry is poured into the container engulfing the master pattern. The silicone rubber slurry is baked at 70°C for 3 hours and upon solidification, a parting line is cut with a scalpel.

The master pattern is removed from the mould, thus forming the tool cavity. The halves of the mould are then firmly taped together. Materials, such as polyurethane, are poured into the silicone tool cavity under vacuum to avoid asperities caused by entrapped air. Further baking at 70°C for 4 hours is carried out to cure the cast polymer part. The vacuum casting process is generally used with such moulds. Each silicone rubber mould can produce up to 20 polyurethane parts before it begins to break apart [8]. These problems are commonly encountered when using hard moulds, making it necessary to have expensive inserts and slides. They can be cumbersome and take a longer time to produce. These are virtually eliminated when the silicone moulding process is used.

AM models can be used as master patterns for creating these silicone rubber moulds. Figures 7.5 (a) to 7.5 (f) describe the typical process of creating a silicone rubber mould and the subsequent urethane-based part.

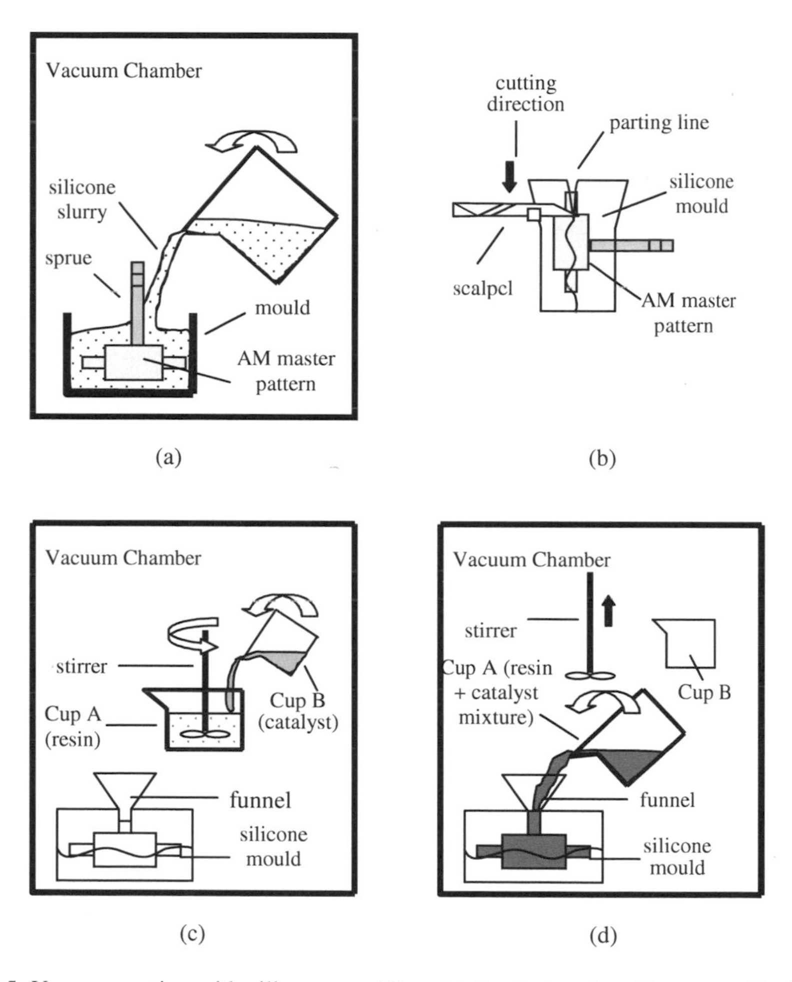

Fig. 7.5. Vacuum casting with silicone moulding. (a) Producing the silicone mould, (b) removing the AM master pattern, (c) mixing the resin and catalyst, (d) casting the polymer mixture.

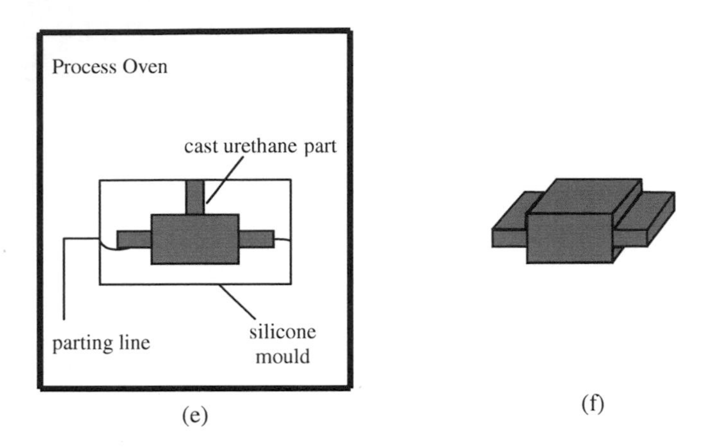

Fig. 7.5. (*Continued*) Vacuum casting with silicone moulding. (e) Cast urethane part cured in a baking oven, (f) the final rapid tooled urethane part.

A variant of this is a process developed by Shonan Design Co Ltd. This process, referred to as the "Temp-less" (temperature-less) process, makes use of similar principles of preparing the silicone mould and casting the liquid polymer, except that no baking is necessary to cure the materials. Instead, ultraviolet rays are used for curing of the silicone mould and urethane parts. The advantages this gives are higher accuracy of replicating the master model because no heat is used. Less equipment and 70% less time are required to produce the parts as compared to standard silicone moulding processes [9].

7.5.2.3 *Spin casting with vulcanised rubber moulds*

Spin casting, as its name implies, applies spinning techniques to produce sufficient centrifugal forces in order to assist in filling the cavities. Circular tooling moulds made from vulcanised rubber are produced in much the same way as in silicone rubber moulding. The tooling cavities are formed closer to the outer parameter of the circular mould to increase centrifugal forces. Polyurethane or zinc-based alloys can be cast using this method [10]. This process is particularly suitable for producing low volumes of small zinc prototypes that will ultimately be mass-produced by die-casting.

7.5.2.4 *Castable resin moulds*

Similar to the silicone rubber moulds, the master pattern is placed in a mould box with the parting line marked out in plasticine [11]. The resin is painted or poured over the master pattern until there is sufficient material for one half of the mould. Different tooling resins may be blended with aluminium powder or pellets so as to provide different mechanical and thermal properties. Such tools are able to typically withstand up to 100 to 200 injection moulding shots.

7.5.2.5 *Castable ceramic moulds*

Ceramic materials that are primarily sand-based can be poured over a master pattern to create the mould [12]. The binder systems can vary with preference of binding properties. For example, in colloidal silicate binders, the water content in the system can be altered to improve shrinkage and castability properties. The ceramic–binder mix can be poured under vacuum conditions and vibrated to improve the packing of the material around the master pattern.

7.5.2.6 *Plaster moulds*

Casting into plaster moulds has been used to produce functional prototypes [13]. A silicone rubber mould is first created from the master pattern and a plaster mould is then made from this. Molten metal is then poured into the plaster mould, which is broken away once the metal has solidified. Silicone rubber is used as an intermediate stage because the pattern can be easily separated from the plaster mould.

7.5.2.7 *Casting*

In the metal casting process, a metal, usually an alloy, is heated until it is in a molten state, whereupon it is poured into a mould or die that contains a cavity. The cavity will contain the shape of the component or casting to be produced. Although there are numerous casting techniques available, three main processes are discussed here: the conventional sand casting, investment casting, and evaporative casting processes. AM models render themselves well to be the master patterns for the creation of these metal dies.

Sand casting moulds are similarly created using AM master patterns. AM patterns are first created and placed appropriately in the sand box. Casting sand is then poured and packed very compactly over the pattern. The box (cope and drag) is then separated and the pattern carefully removed leaving behind the cavity. The box is assembled together again and molten metal is cast into the sand mould. Sand casting is the cheapest and most practical for casting large parts. Figure 7.6 shows a cast metal mould resulting from an AM pattern made by Laminated Object Manufacturing (LOM).

Fig. 7.6. Cast metal (left) and AM pattern for sand casting. (Courtesy of Helysis, Inc.)

Another casting method, the investment casting process, is probably the most important moulding process for casting metal. Investment casting moulds can be made from AM master patterns. The pattern is usually wax, foam, paper or other materials that can be easily melted or vaporised. The pattern is dipped in slurry ceramic compounds to form a relatively strong coating, or investment shell, over it [14]. This is repeated until the shell builds up thickness and strength. The shell is then used for casting, with the pattern being melted away or burned out of the shell, resulting in a ceramic cavity. Molten metal can then be poured into the mould to form the object. The shell is then cracked open

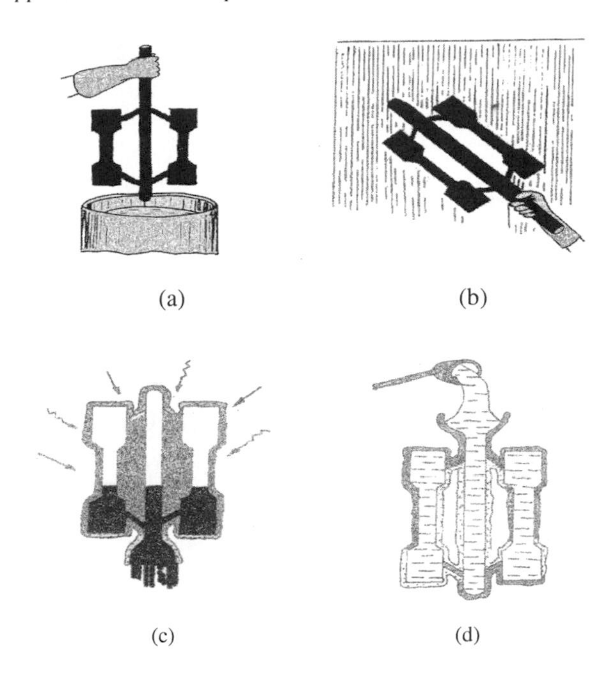

Fig. 7.7. Schematic diagram of the shell investment casting process. (a) Pattern clusters are dipped in ceramic slurry, (b) refractory grain is sifted onto the coated patterns. Steps (a) and (b) are repeated several times to obtain desired shell thickness, (c) after the mould material has set and dried, the patterns are melted out of the mould and (d) hot moulds are filled with metal by gravity, pressure vacuum or centrifugal force.

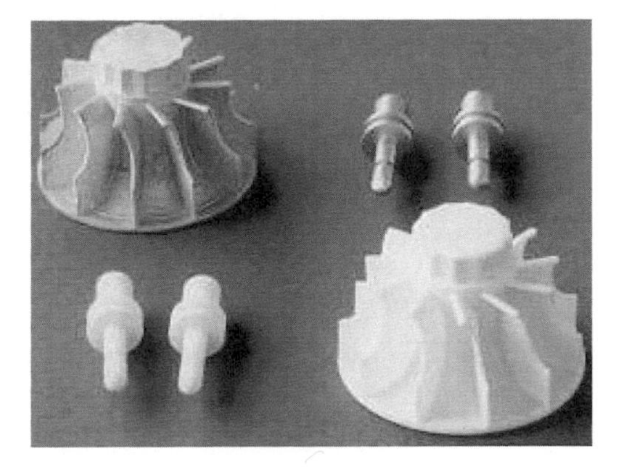

Fig. 7.8. Investment casting of an impeller from AM pattern.

to release the desired object in the mould. The investment casting process is ideal for casting miniature parts with thin sections and complex features. Figure 7.7 shows schematically the investment casting process from an AM-produced wax master pattern while Figure 7.8 shows an investment casting mould resulting from an AM pattern.

The third casting process discussed in this book is the evaporative pattern casting. As its names implies, this uses an evaporative pattern, such as polystyrene foam, as the master pattern. This pattern can be produced using the selective laser sintering (SLS) process along with CastForm™ polystyrene material. The master pattern is attached to sprue, riser and gating systems to form a 'tree'. This polystyrene 'tree' is then surrounded by foundry sand in a container and vacuum compacted to form a mould. Molten steel is then poured into the container through the sprue. As the metal fills the cavity, the polystyrene evaporates with very low ash content [15].

The part is cooled before the casting is removed. A variety of metals, such as titanium, steel, aluminium, magnesium and zinc can be cast using this method. Figure 7.9 shows schematically how an AM master pattern is used with the evaporative pattern casting process.

7.5.3 *Direct Hard Tooling*

Hard tooling produced by AM systems has been a major topic for research in recent years. Although several methods have been demonstrated, much research is still being carried out in this area. The advantages of hard tooling produced by AM methods are fast turnaround times to create highly complex-shaped mould tooling for high volume production. Fast response to modifications in generic designs can be almost immediate. The following are some examples of direct hard tooling methods.

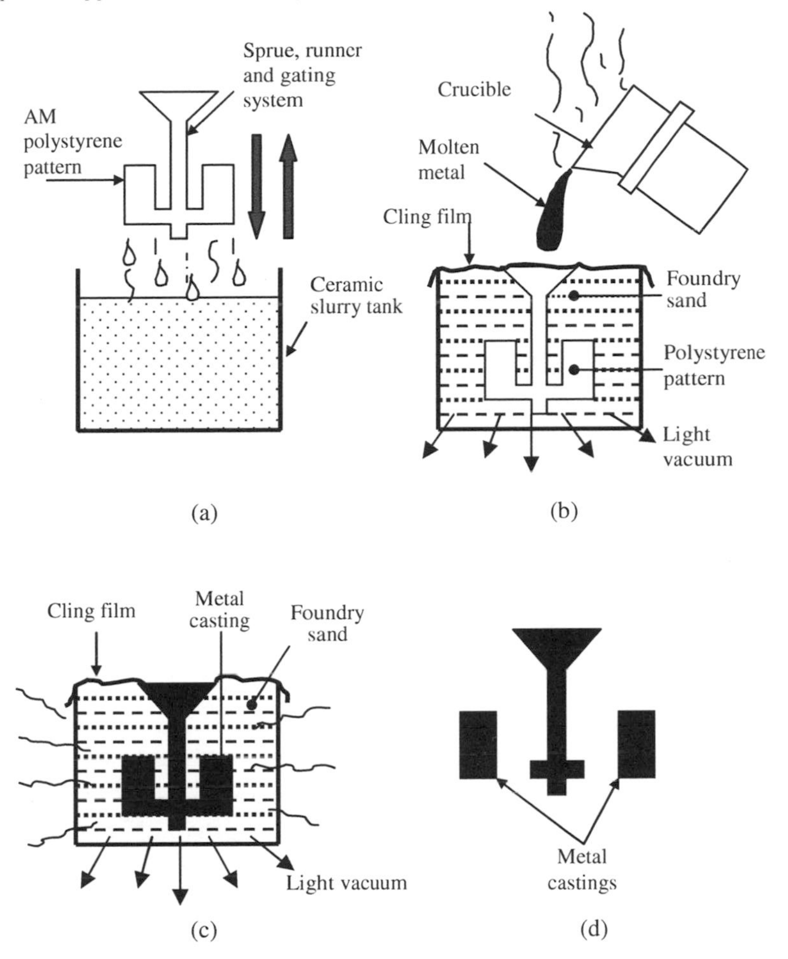

Fig. 7.9. Evaporative pattern casting process. (a) Polystyrene AM pattern 'tree' is coated by dipping with a ceramic slurry and air-dried, (b) coated AM pattern is packed with foundry sand in a container. The container is sealed with cling film and vacuumed to compact the sand further, (c) the polystyrene pattern evaporates as the molten metal is cast into the mould. The casting is then left to cool and (d) after solidification, the final cast parts are removed from the sprue, runner and gating system.

7.5.3.1 *RapidTool™*

RapidTool™ is a technology previously patented by the DTM Corporation to produce metal moulds for plastic injection moulding directly from the SLS Sinterstation. The moulds are capable of being used in the conventional injection moulding machines to mould the final product with the functional material [16]. The CAD data is fed into the Sinterstation™ which bonds polymeric binder coated metal beads together using the SLS process. Next, debinding takes place and the green part is cured and infiltrated with copper to make it solid. The furnace cycle is about 40 hours with the finished part having similar properties equivalent to aluminium. The finished mould can be easily machined. Shrinkage is reported to be no more than 2%, which is compensated for in the software.

Typical time frames allow relatively complex moulds to be produced in 2 weeks as compared to 6 to 12 weeks using conventional techniques. The finished mould is capable of producing up to tens of thousands of injection-moulded parts before breaking down.

7.5.3.2 *Laminated metal tooling*

This is another method that may prove promising for RT applications. The process applies metal laminated sheets with the LOM method. The sheets can be made of steel or any other material which can be cut by the appropriate means, for example by CO_2 laser, water jet or milling, based on the LOM principle [17]. The CAD 3D data provides the sliced 2D information for cutting the sheets layer-by-layer. However, instead of bonding each layer as it is cut, the layers are all assembled after cutting and either bolted or bonded together.

7.5.3.3 *Direct metal laser sintering (DMLS) tooling*

The Direct Metal Laser Sintering (DMLS) technology was developed by EOS. The process uses a very high-powered laser to sinter metal powders directly. The powders available for use by this technology are the bronze-based and steel-based materials. Bronze is used for applications where strength requirements are not crucial. Upon sintering of the bronze powder, an organic resin, such as epoxy, is used to infiltrate the part. For steel powders, the process is capable of producing direct steel parts of up to 95% density so that further infiltration is not required.

Several direct applications are produced with this technology including mould inserts and other metal parts [18].

7.5.3.4 *ProMetal™ rapid tooling*

Based on MIT's Three-Dimensional Printing (3DP) process, the ProMetal™ Rapid Tooling System is capable of creating steel parts for the tooling of plastic injection moulding parts, lost foam patterns and vacuum forming. This technology uses an electrostatic ink jet print head to eject liquid binders onto the powder, selectively hardening slices of an object a layer at a time. A fresh coat of metal powder is spread on top and the process repeats until the part is complete. Loose powders act as supports for the object to be built. The AM part is then infiltrated at furnace temperatures with a secondary metal to achieve full density. Toolings produced by this technology for use in injection moulding have reported withstanding pressures up to 30,000 psi (200 MPa) and survived 100,000 shots of glass-filled nylon [19].

7.5.4 *Indirect Hard Tooling*

There are numerous indirect AM tooling methods that fall under this category and this number continues to grow. However, many of these processes remain largely similar in nature except for small differences e.g. binder system formulations or type of system used. Processes include the Rapid Solidification Process (RSP), Ford's (UK) Sprayform, Cast Kirksite Tooling, CEMCOM's Chemically Bonded Ceramics (CBC) and Swift Technologies Ltd 'SwiftTool,' just to name a few. This section will only cover selected processes that can also be said to generalise all the other methods under this category. In general, indirect methods for producing hard tools for plastic injection moulding generally make use of casting of liquid metals or steel powders in a binder system. For the latter, debinding, sintering and infiltration with a secondary material are usually carried out as post-processes.

7.5.4.1 *3D Keltool*

The 3D Keltool process has been developed by 3D Systems to produce a mould in fused powdered steel [20]. The process uses an SLA model of the tool for the final part that is finished to a high quality by sanding and polishing. The model is placed in a container where silicone rubber is poured around it to make a soft silicone rubber mould that replicates the female cavity of the SLA model. This is then placed in a box and silicone rubber is poured around it to produce a replica copy of the SLA model. This silicone rubber is then placed in a box and a proprietary mixture of metal particles, such as tool steel, and a binder material is poured around it, cured and separated from the silicone rubber model. This is then fired to eliminate the binder and sinter the green metal particles together. The sintered part is about 70% steel. The 30% void is then infiltrated with copper to give a solid mould, which can be used in injection moulding.

An alternative to this process is described as the reverse generation process. This uses a positive SLA master pattern of the mould and requires one step less. This process claims that the CAD solid model to injection-moulded production part can be completed in 4 to 6 weeks. Cost savings of around 25 to 40% can be achieved when compared to that of conventional machined steel tools.

7.5.4.2 *Electro-Discharge Machining (EDM) electrodes*

A method successfully tested in research laboratories but so far not widely applied in the industry is the possible manufacturing of copper electrodes for EDM (Electro-Discharge Machining) processes using AM technology. To create the electrode, the AM-created part is used to create a master for the electrode. An abrading die is created from the master by making a cast using an epoxy resin with an abrasive component. The resulting die is then used to abrade the electrode. A specific advantage of the SLS procedure (see Section 5.1) is the possible usage of other materials. Using copper in the SLS or SLM process, it is possible to generate the electrodes used in EDM quickly and affordably.

7.5.4.3 *Ecotool*

This is a development between the Danish Technological Institute (DTI) in Copenhagen, Denmark, and the TNO Institute of Industrial Technology of Delft in Holland. The process uses a new type of powder material with a binder system to rapidly produce tools from AM models. As its name implies, the binder is friendly to the environment in that it uses a water-soluble base. An AM master pattern is used and a parting line block produced. The metal powder-binder mixture is then poured over the pattern and parting block and left to cure for an hour at room temperature. The process is repeated to produce the second half of the mould in the same way. The pattern is then removed and the mould is baked in a microwave oven.

7.5.4.4 *Copy milling*

Although not broadly applied nowadays, AM master patterns can be provided by manufacturers to their vendors for use in copy milling, especially if the vendor for the required parts is small and does not have the more expensive but accurate CNC machines. In addition, the principle of generating master models only when necessary allows some storage space to be saved. The limitation of this process is that only simple geometrical shapes can be made.

7.6 Aerospace Industry

Aerospace is the leading pioneer industry that other industries look to for a glimpse of what is new on the horizon. The aerospace industry has incorporated AM throughout its product development lifecycle; from the design concept to repairs. With the various advantages that AM technologies promise, it is only natural that aerospace continues to discover and find new applications and invest in research to make them possible. The following are a few examples.

7.6.1 *Prototype and Test*

AM technology has matured into a mainstream rapid prototype fabrication methodology. AM allows designers to skip the fabrication of tools and go straight to the finished parts. Proven to be effective in creating light-weight, durable and complex aerospace parts, AM has been at the forefront of aerospace companies' technology adoption. SelectTech Geospatial, which built the first 3D-printed Unmanned Aerial System (UAS) to take off and land on its gear, uses AM to build prototypes. SelectTech refined the entire airframe through discoveries from physical prototypes. Through a trial-and-error approach, parts were printed for testing and designs were modified to correct problems that arose, learning from each iteration. Using FDM method (See Section 4.1), the entire models were built using Stratasys' Dimension 3D Printer [26]. Eliminating the need for expensive tooling during the design process allowed SelectTech designers to focus on creating lighter weight structures, boosting efficiencies, strengthening parts and speeding up the overall development process. The success of this project proved that complicated and sophisticated products can be made, tested and manufactured through rapid-response model using AM without committing to a final design and manufacturing tools.

When it comes to building highly customised vehicles and testing them in punishing environments, NASA knew only AM offered the design flexibility and quick turnaround process. The NASA team used 3D printing to build a Mars rover with about 70 parts built digitally and directly from computer designs in the heated chamber of a Stratasys 3D Printer [27]. Using FDM, the process created complex shapes such as an ear-shaped exterior housing, which is so deep and twisted that it would have been impossible or even financially infeasible to build. NASA engineers used AM to build prototypes to test the form, fit, and function of parts that they will eventually make in other materials. 3D printing allowed designers to solve difficult challenges before committing to expensive tooling, ensuring machined parts are based on the best possible design.

7.6.2 *General Electric (GE) and European Aeronautic Defence (EADS) to 3D Print Better and Lighter Parts for Aeroplanes*

In 2011, GE started a new lab at its global research headquarters in New York, USA devoted to turning AM technology into a viable means of manufacturing functional parts for its businesses. AM technology has vastly improved to the point that AM systems are able to produce intricate objects out of durable materials, including metals such as titanium and aluminium. 3 years later, GE Aviation announced that it will open a new $100 million assembly plant in Indiana, USA to build the world's first aeroplane with 3D printed fuel nozzles. EADS or European Aeronautic Defence and Space Company used DMLS to print metal hinges for engine covers. The parts have intricate shapes that maintain strength while cutting the weight in half. These 3D printed hinges have also been tested to meet performance requirements. With AM, up to 75% of raw materials can be reduced as compared to the conventional production techniques. Furthermore, lighter parts translate to weight savings which are critical in the aerospace industry. According to EADS, reducing the weight of the plane by just one kilogram can result in fuel savings of $3,000 per year, or $100,000 over 30 years, the typical life of an aeroplane.

7.6.3 *Production*

The commercial aviation industry has high performance standards to comply with and has now accepted AM in the manufacturing of parts such as air grates, panel covers, heating, ventilation, and air conditioning (HVAC) ducts, power distribution panes, some interior parts as well as lots of mounting and attachment hardware.

An example of a company which utilises FDM is Taylor-Deal Automation. They use AM for prototyping through production for its engineering and modifications of specialty fluid and air handling parts. Even with low-quantity production, AM has enabled them to reduce cost and weight, and improved lead times as well as flexibility in designs.

The world's largest manufacturer of general aviation instruments is Kelly Manufacturing Company (KMC). They are also the manufacturer of the popular R.C. Allen line of aircraft instrumentation. Through the implementation of FDM they are now able to produce 500 pieces of toroid housings using one overnight run as opposed to their traditional method of using a urethane moulded technique which took 4 weeks [26]. This has helped save them cost and time.

Another industry that has been growing rapidly in its production through AM is Unmanned Aerial Systems (UAS). The highly complex, low-volume, rapid design iterations and structural complexity put AM technology at the forefront of its production, and having no passenger safety regulations helped with the vast adoption.

Aurora Flight Sciences is a manufacturer of advanced unmanned systems and aerospace vehicles. The company successfully created and flew a 62-inch wingspan aircraft using AM technology [26]. AM manufacturing solves the problems of design constraints that traditional fabrication techniques often pose.

Another emerging application of 3D printing is ‘smart parts’, which are hybrid parts that include 3D-printed structures and electronics. Stratasys and Optomec helped Aurora combined FDM and Aerosol Jet Electronics printing to produce wings integrated with electronic components. This is truly a game changer for design and manufacturing as a whole, as the production of goods can be streamlined to require less materials and steps to bring a product to market.

7.6.4 *Making Parts for Jet Engines*

GE Aviation, part of the world’s biggest manufacturing group, has made significant investments in additive manufacturing to build more than 85,000 fuel nozzles for its new Leap jet engines. Previously, the nozzles were assembled using 20 different parts, a labour-intensive process in which a lot of materials were wasted [28]. With AM, complex and intricately designed parts can be produced without any waste.

Using the selective laser sintering process (SLS), thin layers of metallic powder are spread onto a build platform and fused together with a laser beam. The process is repeated until an object emerges. SLS is capable of creating all kinds of metal parts, including components made from aerospace-grade titanium. Aside from the cost and time-saving benefits of using AM, the 3D printed fuel nozzles will each be 25 percent lighter and 5 times more efficient than conventionally manufactured nozzles.

While there is still some way to go before AM can be a fully practical alternative to traditional production methods, the potential AM has to offer is enormous. AM has certainly made its marks in designing, testing, tooling and production in the aerospace industry, and the versatility of this technology will extend beyond aerospace applications. AM has the potential to become a transformational technology, one which could take manufacturing in entirely different directions and help launch new enterprises and business models.

7.7 Automotive Industry

As early adopters of rapid prototyping (later called additive manufacturing), the automotive industry has more 3D printing applications than any other industry. The automotive industry has always been a technologically and innovation-driven industry, where consumers' demands for design, safety, comfort and environmentally friendly performance vehicles create various challenges for companies to compete to meet these demands.

7.7.1 *Multi-Material Capability and the Speed of Additive Manufacturing*

Bentley Motors Ltd, a name synonymous with luxury and quality, has utilised the patented PolyJet technology to speed up the production of small-scale models as well as full-size parts for assessment and testing prior to production on the assembly line. Every part of the car is prototyped in miniature. The accuracy of the Objet30 3D Printer enabled the operations and projects team to emulate a full-size part and scale it down to a one-tenth-scale model [29]. Once the miniature model is approved, some full-sized parts and parts that combine different materials can be produced even without assembly.

Stratasys' 3D Printer Objet500 Connex has given the design studio team the power to combine a variety of material properties using 3D printing technology. A single prototype can combine rigid and rubber-like, clear and opaque materials with no assembly necessary by 3D printing, for example, a rubber tyre on a wheel rim. Polyjet's rubber-like material also enables Bentley to simulate rubber according to different levels of hardness and tear resistance.

Using traditional methods, such a prototype would have been costly and time-consuming to build, and it might not have been always possible to include all fine details of the design. Fabrication of the model based on drawings was often subject to human interpretation and consequently error-prone, thus further complicating the prototyping process. All these difficulties were avoided by using AM technology, as the production of the model was based entirely on the CAD model that was created before it was 3D printed.

AM technology has shrunk the prototype development time from months to a couple of weeks, and considerable time and cost savings were achieved. The ability to print parts with different materials combined has also revolutionised the entire Bentley Motors design process.

7.7.2 *Direct Digital Manufacturing*

While additive manufacturing has become an integral part of product development in automotive manufacturing, German automaker BMW is extending FDM's application to other areas, including Direct Digital Manufacturing. Stratasys' 3D Production System is used to build ergonomically designed assembly line hand tools which perform better than conventional tools. These tools are easier and more comfortable to use for repetitive processes, which ultimately improves productivity. For example, a hand-held device's weight was reduced by 72% with a sparse-fill build technique using 3D printing, taking 1.3kg off the device. The difference of the weight may not seem significant but it does for a worker who uses the tool hundreds of times during a shift. The FDM process has turned out to be a cost saving alternative to conventional manufacturing methods, with cost reductions in engineering documentation, warehousing and manufacturing. BMW has saved 58% in costs and 92% in time by using additive manufacturing [30]. The layered FDM manufacturing process is also suitable for the production of complex bodies compared with conventional metal-cutting processes which would be difficult and expensive to produce. For example, a tool used for attaching bumper supports is a rather convoluted tube that negotiates around obstructions and attaches fixturing magnets exactly where needed. Without FDM, this tool would have been next to impossible to produce.

FDM enable designers to create devices that utilise the advantages of additive manufacturing. This helps enterprises with their product development as well as offers an alternative method for manufacturing components in small numbers.

7.7.3 *More Creativity and Faster Prototypes*

Ford has been at the forefront of AM for 25 years and was involved with the invention of 3D printing in the 1980s. Using various systems such as SLS, FDM and SLA, Ford uses AM to speed up the production of prototype parts, shaving months off the development time for individual components used in all of its vehicles, such as cylinder heads, intake manifolds and air vents. With the conventional method of using custom machine tools, an engineer would create a computer model of an intake manifold, the most complicated engine part, and wait four months for one prototype at a cost of $500,000. With AM technology, Ford can print the same part in four days including multiple iterations and no tooling limits at a cost of $3,000 [31]. This is a huge saving for Ford and for their customers. AM allows significant improvement in the product that can be weight-optimised to improve fuel efficiency.

Without the need for special tooling or a dedicated mould, 3D printing has helped Ford save millions of dollars in product development. The technology allows engineers to be more creative and experiment with more radical, innovative part designs that would not have been explored using the extremely costly, long and tedious traditional methods. Ford is showing its commitment to 3D printing by working with their suppliers to bring many of those new technologies to the market and equipping all Ford engineers with 3D printers at their workstations [32].

7.7.4 *World's First Energy-Efficient Prototype Car Body Built with 3D Printing*

KOR EcoLogic Inc, a design and engineering consultancy firm based in Winnipeg, Canada is responsible for the world's first prototype car that has its entire body 3D printed including its glass panel prototypes. The car, code-named URBEE which stands for urban electric, has been designed from the ground up with environmentally sustainable principles in the mind of its founder, Jim Kor. The electric and liquid fuel vehicle offers some impressive fuel efficiency numbers while achieving more than 200mpg (1.2l/100km) on the highway and 100mpg (2.3l/100km) in city driving. [33]

The traditional prototyping method would have taken months, if not years, of iterating designs, tooling and materials. The original plan was to use a clay model that took the team 3 months to make and to build the vehicle from fibre-reinforced polymer (FRP) or fibreglass panels which would have taken another few months of labour-intensive work. With 3D printing, designers can tweak and make design adjustments on the CAD file and the parts can be printed within a few days. Before printing the full-sized body panels, the computer files were first verified to be correct by printing a 1/6th scale model (See Fig. 7.10). KOR EcoLogic worked closely with Stratasys in Minneapolis which is responsible for printing all the vehicle's exterior components. The body was produced in ABS plastic using FDM, by applying the thermoplastics in layers from the bottom up. This allows the elimination of tooling and machining, bringing improved efficiency when a design change is needed. The FDM process allowed the team to fabricate an extremely precise, light and strong body for the car. The 3D printers eject polymers right down to a microscopic layer, creating the entire car in 2,500 hours continuously even after the engineers have gone home. Unlike traditional cars made from sheet metal, the URBEE car framework did not require connecters, nuts and bolts and was constructed out of large single pieces of plastic. The entire finished 3D printed product is shown in Figure 7.11.

Fig. 7.10. The 1/6th scale model of URBEE. (Courtesy of Dana McFarlane and KOR EcoLogic)

At the time of writing, the company has moved on to the next project of producing URBEE 2, which would take the concept of the original URBEE to a higher level by featuring fully functioning heaters, windscreen wipers and even mirrors. The KOR EcoLogic team will fully design URBEE 2 in CAD files and then collaborate with Stratasys for printing through the FDM process. While the original URBEE only used 3D printing for the body, URBEE 2 will contain interior and internal 3D printed parts. The firm hopes to push fuel efficiency even further with URBEE 2 and hopes to travel across the United States on only 10 gallons of bio-fuel. [34]

Fig. 7.11. The URBEE's entire body was 3D printed. (Courtesy of KOR EcoLogic)

3D printing is still very much in its early stage for manufacturing and may not fully yet replace the automotive industry's current high speed and high volume direct production process. However, the technology is definitely making its impact in the product and testing development requiring niche product applications, which go through frequent development changes.

7.8 Jewellery Industry

The jewellery industry has traditionally been regarded as one which is heavily craft-based, and automation is generally restricted to the use of machines in the various individual stages of jewellery manufacturing. The use of AM technology in jewellery design and manufacture offers a significant breakthrough in this industry. In an experimental computer-aided jewellery design and manufacturing system jointly developed by Nanyang Technological University and Gintic Institute of Manufacturing Technology in Singapore, the SLA (from 3D Systems) was used successfully to create fine jewellery models [35]. These were used as master patterns to create the rubber moulds for making wax patterns that were later used in investment casting of the precious metal end products (see Figure 7.12). In an experiment with the design of rings, the overall quality of the SLA models were found to be promising, especially in the generation of intricate details in the design. However, due to the nature of the step-wise building of the model, steps at the 'gentler' slope of the model were visible. With the use of better resin and finer layer thickness, this problem was reduced but not fully eliminated. Further processing was found to be necessary, and abrasive jet deburring was identified to be most suitable [36].

Though postprocessing of SLA models is necessary in the manufacture of jewellery, the ability to create models quickly (a few hours compared to days or even weeks, depending on the complexity of the design) and its suitability for use in the manufacturing process offer great promise to improve design and manufacture in the jewellery industry.

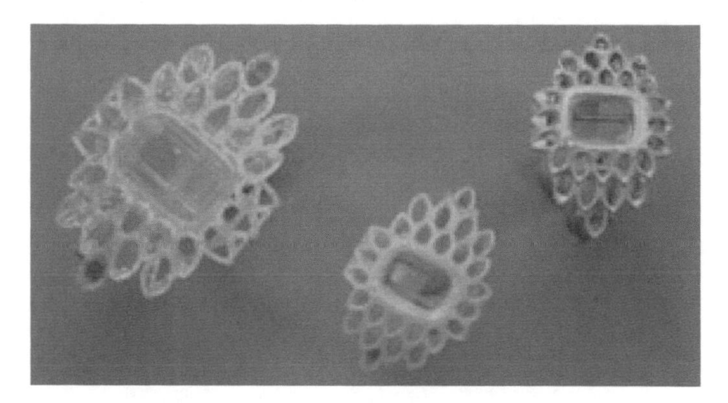

Fig. 7.12. A two-times scaled SLA model to aid visualisation (left), full-scale wax pattern produced from the silicone rubber moulding (centre), and an investment cast silver alloy prototype of a brooch (right).

7.9 Coin Industry

Similar to the jewellery industry, the mint industry has traditionally been regarded as very labour-intensive and craft-based. It relies primarily on the skills of trained craftsmen in generating "embossed" or relief designs on coins and other related products. In another experimental coin manufacturing system using CAD/CAM, CNC and AM technologies developed by Nanyang Technological University and Gintic Institute of Manufacturing Technology in Singapore, the SLA (from 3D Systems) was used successfully with Relief Creation Software to create tools for coin manufacture [37]. In the system involving AM technology, its working methodology consists of several steps.

Firstly, 2-D artwork is read into ArtCAM, the CAD/CAM system used in the system, utilising a Sharp JX A4 scanner. Figure 7.13 shows the 2-D artwork of a series of Chinese characters and a roaring dragon. In the ArtCAM environment, the scanned image is reduced from a colour image to a monochrome image with the fully automatic "Grey Scale" function. Alternatively, the number of colours in the image can be reduced using the "Reduce Colour" function. A colour palette is provided for colour selection and the various areas of the images are coloured, either using different sizes and types of brushes or the automatic flood fill function.

Fig. 7.13. 2-dimensional artwork of a series of Chinese characters and a roaring dragon.

The second step is the generation of surfaces. The shape of a coin is generated to the required size in the CAD system for model building. A triangular mesh file is produced automatically from the 3D model. This is used as a base onto which the relief data is wrapped and later combined with the relief model to form the finished part.

The third step is the generation of the relief. In creating the 3-D relief, each colour in the image is assigned a shape profile. There are various fields that control the shape profile of the selected coloured region, namely, the overall general shape for the region, the curvatures of the profile (convex or concave), the maximum height, base height, angle and scale. The relief detail generated can be examined in a dynamic Graphic Window within the ArtCAM environment itself. Figure 7.14 illustrates the 3-D relief of the roaring dragon artwork.

Fig. 7.14. 3-dimensional relief of artwork of the roaring dragon.

The fourth step is the wrapping of the 3-D relief onto the coin surface. This is done by wrapping the 3-dimensional relief onto the triangular mesh file generated from the coin surfaces. This is a true surface wrap and not a simple projection. The wrapped relief is also converted into triangular mesh files. The triangular mesh files can be used to produce a 3-D model suitable for colour shading and machining. The two sets of triangular mesh files, of the relief and the coin shape, are automatically combined. The resultant model file can be colour-shaded and used by the SLA to build the prototype.

The fifth step is to convert the triangular mesh files into the .STL file format. This is to be used for building the AM model. After the conversion, the .STL file is sent to the SLA to create the 3-D coin pattern, which will be used for proofing of design [33].

7.10 Tableware Industry

In another application to a traditional industry, the tableware industry, CAD and AM technologies are used in an integrated system to create better designs in a faster and more accurate manner. The general methodology used is similar to that used in the jewellery and coin industries. Additional computer tools with special programs developed to adapt decorative patterns to different variations of size and shape of tableware are needed for this particular industry [38]. Also, a method for generating motifs along a circular arc has been developed to supplement the capability of such a system [39].

The general steps involved in the art to part process for the tableware include the following:

(1) Scanning of the 2-D artwork.
(2) Generation of surfaces.
(3) Generation of 3-D decoration reliefs.
(4) Wrapping of reliefs on surfaces.
(5) Converting triangular mesh files to STL file.
(6) Building of model by the AM system.

Two AM systems are selected for experimentation in the tableware system. One is 3D Systems' SLA, and the other is Helysis' LOM. The SLA has the advantages of being a pioneer and a proven technology with many excellent case studies available. It is also advantageous to use in tableware design as the material is translucent and thus allows designers to view the internal structure and details of tableware items like tea pots and gravy bowls. On the other hand, the use of LOM has its own distinct advantages. Its material cost is much lower and because it does not need support in its process (unlike the SLA); it saves a lot of time in both preprocessing (deciding where and what supports to use) and postprocessing (removing the supports). Examples of dinner plates produced using the systems are shown in Figure 7.15.

Fig. 7.15. Dinner plate prototype built using SLA (lcft) and LOM (right).

In an evaluation test of making the dinner plate prototype, it was found that the LOM prototype was able to recreate the floral details more accurately. The dimensional accuracy was slightly better in the LOM prototype. In terms of build-time, including pre- and postprocessing, the SLA was about 20% faster than the LOM process. However, with sanding and varnishing, the LOM prototype appeared to be a better model which could be used later to create the plaster of Paris moulds for the moulding of ceramic tableware (see Figure 7.16 for a tea pot built using LOM). Apart from these technical issues, the initial investment, operating and maintenance costs of the SLA were considerably higher than that of the LOM, estimated to be about 50 to 100% more.

Fig. 7.16. LOM model of a tea pot. (Courtesy of Champion Machine Tools, Singapore)

In the ceramic tableware production process, the LOM model can be used directly as a master pattern to produce the block mould. The mould is made of plaster of Paris. The result of this trial is shown in Figure 7.17. The trials highlighted the fact that plaster of Paris is an extremely good material for detailed reproduction. Even slight imperfections left after hand finishing the LOM model are faithfully reproduced in the block mould and pieces cast from these moulds.

Whichever AM technology is adopted, such a system saves time in designing and developing tableware, particularly in building a physical prototype. It can also improve designs by simply amending the CAD model, and the overall system is easy and friendly to use.

Fig. 7.17. Block mould cast from the LOM model of the dinner plate. (Courtesy of Oriental Ceramics Sdn Bhd, Malaysia)

7.11 Geographic Information System (GIS) Applications

AM has been applied to create physical models of 3D Geographic Information System (GIS) objects to replace 2D representations of geographical information. Figure 7.18 shows a three-dimensional physical model of a city [40]. To be able to do this, CONTEX Scanning Technology introduced its PUMA HS 36 colour scanner. It features iJET Technology and the 18 flatbed colour scanner, which is designed to scan all types of originals, including rare and valuable documents up to A2/C size.

Fig. 7.18. A physical model of a city. (Courtesy of CONTEX Scanning Technology A/S)

7.11.1 *3D Physical Map*

The first experiences concerning the reception of the models were very encouraging and surprising: nearly everybody tried to touch the models with his hands, to get a feel in the literal sense [40]. The sensual capability creates a possibility to produce maps for blind and visually impaired persons. The haptic experience could be utilised as an additional stimulus for transmitting a cartographic message and to induce insight.

7.11.2 *3D Representation of Land Prices*

A three-dimensional map visualising the average cost of land in a country or city could be plotted. The height of the prisms would be proportional to the average price of the land. In contrast to a 2D choropleth map depicting value classes, the absolute differences in height could be perceived immediately [40].

7.11.3 *3D Representation of Population Data*

The volume of the prisms could proportional to the number of inhabitants in each unit area. The height of the prisms would then be proportional to the population density (see Figure 7.19).

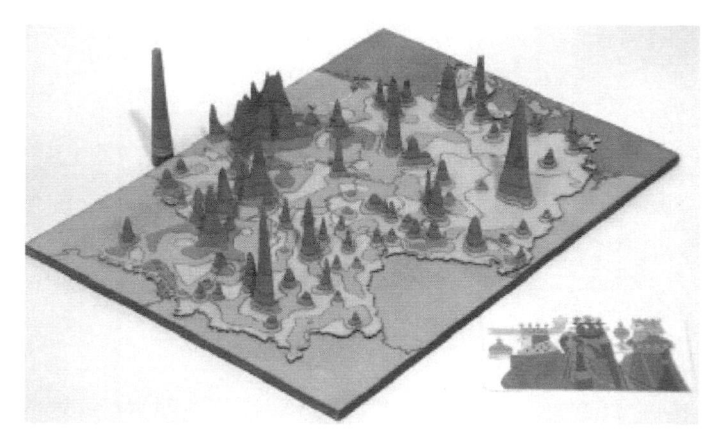

Fig. 7.19. Smooth surface depicting population density. (Courtesy of Federal Office for Building and Spatial Planning (BBR))

7.12 Arts and Architecture

7.12.1 *Architecture*

Architectural models are used for visualising the perception and elements of a building or a structure are considered most important to decision-makers (see Figure 7.20). Architectural model making has been revolutionised recently in the past few years. AM technology has been considered to replace traditional techniques as it has the following advantages: (a) The ability to create scaled models directly from 3D CAD data in a few short hours. It is currently used by large architectural firms for the creation of proposed and as-built designs. Design changes in particular segments of the construction can be readily rebuilt based on the reproducibility of the AM models, and (b) the ability to create multiple models at reasonable cost. For larger projects involving more than one key main contractor, it is advantageous to build multiple copies of the model for co-ordination and communication.

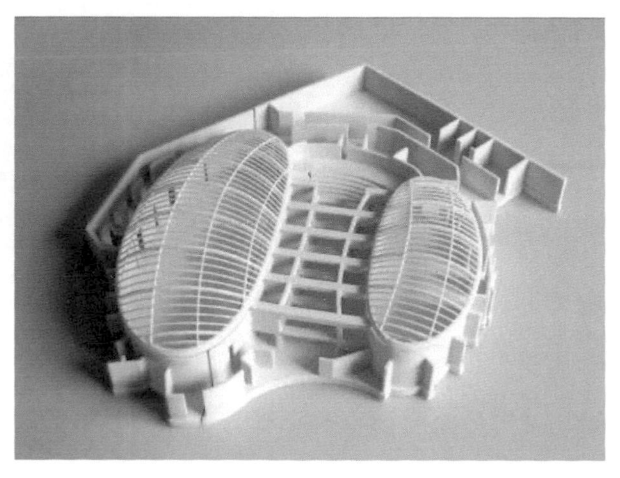

Fig. 7.20. An architecture model built from 3DP. (Courtesy of CONTEX Scanning Technology A/S)

7.12.2 *Arts*

While AM technology has been widely used to generate replicas of natural and man-made structures, Neri Oxman, an architect and designer named by Fast Company as one of the "100 Most Creative People" in 2009, is discovering new design and engineering principles through 3D printing to produce complex structures impossible by other fabrication techniques [41]. With 3D printers' capabilities of printing multi-materials, Oxman believes the technology will revolutionise architecture and enable new forms, design freedom and possibilities. Her work has been featured in museums and one of her best known works, the Beast, is a sensually curvy chair that can adjust its shape, flexibility and softness to fit each person who sits in it (See Fig 7.21).

Fig. 7.21. Beast, a chair designed by Neri Oxman using 3D printing that responds to the individual's body weight. (Courtesy of Neri Oxman)

Today, AM technology is being used by an increasing number of artists to build a wide variety of sculptural objects. Some of these works are visually realistic and representational while others are abstract and even obscured. The abstract objects can be the result of pure imagination and artistic free will, or may be derived solely from mathematics or computation. Some of the works created with AM may not have been possible to be made any other way [42].

A Trefoil Torus Knot with a computer-generated texture map was rendered in 3D Printing. Colour 3D printers now build computer-specified colour textures that are directly modelled into the 3D part (see Figure 7.22). It is possible to reproduce surface colouration which defies reproduction by hand, i.e. spaces where a paint brush cannot fit.

Fig. 7.22. Torus Knot. (Courtesy of Stewart Dickson)

7.13 Construction

AM has typically been used extensively in building architectural models but recently the focus has shifted to the delivery of full-scale architectural components and elements of buildings such as walls and facades as well.

A group of researchers from Loughborough University has led the project known as 3D Concrete Printing and Freeform Construction, an additive manufacturing process capable of producing full-scale building components [43]. The team led by Professor Richard Buswell utilised concrete as its main construction material for the entire AM process, where layers and layers of concrete are deposited in great precision under computer control. Figure 7.23 shows the one-tonne reinforced concrete bench printed using the AM process, and a two square metre curvy panel being featured in an exhibition (Fig 7.24).

Fig. 7.23. A wall-like artefact named Wonder Bench built out of concrete using the AM process. (Courtesy of Loughborough University)

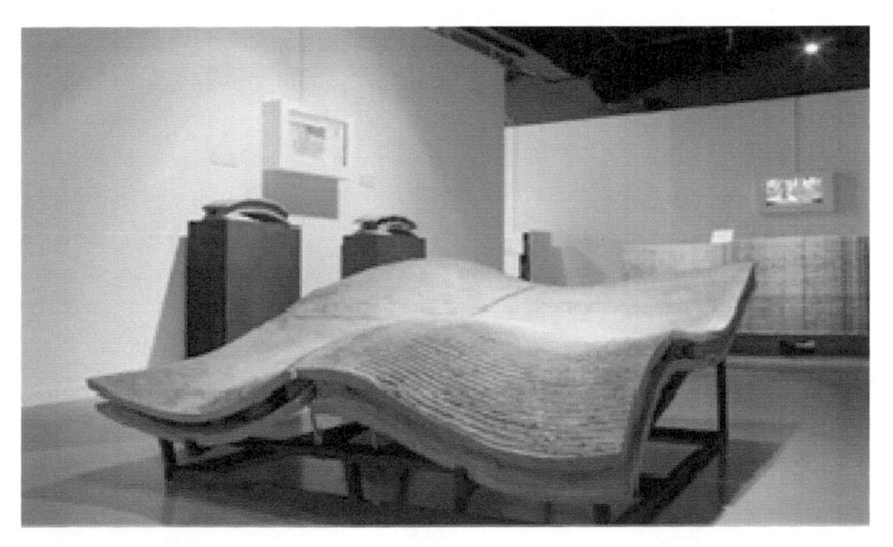

Fig. 7.24. 3D Printing used to create a double curved concrete panel. (Courtesy of Agnese Sanvito)

Concrete printing is based on highly controlled extrusion of cement-based mortar, which is precisely positioned according to a 3D computer-aided-design (CAD) model. Using this technology allowed complicated sections of the building to be created, which may traditionally have been too costly to build using the injection moulding method. The process consists of data preparation typical to most AM processes: material preparation and printing using cement-based mortar.

In construction, every part and structure is unique in size and dimension, which means traditional *reductive manufacturing* would require standard sized materials to be removed to fit; *formative manufacturing* would require a modified mould to shape each component [44]. AM processes have advantages over conventional manufacturing processes because: (a) cost-based opportunity to save time and materials by eliminating waste and the need for bespoke moulds. (b) The computational design ability provides the freedom to design around individuals and the environment, making each part unique. (c) AM allows form optimisation, which significantly reduces the quantity of materials needed. (d) It has the potential of building additional functionality into structures [45].

To bring automation into the construction industry in the form of AM, researchers have partnered with major design and construction firms who provide first-hand industrial knowledge and assess its commercial viability in a conservative market that emphasises efficiency and value over money.

7.14 Fashion and Textile

The fashion industry is starting to embrace 3D printing. AM's ability to produce complex, currently impossible to manufacture forms, coupled with the ease with which customisation can be achieved using it, is transforming fashion both on the high street and on the runway.

One can now order a 3D printed bikini from Continuum's Shapeways Shop and its creators hope to be able to one day adapt the patterning algorithm used to any surface, allowing fully customised clothing to be made using body scans [46].

Fashion designers Gabi Asfour, Angela Donhauser and Adi Gil who run the fashion company ThreeASFOUR created their "Mer Ka Ba" collection inspired by religion and geometry. Each of the three syllables of the collection name represents a sacred religious concept and the geometric patterns included in their designs were drawn from religious architecture. AM technology allowed the trio to create complex and intricately designed geometric patterns [47].

In contrast to the rigid geometric shapes of traditional laser cutting technology, 3D printing gives a softer and delicate cocoon effect. Figure 7.25 and Figure 7.26 both show 3D printed dresses and shoes designed by ThreeASFOUR.

Fig. 7.25. 3D printed dress and shoes on display. (Courtesy of ThreeASFOUR)

Fig. 7.26. A model displaying a 3D printed dress collection. (Courtesy of ThreeASFOUR)

The physical printing of the dresses was outsourced to Materialise, a company with a solid history of 3D printing haute couture for the catwalk.

Singapore's Nanyang Technological University's School of Mechanical and Aerospace Engineering (MAE) held its first 3D printing festival on 17 December 2013, showcasing fashion pieces made using 3D printers rather than the traditional needle and thread. The international 3D printing competitions attracted over 30 entries from 7 countries where participants created unique and complex designs made possible with this groundbreaking technology. The winning entry for the fashion category came from Australia and the XYZ designers created a fashion piece called the "Hydro-shift top". The network of open spheres created a lace-like fabric which showed a distinct contrast between solid and transparent pieces. The silhouette of the design echoes a traditional Chinese cheongsam, creating a masterpiece combining both tradition and technology. Fig 7.27 shows the winning 3D printed design by XYZ designers.

Fig. 7.27. 3D printed winning design "Hydro-shift top". (Courtesy of NTU Additive Manufacturing Centre (NAMC))

The commendation prize went to a group of NTU students comprising 3 Mechanical and Aerospace students and 4 Art, Design and Media students. Their design cleverly combined both art and technology to form a flowing 'chainmail' inspired by the Chinese word for water (水). There

Fig. 7.28. 3D printed fashion created by 7 NTU students. (Courtesy of NTU Additive Manufacturing Centre (NAMC))

are also patterns with large drops of water at the shoulders trickling to smaller drops of water as it unfolds from top to bottom (Fig 7.28).

Aside from clothing, AM has infiltrated the accessories and footwear sectors, allowing designers to create dramatic, one-off pieces quickly and fairly cheaply with shorter lead time. Designer Kimberly Ovitz, who displayed a small range of Shapeways-printed nylon jewellery with her ready to wear collection at New York Fashion Week, believes that 3D printing will revolutionise her production timetable as consumers could order and receive the jewellery two weeks after the show [48].

It has also allowed sportswear companies such as Nike and adidas to rapidly prototype new shoe designs, resulting in reduced labour hours and slashing the amount of time needed to evaluate a new product. With traditional injection moulding techniques, Nike would typically update complex product parts every couple of years but by leveraging AM, many rounds of prototype iterations can be fully tested within 6 months [49]. While 3D printing is accelerating the speed of product prototyping, many wonder when and if the technology will be used to create the finished products as for now, the process is too tedious and time-consuming.

While AM has been the forefront of fashion and accessories, the latest invention by Tamar Giloh has been created to solve the problem of heavy menstruation for women worldwide. She and her husband set up a fabric printing company called Tamicare using AM technology to manufacture biodegradable underwear [50]. Their proprietary production line machinery sprays fibres and polymers in a computerised manner to create stretchable biodegradable fabric. In comparison with traditional underwear production, this cutting edge technology requires no cutting therefore no waste is produced. Once perfected, this fabric printing process will allow for instantly created finished products from raw materials and automated mass production.

Retail, distribution and manufacturing could very well be on the verge of a new era due to these types of revolutionary technologies. Eventually, customers might be able to co-design and print out clothes from the comfort of their homes.

7.15 Weapons

With the maturity and increasing popularity of additive manufacturing technology, it offers unlimited possibilities to its users. The application has extended to an area where there has been much controversy – weapons, or more specifically, 3D printed guns.

The world's first 3D printed gun was introduced in May 2013. Its creator, Texan law student Cody Wilson from Defense Distributed, took 8 months to create the gun and made headlines all over the world when he successfully demonstrated a working 3D printed gun and uploaded its blueprint on the Internet [51]. The 3D printed gun, also known as "the Liberator", is shown in Figure 7.29.

Fig. 7.29. The world's first 3D printed gun, 'the Liberator'. (Courtesy of Defense Distributed)

All 16 pieces except for one firing pin of the Liberator's prototype were printed in ABSplus thermoplastic using Stratasys' Dimension SST 3D Printer (see Figs. 7.30 (a) and 7.30 (b)). The gun is able to fire standard handgun rounds [52].

Fig. 7.30 (a). All 16 pieces except one firing pin printed using ABSplus thermoplastic. (Courtesy of Forbes Collection/Michael Thad Carter)

Using Fused Deposition Modelling (FDM) technology, plastic filament flows through a tube to the print head, gets heated to a semi-liquid state and is extruded in thin, accurate layers to create precisely shaped solid objects. The CAD files for the gun designs were downloaded over 100,000 times before the US government intervened to get the files removed from public access.

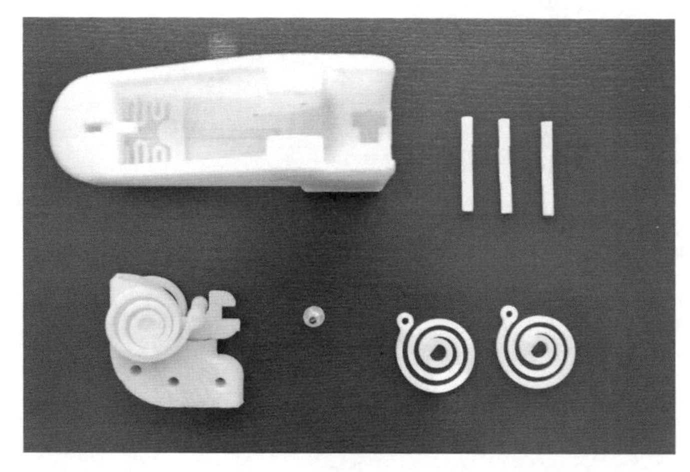

Fig. 7.30 (b). Parts of the 3D printed gun. (Courtesy of Defense Distributed)

A few months after 'the Liberator', which was made from plastic, was introduced, another US based company called Solid Concepts successfully manufactured and fired the world's first 3D-printed metal gun [53]. Using another technique, every part of the gun except for the spring was made using DMLS, which blasts the powdered metal with a laser. The company made the gun as a testament to prove that 3D Printing is a viable solution that can be used in modern manufacturing.

On one hand, the success of 3D printed guns represents an advance in the hyping AM technology. On the other hand, the controversies surrounding how easy it is to print a gun from home using only a 3D printer, a computer and Internet connection have shown the dangerous aspect of 3D printing, which until then, had many marvelling at the many opportunities and benefits presented by this technology. Law enforcement agencies worldwide are scrambling trying to come up with new legislations to curb such 'undetectable weapons', which are made entirely from ABS plastic. With the significant advances of 3D printing capabilities and accessibility, 3D printed firearms will pose an interesting security threat to public safety.

7.16 Musical Instruments

The art of crafting musical instruments is a skill that takes years to master. A craftsman would design an instrument to create a certain sound tone based on his knowledge and experience using various materials, especially woods of different ages. As the skills and experience of the craftsman improve, so does the quality of his instruments.

The controversial manufacturing of traditional musical instruments using 3D printing has opened up a new horizon for the industry. Traditional designs are now being challenged with new shapes which integrate the craftsman's (now a computer engineer) complex artwork into the framework of the instruments.

The construction of musical instruments has traditionally been a precise craft, taking many years to master. The strict conformation to traditional standards has also led to there being little innovation in the design of orchestral instruments. AM, with its ability to manufacture complex forms, can therefore be possibly applied to manufacture intricately shaped instruments, potentially allowing instruments to be mass-produced at a lower cost in the future. In addition, it will allow designers to produce intricate forms not previously possible, leading to more possibilities for tone production [54].

The technology is already being used to produce novelty electric guitars. Started by Olaf Diegel, a design engineer and professor of mechatronics at Massey University, ODD guitars are a range of electric guitars featuring beautifully intricate, customisable bodies [55]. Shown in Figure 7.31 is the Scarab guitar from the range.

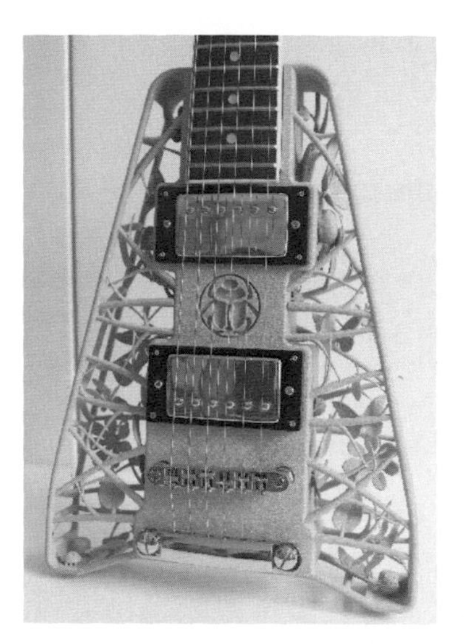

Fig. 7.31. The Scarab 3D printed guitar. (Courtesy of Olaf Diegel)

Other musical instruments that have been produced using AM include a 3D printed flute and violins. Designed by Amit Zoran, a PhD candidate at MIT, the concert flute in Figure 7.32 made using Stratasys' PolyJet technology was a proof of concept piece used to study the application of AM to the fabrication of musical instruments [54].

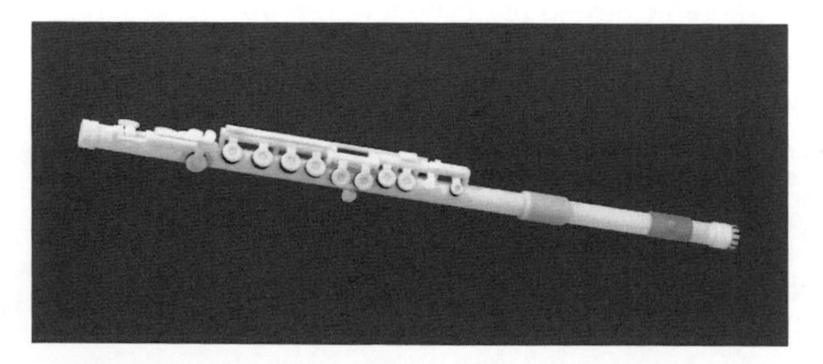

Fig. 7.32. A 3D printed flute. (Courtesy of Amit Zoran)

EOS produced a replica of a Stradivarius violin body using their technology (see Figure 7.33). The body is made of EOS PEEK HP3, a high performance polymer, instead of wood. The violin was then assembled and set up by a luthier with strings, peg box and chin rests etc. [56]. It is fully functional. The production of a complex instrument, usually made labouriously in a traditional way with traditional materials, via AM is almost blasphemy but it illustrates perfectly AM technology's potential to change everything.

Fig. 7.33. Stradivarius replica by EOS. (Courtesy of EOS GmbH)

Eventually, the technology may be used to produce extraordinary instruments that are impossible to fabricate using traditional manufacturing methods, such as the multi-tube trumpet shown in Figure 7.34.

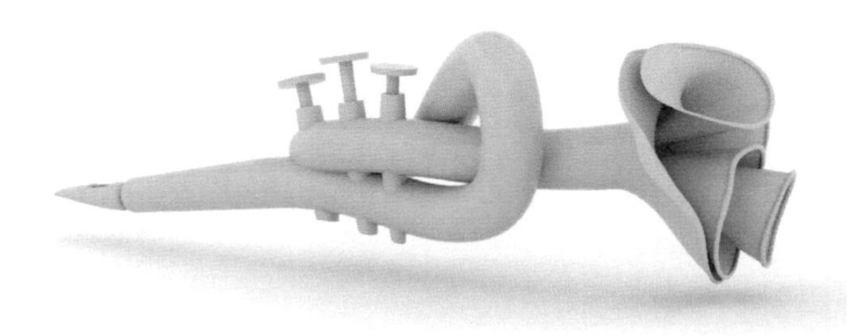

Fig. 7.34. Rendering of conceptual multi-tube trumpet. (Courtesy of Amit Zoran)

7.17 Food

NASA recently funded the development of a 3D food printer to feed astronauts in space. The agency awarded $125,000 to Systems and Materials Research Consultancy, a Texas-based company, to study how to make high nutrient food using AM technology [57].

The existing food system does not meet the food requirements of an astronaut crew on a mission to Mars or other longer missions. The project, which has already made its first 3D Printed Space Pizza, is set to explore the provision of food nutrition stability while minimising waste and time for space crews.

Researchers at Cornell University pioneered work in AM food printing using FDM called the Fab@Home Lab. They have partnered with the French Culinary Institute in Manhattan to 3D print personalised chocolate and cheese, cookies and even tiny spaceships made of deep fried scallops (see Fig. 7.35). Using AM technology, it allows for the fabrication of multi-material objects with high geometric complexity. Potential future applications may include cakes with 3D letters embedded in them and upon cutting the cake, the message is revealed.

Fig. 7.35. 3D printed deep fried scallop in space shuttle form. (Courtesy of Dan Cohen)

The professional culinary community stands to benefit from AM technology as it would enable mass-customisation in the industrial culinary sector. Currently, industrial food producers depend heavily on high-throughput processes such as moulding, extrusion and die-cutting. These processes are not amenable to mass-customisation and as such would require substantial custom-tooling; consequently, producing custom output for low-quantity runs is simply unfeasible. This is where AM's inherent strengths can be leveraged: producing food with custom, complex geometries while maintaining cost-effectiveness. The cost-effectiveness is enabled by the fact that AM does not require custom-tooling or extensive manual labour. One potential future application is the custom production of edible giveaways, for example, as marketing collateral for small corporate events [58].

Jeffrey Lipton, a PhD candidate at Cornell University and also the head of the Cornell Creative Machines Lab project, believes that in the future, food can be 3D printed according to the individual's dietary requirements. The lab successfully printed what Lipton called "data-driven cookies" where cookies are 3D printed based on the data compiled about the individual's height, weight, body mass index, and the caloric deficit for the day [59]. The cookies are of the same size but they are made up of ingredients based on the individual's nutritional requirements.

Lipton has also developed a new printing technique called stochastic printing that allows the printer to change the texture of the food being printed. Using corn masa dough, the team printed a new form of corn chip in the shape of a gear (see Fig. 7.36), that could be deep fried evenly. Instead of a solid form, the technique allows for a porous structure where absorbent quality can be controlled.

Lead scientist Dr Liang Hao from the University of Exeter established Choc Edge after he managed to develop a chocolate 3D printer that can print custom-made chocolate candy layer-by-layer without any moulding tools [60]. A number of chocolate retailers and manufacturers all over the world have expressed interest in acquiring the 3D chocolate printer when the prototype first came out.

Another amazing way 3D printing can revolutionise the way we live is getting meat on our table without slaughtering any animals. A company in the United States called Modern Meadow has set out to achieve this unimaginable feat by making artificial raw meat using a 3D bioprinter [61]. This is a more challenging process than any other 3D printing processes, as the cells are alive as the printing is being done. The 3D printer contains bioink which is made up of live cells' particles taken from living animals to form living tissue that can be assembled into meat. The commercialisation of this process will bring enormous market potential and may render the current agricultural method of livestock production irrelevant in the future.

Fig. 7.36. Corn masa dough 3D printed into a gear form. (Courtesy of Jeff Lipton)

7.18 Movies

3D printing has helped the movie industry through the manufacturing of props for use in movies. When filmmakers wanted to blow up an iconic Aston Martin DB5 which appeared in the James Bond film, *Skyfall*, they looked to Propshop Modelmakers, a British company that specialises in the production of film props using a 3D printer by Voxeljet [62]. The car that was blown up was in fact a 3D printed model, and the priceless car was spared.

Computer files with the design data for all components were given to Voxeljet to begin the 3D printing process. The models are produced with the layer-by-layer application of particle material that is glued together with a binding agent. As each layer is finished, another is printed on top to build up a 3D model. A total of 18 individual components for each of the three vehicle models were cleaned and then meticulously assembled and painted to look like the real Aston Martin DB5. This elaborately constructed model was then decimated in flames in the film.

Another film that could not have possibly been made as good as it was without AM technology is Kathryn Bigelow's film *Zero Dark Thirty*. Propshop was approached to create several important pieces for the movie, which includes US military night vision goggles for the actors [63]. As the goggles were a military issue, Propshop was unable to gain access to the real gadgets. The CG department had to make use of a small number of low resolution images to create a near perfect 3D model replica. At the conceptual design stage, a 3D printer was used to rapidly prototype the parts in order to improve the model in the shortest possible time frame. Many versions were created during the design stage but once approved, 3D printing production on the final model began.

In *Iron Man 2*, 3D printing allowed film production company Legacy Effects to create flexible gauntlets no thicker than a dime and it allowed the gloves to be worn for a long time without the heat getting onto Robert Downey Jr.'s hands. First, the actor's hands were scanned, and then the metal gloves were 3D printed through DMLS, using an inkjet cartridge to print a layer of powdered plastic which is then fused with UV light [64]. All these from design to printing can be done in a matter of hours. Creating the models by hand would have been too time-consuming but 3D printing allows for an extremely fast way of sculpting and creating concepts. Any changes to be made can be done on the CAD program, keeping the model in a digital format.

The main advantage of AM, besides speed, is that one can print out custom-fit costumes, allowing actors to be more comfortable while acting their roles. Downey had previously complained that the original Iron Man suits, which were not 3D printed, were too clunky and uncomfortable to act in. 3D printing is slowly finding its place in the movie industry which has seen the dominance of CGI in creating special effects.

References

[1] Mueller, T. (1991). Applications of stereolithography in injection molding, *Proceedings of the Second International Conference on Rapid Prototyping*, Dayton, USA, June 23-26:323-329.

[2] Wang, J. H., Lim, C. S., & Yeo, J. H. (2000). CFD Investigations of Steady Flow in Bi-Leaflet Heart Valve. In *Critical Reviews in Biomedical Engineering*, **28**(1-2): 61 – 68.

[3] Chua, C. K., Hong, K. H., & Ho, S. L. (1999). Rapid tooling technology (Part 1) – A comparative study. In *International Journal of Advanced Manufacturing Technology*, **15**(8): 604-608.

[4] Wilkening, C. (1996). Fast production of rapid prototypes using direct laser sintering of metals and foundry sand. In *Second National Conference on Rapid Prototyping and Tooling Research*, UK, November 18-19: 153-160.

[5] Tsang, H. B., & Bennett, G. (1995). Rapid tooling – direct use of SLA moulds for investment casting. In *First National Conference on Rapid Prototyping and Tooling Research*, November 6-7: 237-247. U.K.

[6] Atkinson, D. (1997). *Rapid Prototyping and Tooling: A practical guide, Strategy Publications*. U.K.

[7] Chua, C. K., Hong, K. H., & Ho, S. L. (1999). Rapid Tooling Technology (Part 2) – A case study using arc spray metal tooling. In *International Journal of Advanced Manufacturing Technology*, **15**(8): 609-614.

[8] Venus, A. D., Crommert, S. J., & Hagan, S. O. (1996). The feasibility of silicone rubber as an injection mould tooling process using rapid prototyped pattern. In *Second National Conference on Rapid Prototyping and Tooling Research*, UK, November 18-19: 105-110.

[9] Shonan Design Co. Ltd. (1996). *Temp-Less 3.4.3 – UV RTV process for quick development and fast to market.*

[10] Schaer, L. (1995). Spin casting fully functional metal and plastic parts from stereolithography models. In *The Sixth International Conference on Rapid Prototyping,* Dayton, USA, June: 217-236.

[11] Male, J. C., Lewis, N. A. B., & Bennett, G. R. (1996). The accuracy and surface roughness of wax investment casting patterns from resin and silicone rubber tooling using a stereolithography master. In *Second National Conference on Rapid Prototyping and Tooling Research*, UK, November 18-19: 43-52.

[12] Bettay, J. S., & Cobb, R. C. (1995). A rapid ceramic tooling system for prototyping plastic injection moldings. In *First National Conference on Rapid Prototyping and Tooling Research*, UK, November 6-7: 201-210.

[13] Warner, M. C. (1993). Rapid prototyping methods to manufacture functional metal and plastic parts. *Rapid prototyping systems: Fast track to product realization*: 137-144.

[14] Lim, C. S., Siaminwe, L., & Clegg, A. J. (1996). Mechanical Property Enhancement in an Investment Cast Aluminium Alloy and Metal-Matrix Composite. In *9th World Conference on Investment Casting*, Oct 1996, San Francisco, USA, 17:1 – 18.

[15] Clegg, A. J. (1991). *Precision Casting Processes.* Oxford, England: Pergamon Press Plc.

[16] Venus A. D., & Crommert, S. J. (1996). Direct SLS nylon injection. In *Second National Conference on Rapid Prototyping and Tooling Research*, UK, November 18-19: 111-118.

[17] Soar, R. C., Arthur, A., & Dickens, P. M. (1996). Processing and application of rapid prototyped laminate production tooling. In *Second National Conference on Rapid Prototyping and Tooling Research,* UK, November 18-19: 65-76.

[18] Industrial Technology. (Jun 2002). *Laser sintering for rapid production*. Retrieved from http://www.industrialtechnology.co.uk/2001/may/eos.html

[19] Rapid Prototyping and Tooling State of the Industry. (2000). In *Wohlers Report 2000*. USA: Wohlers Associates Inc.

[20] Eyerer, P. (1996). Rapid tooling – manufacturing of technical prototypes and small series. In *Mechanical Engineering*: 45-47.

[21] Sundstrand Aerospace uses laminated object manufacturing to verify large complex assembly. (1995). In *Rapid Prototyping Report,* **5**(6), June 1995:1-2. CAD/CAM Publishing Inc.

[22] 3D Systems. (1993). *User Focus: AlliedSignal Aerospace, Stereolithography and QuickCastTM provide the winning combination for meeting critical deadline in AlliedSignal's development of the TFE 731-20 Turbo Fanjet Engine.*

[23] Sundstrand Power Systems uses selective laser sintering to create large investment casting patterns. (1995). In *Rapid Prototyping Report,* **5**(1), January 1995:1-2. CAD/CAM Publishing Inc.

[24] DTM Corporation Press Release. (1994). *SLS-generated Polycarbonate Patterns Speed Casting of Intricate Aircraft Engine Part for Sundstrand Power Systems.* DTM Corporation.

[25] Bell Helicopter uses QuickCast to fabricate flight-certified production casting. (1995). In *Rapid Prototyping Report*, **5**(7), July 1995:1-2. CAD/CAM Publishing Inc.

[26] Stratasys. (2013). *Additive Manufacturing Trends in Aerospace: Leading The Way*. Retrieved from http://www.mcad.com/3d-printing/3d-printing-whitepaper-aerospace/

[27] Newman, J. (Aug 2012). *NASA Turns to 3D Printing for Space Exploration Vehicle*. Rapid Ready Technology. Retrieved from http://www.rapidreadytech.com/2012/08/nasa-turns-to-3d-printing-for-space-exploration-vehicle/

[28] Catts, T. (Nov 2013). *GE Turns to 3D Printers for Plane Parts*. Bloomberg Business Week. Retrieved from http://www.businessweek.com/articles/2013-11-27/general-electric-turns-to-3d-printers-for-plane-parts/

[29] Halterman, T. (Sep 2013). *Bentley Motors – Automotive Perfection Via 3D Printing*. 3D Printer World. Retrieved from http://www.3dprinterworld.com/article/bentley-motors-automotive-perfection-via-3d-printing/

[30] Chatterjee, R. (Mar 2012). *3D Printing: The technology can revolutionize manufacturing & healthcare industry*. The Economic Times. Retrieved from http://articles.economictimes.indiatimes.com/2012-03-15/news/31197145_1_body-parts-3d-scooter/

[31] *Ford's 3D-printed auto parts save millions of dollars: Technology improves quality in vehicles*. (Dec 2013). Arab Times English Daily. Retrieved from http://www.arabtimesonline.com/NewsDetails/tabid/96/smid/414/ArticleID/202365/reftab/36/Default.aspx

[32] Higginbotham, S. (Dec 2012). *Ford's Gift to Engineers: MakerBot 3D Printers.* Bloomberg Business Week. Retrieved from http://www.businessweek.com/articles/2012-12-21/fords-gift-to-engineers-makerbot-3d-printers

[33] Beck, J. (2010). *Urbee is the First Car Made by a 3D Printer*. Popular Science. Retrieved from http://www.popsci.com/cars/article/2010-11/hybrid-car-created-completely-3d-printing/

[34] Maxey, K. (Nov 2013). *Urbee 2 to Cross the US on 10 Gallons of Fuel.* Retrieved from http://www.engineering.com/3DPrinting/3DPrinting Articles/ArticleID/6661/Urbee-2-to-Cross-the-US-on-10-Gallons-of-Fuel.aspx

[35] Lee, H. B., Ko, M. S. H., Gay, R. K. L., Leong, K. F., & Chua, C. K. (1992). Using Computer-based Tools and Technology to Improve Jewellery Design and Manufacturing. In *International Journal of Computer Applications in Technology*, **5**(1):72-88.

[36] Leong, K. F., Chua, C. K., & Lee, H. B. (1994). Finishing Techniques for Jewellery Models Built using the Stereolithography Apparatus. In *Journal of the Institution of Engineers, Singapore,* **34**(4):54-59.

[37] Chua, C. K., Gay, R. K. L., Cheong, S. K. F., Chong, L. L., & Lee, H. B. (1995). Coin Manufacturing Using CAD/CAM, CNC and Rapid Prototyping Technologies. In *International Journal of Computer Applications in Technology*, **8**(5/6):344-354.

[38] Chua, C. K., Hoheisel, W., Keller, G., & Werling, E. (1993). Adapting Decorative Patterns for Ceramic Tableware. In *Computing and Control Engineering Journal*, **4**(5):209-217.

[39] Chua, C. K., Gay, R. K. L., & Hoheisel, W. (1994). A Method of Generating Motifs Aligned Along a Circular Arc. In *Computers & Graphics: An International Journal of Systems and Applications in Computer Graphics*, **18**(3):353-362.

[40] Rase, W-D. (2002). Physical Models of GIS Objects by Rapid Prototyping. In *Symposium on Geospatial Theory, Processing and Applications*. Ottawa.

[41] Hill, D. J. (2012). *3D Printing is the Future of Manufacturing and Neri Oxman Shows How Beautiful It Can Be.* SingularityHUB. Retrieved from http://singularityhub.com/2012/06/04/3d-printing-is-thc-futurc-of-manufacturing-and-neri-oxman-shows-how-beautiful-it-can-be/

[42] Dickson, S. (2003). *Computer-aided Rapid Mechanical Prototyping or Automatic Fabrication*. Retrieved from http://www.emsh.calarts.edu/-mathart/R_Proto_ref.html

[43] Lim, S., & et al. (2012). Developments in construction-scale additive manufacturing processes. In *Automation in Construction*, vol. 21 (1), pp. 262-268. Retrieved from https://dspace.lboro.ac.uk/2134/9176

[44] Buswell, R. A., & et al. (2007). Freeform Construction: Mega-Scale Rapid Manufacturing for Construction. In *Automation in Construction*, vol. 16 (2), pp. 224-231. Retrieved from https://dspace.lboro.ac.uk/dspace-jspui/handle/2134/9925

[45] Lim, S., & et al. (2011). Development of a Viable Concrete Printing Process. In *The 28th International Symposium on Automation and Robotics in Construction (ISARC2011)*. Retrieved from http://www.iaarc.org/publications/fulltext/S20-3.pdf

[46] Shapeways. (Jul 2013). *N12: 3D Printed Bikini*. Retrieved from http://www.shapeways.com/n12_bikini

[47] Brooke, K. (Sep 2013). *Threesfour and Materialise Hit the Catwalk.* 3D Printer World. Retrieved from http://www.3dprinterworld.com/article/threeasfour-and-materialise-hit-catwalk/

[48] Brooke, E. (Jul 2013). *Why 3D Printing Will Work in Fashion.* Techcrunch. Retrieved from http://techcrunch.com/2013/07/20/why-3d-printing-will-work-in-fashion/

[49] Maxey, K. (Jun 2013). *Nike and Adidas Use 3D Printing to Speed Up Prototyping.* Engineering.com. Retrieved from http://www.engineering.com/3DPrinting/3DPrintingArticles/ArticleID/5847/Nike-and-Adidas-Use-3D-Printing-to-Speed-Up-Prototyping.aspx

[50] Molitch-Hou, M. (Nov 2013). *Disposable Panties Yield 3D Printing Fabric Innovation.* 3D Printing Industry. Retrieved from http://3dprintingindustry.com/2013/11/13/disposable-panties-yield-3d-printing-fabric-innovation/

[51] Hadhazy, A. (May 2013). *Why you should, and shouldn't, worry about the 3D-Printed gun.* Popular Mechanics. Retrieved from http://www.popularmechanics.com/technology/military/weapons/why-you-should-and-shouldnt-worry-about-the-3D-printed-gun-15450141

[52] Greenberg, A. (May 2013). *This is the world's first 3D-Printed gun.* Forbes. Retrieved from http://www.forbes.com/sites/andygreenberg/2013/05/03/this-is-the-worlds-first-entirely-3d-printed-gun-photos/

[53] Griffiths, S. (Nov 2013). *World's first 3D-printed gun made from 30 metal components is created.* Daily Mail UK. Retrieved from http://www.dailymail.co.uk/sciencetech/article-2492505/Worlds-3D-printed-gun-30-METAL-components-created.html?ico=sciencetech%5Eheadlines

[54] Zoran, A. (2011). The 3D Printed Flute: Digital Fabrication and Design of Musical Instruments. In *Journal of New Music Research,* vol. 40, pp. 379-387, 2011/12/01.

[55] Odd Guitars. (2011). *About.* Retrieved from http://www.odd.org.nz/about.html

[56] Lanxon, N. (2011). *Hands-on with the EOS 3D-printed Stradivarius violin.* Retrieved from http://www.wired.co.uk/news/archive/2011-09/20/3d-printed-stradivarius-violin-eos

[57] Gannon, M. (May 2013). *How 3D Printers Could Reinvent NASA Space Food.* Space.com. Retrieved from http://www.space.com/21308-3d-printing-nasa-space-food.html

[58] Periard, D., Schaal, N., Schaal, M., Malone, E., & Lipson, H. (2007). Printing Food. In *Proceedings of the 18th Solid Freeform Fabrication Symposium, Austin TX, Aug 2007*, pp.564-574.

[59] Bosker, B. (Apr 2013). *3D Printers Could Actually Make Donuts Healthy.* Retrieved from http://www.huffingtonpost.com/2013/04/24/3d-printed-food_n_3148598.html

[60] *Chocolate printer to go on sale after Easter* (Apr 2012). BBC News. Retrieved from http://www.huffingtonpost.com/2013/04/24/3d-printed-food_n_3148598.html

[61] McCullagh, D. (Aug 2012). *3D printed meat: It's what's for dinner.* Cnet.com. Retrieved from http://news.cnet.com/8301-11386_3-57493377-76/3d-printed-meat-its-whats-for-dinner/

[62] *Voxeljet builds Aston Martin models for James Bond film Skyfall.* (2012). Retrieved from http://www.3ders.org/articles/20121107-voxeljet-builds-aston-martin-models-for-james-bond-film-skyfall.html

[63] Darby, A. (Apr 2013). *Gearing up Kathryn Bigelow's Latest Film Zero Dark Thirty.* Retrieved from http://www.propshop.co.uk/zero-dark-thirty/

[64] Kluang, C. (2010). *Iron Man 2's Secret Sauce: 3-D Printing*. Fast Company. Retrieved from http://www.fastcompany.com/1640497/iron-man-2s-secret-sauce-3-d-printing

Problems

1. How is the application of AM models related to the purpose of prototyping? How does it also relate to the materials used for prototyping?

2. List the types of industries that AM can be used in. List specific industrial applications.

3. What are the finishing processes that are used for AM models and why are they necessary?

4. What are the typical AM applications in design? Briefly describe each of these applications and illustrate them with examples.

5. What are the typical AM applications in engineering and analysis? Briefly describe each of them and illustrate them with examples.

6. How would you differentiate between the following types of rapid tooling processes: (a) direct soft tooling, (b) indirect soft tooling, (c) direct hard tooling and (d) indirect hard tooling?

7. Explain how an AM pattern can be used for vacuum casting with silicone moulding. Use appropriate examples to illustrate your answer.

8. What are the ways an AM pattern can be used to create injection moulds for plastic parts? Briefly describe the processes.

9. Compare and contrast the use of AM patterns for the following:

   i. casting of die inserts,

   ii. sand casting, and

   iii. investment casting

10. What are the AM systems that are suitable for sand casting? Briefly explain why and how they are suitable for sand casting.

11. Compare the relative merits of using LOM parts with SLA parts for investment casting.

12. Explain whether AM technology is more suitable for 'high technology' industries like aerospace than it is for consumer product industries like electronic appliances. Give examples to substantiate your answer.

13. Explain how AM systems can be applied to traditional industries like jewellery, coin and tableware.

14. Briefly describe how AM systems can be applied in GIS.

15. What the three advantages of employing AM technology in the field of architecture?

16. Briefly describe how AM systems can be applied to fashion and textile.

17. Explain the advantages of using AM systems in the movie industry.

Chapter 8

MEDICAL AND BIOENGINEERING APPLICATIONS

8.1 Planning and Simulation of Complex Surgery

When facing complex operations, such as craniofacial and maxillofacial surgeries, surgeons usually have difficulty figuring out visually the exact location of a tumour or the precise profile of a defect. The precision and speed of a surgical operation depend significantly on the surgeon's prior knowledge of the case and experience. Additive manufacturing enables surgeons to practise on a precise model and master the essential details before the actual operation. The hands-on experience helps surgeons achieve better success rates. Hence AM models are frequently used and are vital in surgical planning. The following sections describe some of these applications.

8.1.1 *Cranioplasty of Large Cranial Defects*

Cranioplasty is a surgical correction of a defect in cranial bone by implanting a metal or plastic replacement to restore the missing part [1]. In Figure 8.1, a patient was severely impacted on the right side of his head. A portion of the skull was crushed and had to be removed. To cover the large hole left in his skull, an implantable prosthesis with a good fit was required. Surgeons first performed computed tomography (CT) scans on the patient, and then transformed the data to an STL file, which was subsequently sent to an AM machine where the wax prosthesis was produced. The wax pattern was then used as a mould and a biocompatible material (e.g. poly(methyl-methacrylate) (PMMA)) cast into it. After sterilisation, the prosthesis was implanted into the patient. The surgery took very little time as the prosthesis exactly fitted the defect. To the experienced surgeons, it was similar to a "plug and play." A good fit has two main advantages: first, the smoothness continues across the defect outline and gives the patient an improved symmetrical cosmesis; second, it reduces redundant or improper protrusions that may damage brain tissue severely, thus preventing post-surgery injury.

8.1.2 *Congenital Malformation of Facial Bones*

Restoration of facial anatomy is important in cases of congenital abnormalities, trauma or post cancer reconstruction. In one case, the patient had a deformed jaw at birth, and a surgical operation was necessary to cut out the shorter side of the jaw and alter its position [2]. The difficult part of the operation was the evasion of the nerve canal that ran inside the jawbone. Such an operation was impossible in the conventional procedure because there was no way of visualising the inner nerve canal.

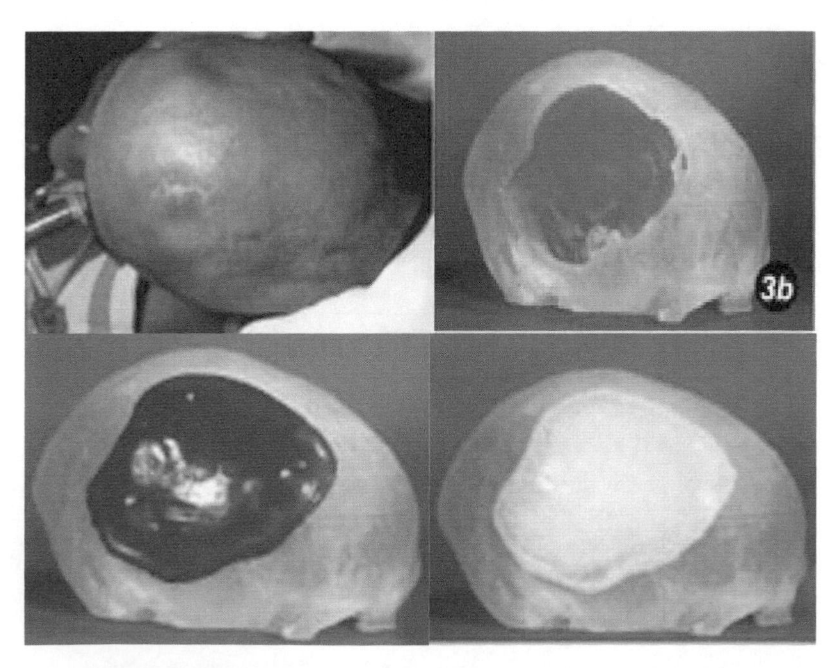

Fig. 8.1. A large cranial prosthesis was modelled using AM technology to close up a defect in the cranium. (Courtesy of Materialise)

Using a CAD model reconstructed from the CT images, the position of the canal was identified and a simulation of the amputating process was carried out to determine the actual line of the cut. Furthermore, the use of a semi-transparent resin prototype of the jawbone allowed the visualisation of the internal nerve canal and facilitated the determination of the amputation line prior to the surgery. The end result was a more efficient surgery and vastly improved post-surgery results.

In another case study, a laser digitiser was used instead of CT [3] to capture the external surface profile of a patient with a harelip problem. The triangulated surface of the patient's face was reconstructed in CAD, as seen in Figure 8.2. Figure 8.3 shows the SLA prototype derived from the CAD data. In this case study, the prototype model provided the validation for the laser scan measurements. In addition, it facilitated an accurate predictive surgical outcome and postoperative assessment of changes in the facial surgery.

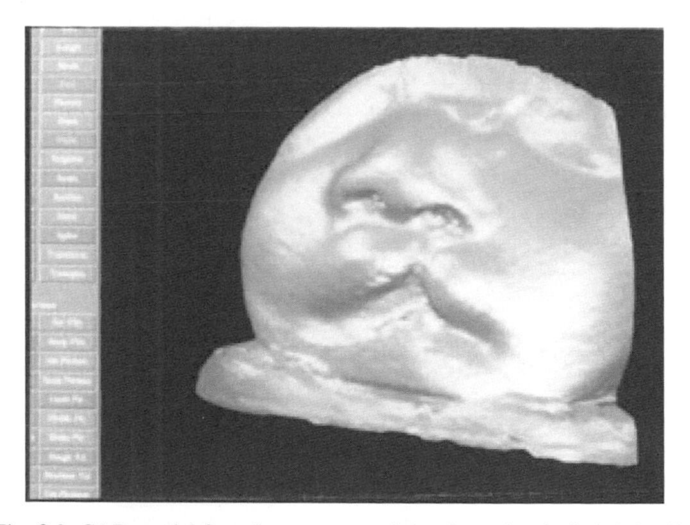

Fig. 8.2. CAD model from laser scanner data of a patient's facial details.

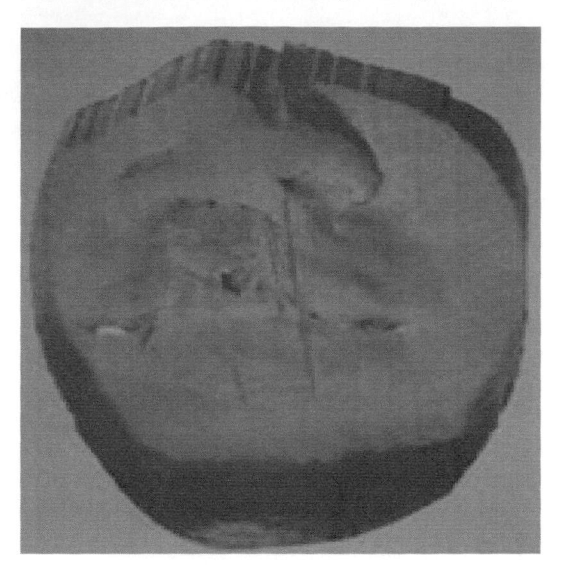

Fig. 8.3. SLA model of a patient's facial details.

8.1.3 *Cosmetic Facial Reconstruction*

Due to a traffic accident, a patient had a serious bone fracture of the upper and lateral orbital rim in the skull [4]. In the first reconstructive surgery, the damaged part of the skull was transplanted with the shoulder bone. However, shortly after the surgery, the transplanted bone dissolved. Another surgery was required to re-transplant another artificial bone that would not dissolve this time. The conventional procedure of such a surgery would be for surgeons to manually carve the transplanted bone during the operation virtually by "trial and error" until it fitted properly. This operation would have required a lot of time due to the difficulty in carving the bone, which has to be done during the surgery. Using AM, a SLA prototype of the patient's skull was made and then used to prepare an artificial bone that would fit the hole caused by the dissolution. This preparation not only greatly reduced the time required for the surgery, but also improved its accuracy.

In another case at Keio University Hospital of Japan [5], a five-month-old baby had a symptom of scaphocephary in the skull. This is a condition that can lead to serious brain damage because it would only permit the skull to grow in the front and rear directions. The procedure required was to take the upper half of the skull apart and surgically reconstruct it completely, so that the skull would not suppress the baby's brain as it grows. Careful planning was essential for the success of such a complex operation. This was aided by producing a replica skull prototype with the amputation lines drawn into the model. Next, a surgical rehearsal was carried out on the model with the amputation of the skull prototype according to the drawn lines followed by the reconstruction of the amputated part. In this case, the AM model of the skull provided the surgical procedure with: (1) good three-dimensional visualisation support for the planning process, (2) an application as training material, and (3) a guide for the real surgery.

8.1.4 *Separation of Conjoined Twins*

On 24th July 2001, two twin sisters were born in a rural village in Guatemala, healthy in every way except for the fact that they were joined at the head [6] (see Figure 8.4). X-ray pictures showed that the two brains

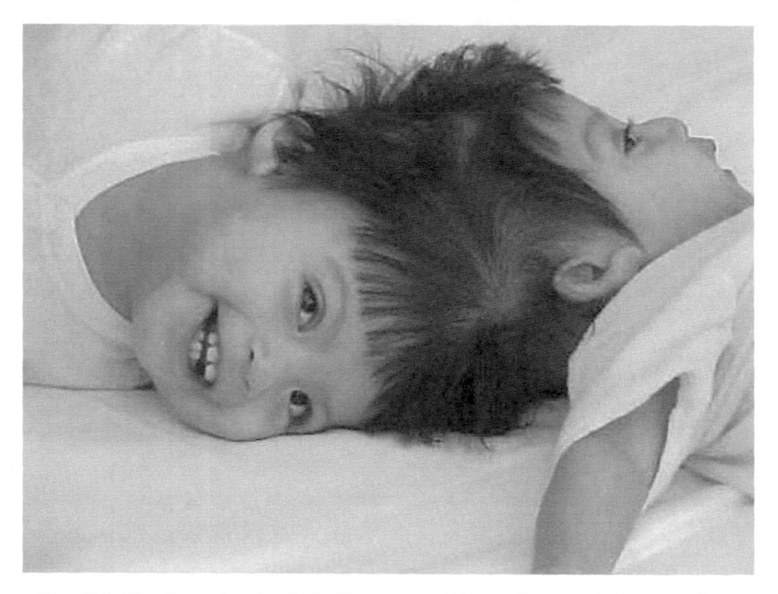

Fig. 8.4. Conjoined twins [6]. (Courtesy of Cyon Research Corporation)

were separated by a membrane and were otherwise normal in size and had a complete structure. This meant that no brain tissue would have to be cut through during the separation surgery. The arteries that carried oxygenated blood to their brains were also separate, but the veins that drained the blood were interwoven and fed into each other's circulatory systems. The most complex part of the operation was to sort out these veins and reroute each girl's blood supply correspondingly. In this situation, AM had an essential role to play.

A series of CT scans of the two girls were taken. Complicating the task was the fact that the two girls, while connected, could not be arranged in the CT system so that a single scan of their heads could be made. Instead, three sets of scanned data were collected at different angles and then combined into a single three-dimensional model. It took about three days to process the CT data and create STL files for AM. Objet's Tempo AM system, made by the Israeli company Objet Geometries Ltd and since acquired by Stratasys, was used to construct the skull models of each girl

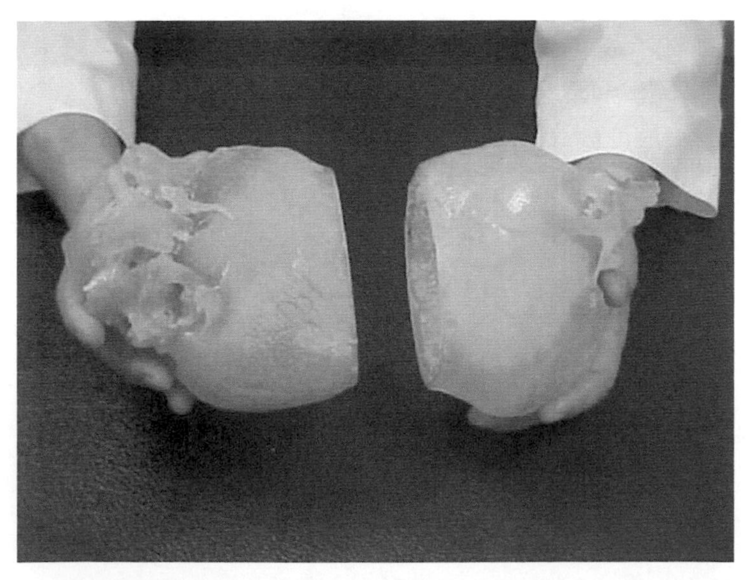

Fig. 8.5. Models of each girl's skull [6]. (Courtesy of Cyon Research Corporation)

(Figure 8.5). The Tempo built the parts by selectively jetting tiny droplets of acrylate photopolymer and then curing the drops, layer-by-layer, with light. It required semi-hardened supports (like a gel) which could be removed simply by jetting water on it. The model of the intersection of the two skulls helped surgeons plan how they would reroute the necessary blood vessels (Figure 8.6). The operation took about 22 hours to complete. Similar procedures in the past took as long as 97 hours. This significant reduction in time was undoubtedly attributed to the AM model.

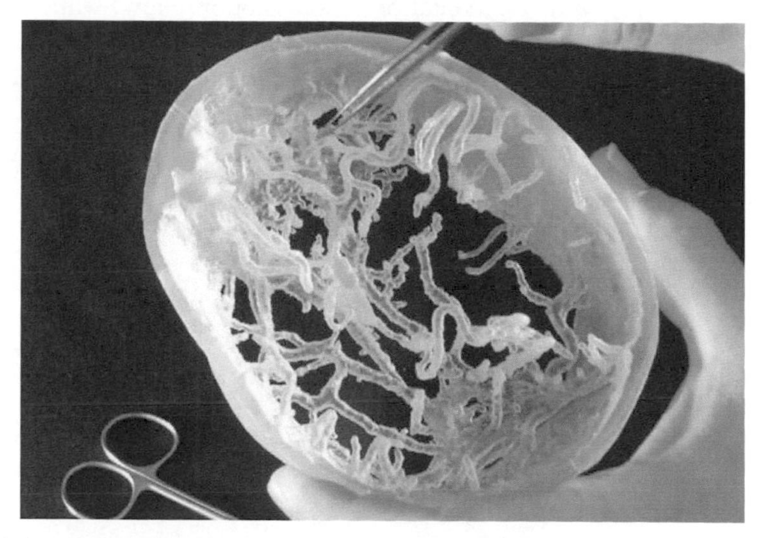

Fig. 8.6. A model of the maze of blood vessels where the two skulls connected [6]. (Courtesy of Cyon Research Corporation)

8.1.5 *Tumour in the Jaw*

Ameloblastoma is a rare, benign tumour that appears in the jaw. Though these tumours are rarely malignant or metastatic (that is, they rarely spread to other parts of the body), the resulting lesions can cause severe abnormalities in the face and jaw. Additionally, because abnormal cell growth easily infiltrates and destroys surrounding bony tissues, wide surgical excision is required to treat this disorder.

A 51-year-old male was seen in the university clinic of Kiev Medical Academy with ameloblastoma of the posterior part of the left alveolar ridge of the mandible [7]. A series of CT scans were performed and a virtual model was built using a computer. Figure 8.7 (a) shows the front view of the 3D-reconstruction of CT dataset. Figure 8.7 (b) shows a zoomed in picture of the 3D-reconstruction. An ovoid tumour cavity with precise contours is clearly seen. X-ray density is about -41/73 HU, dimensions: 12 x 25 x 22 mm. Based on the CT dataset, a STL file was created and an AM model was subsequently built. Figure 8.7 (c) shows the model created from segmentation masks (bottom view) while Figure 8.7 (d) shows the view from the top (plan view). The size of the tumour and the exact location were identified from the AM model, and the surgical procedure was planned in detail. The patient was successfully operated on by Professor A. Timofeev in the CMF Surgery Clinic, Kiev Medical Academy of Postgraduate Education, Kiev, in Ukraine.

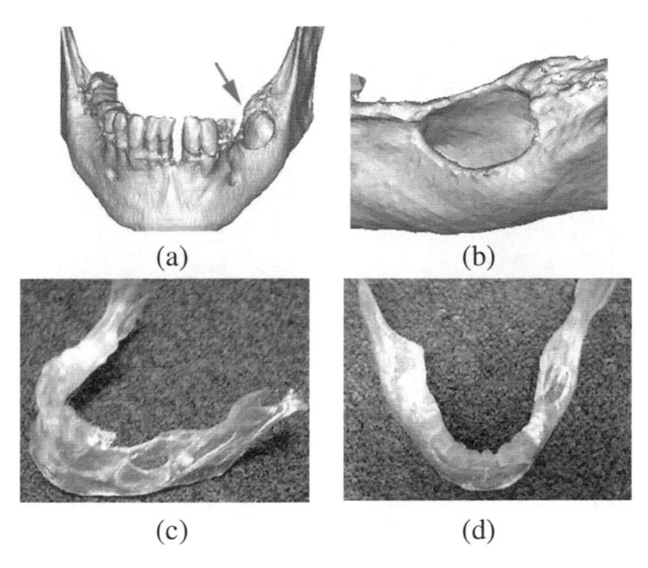

Fig. 8.7. Ameloblastoma of left mandible using SLA. (Courtesy of Materialise)

The planning for removing of the facial bone tumour was enhanced by merging different procedures: 3D-reconstruction, advanced segmentation with different thresholds and direct control on stereolithography models. This procedure increased the quality of the planning, allowing surgeons to practise for the real situation beforehand and to prepare the exact shape of the bone graft.

8.1.6 *Cancerous Brain*

In another case, a patient had a cancerous bone tumour in his temple area, requiring the surgeon to have to access the growth via the front through the right eye socket. The operation was highly dangerous as damage to the brain was likely, which would result in the impairment of some motor functions. In any case, the patient would have lost the function of the right eye [8]. However, before proceeding with the surgery, the surgeon wanted another examination of the tumour location, but this time using a 3-dimensional plastic replica of the patient's skull. By studying the model, the surgeon realised that he could reroute his entry through the patient's jawbone, thus avoiding the risk of harming the eye and motor functions. Eventually, the patient lost only one tooth and of course, the tumour. The plastic AM model used by the surgeon was fabricated by SLA from a series of CT scans of the patient's skull.

Other case studies relating to bone tumours have also been reported to have improved success [9, 10]. In all cases, not only was the patient spared physical disability as well as the emotional and financial price tags associated with that, but the surgeon also gained valuable insights into his patient using the AM model. From here, non-intuitive alternative improved strategies for the surgery were created by the enhancement to the surgeon's pre-surgical planning stage.

8.1.7 *Dental Precision Planning*

Treating an impacted maxillary canine requires identifying its exact position; this can pose a challenge to both orthodontists and oral surgeons. Relying only on the conventional method of using 2-D slices of CT images would pose great difficulty in determining the exact location of an impacted tooth.

One case presented a new method for diagnosis and treatment planning of maxillary canine impaction by using CT combined with AM [11]. The CT image files of a patient with tooth-13 impaction were edited to produce, by means of AM, an anatomic model of the maxillary teeth and a single attachment model that was to be used later on to fabricate a metal attachment for bonding to the impacted tooth. The dental model was used in the diagnosis and orthodontic treatment planning, and to communicate with the patient and his parents. The model showed the exact anatomical relationship between the impacted tooth and the other teeth; it was the main aid in the intraoperative navigation during surgery to expose the tooth. The metal attachment built from the prototype was bonded to tooth 13 during surgery. Thus, AM is also an important tool for fabricating brackets and other precision accessories for specific dental needs. In this case, a series of dental models made with AM was the diagnostic procedure of choice for evaluating impacted maxillary canines. Several AM manufacturers have introduced dedicated AM machines for the dental industry e.g. 3D Systems' ProJet® 3510 DP and MP Professional 3D Printer System, and EnvisionTEC GmbH's 3Dent™.

8.1.8 *Biomodelling as an Aid to Spinal Instrumentation*

Stereotactic surgery or stereotaxy is a minimally invasive form of surgical intervention which makes use of a three-dimensional coordinates system to locate small targets inside the body and to perform actions such as removal, biopsy, lesion, injection, stimulation and implantation. Previously, frameless stereotaxy was used in spinal surgery, but this has significant limitations. In one study, a novel stereotactic technique using biomodels was developed [12]. Biomodelling was found to be helpful for complex skeletal surgery and has advantages over frameless stereotaxy. In that study, 20 patients with complex spinal disorders requiring instrumentation were recruited. 3D CT scans of their spines were performed, and the data was used to generate acrylate biomodels of each spine using AM. The biomodels were used to simulate the surgery. Simulation was performed using a standard power drill to place trajectory pins into the spinal biomodel. Acrylate drill guides were manufactured using the biomodels and trajectory pins as templates. The biomodels were found to be highly accurate and of great assistance in the planning and execution of the surgery. The ability to drill optimum screw trajectories into the biomodel and then accurately replicate the trajectory was judged to be especially helpful. Accurate screw placements were confirmed with postoperative CT scanning. The design of the first two templates was suboptimal as the contact surface geometry was too complex. Approximately 20 minutes was spent before surgery preparing each biomodel and template. Operating time was reduced, as less reliance on intraoperative radiograph was necessary.

8.2 Customised Implants and Prostheses

For hip replacements and other similar surgeries, these were previously carried out using standardised replacement parts selected from a set range provided by manufacturers based on available anthropomorphic data and the market needs. This works satisfactorily for some types of procedures and patients, but not all. For those patients outside the standard range, in-between sizes, or with special requirements caused by disease or genetics, the surgical procedure may become significantly more complex and expensive. For imperfect fits, these implants may even cause poor gait outcomes and further wear or injury to other joints. AM has made it possible to manufacture a custom prosthesis that precisely fits a patient at a reasonable cost. The following application examples are drawn from a range of applications that span the human anatomy.

8.2.1 *Cranium Implant*

A patient suffered from a large frontal cranium defect after complications from a previous meningioma tumour surgery. This left the patient with a missing cranial section, which caused the geometry of the head to look deformed. Conventionally, a titanium-mesh plate would be hand-formed during the operation by the surgeon. This often resulted in inaccuracies and time wasted on trial and error. Using AM, standard preparations of the patient were made and a CT scan of the affected area and surrounding regions was taken during the pre-operation stage. The 3D CT data file was transferred to a CAD system and the missing section of the cranium topography was generated. After some software repair and cleaning up were carried out on the newly generated section, an inverted mould was produced on CAD. This 3D solid model of the mould was saved in .STL format and transferred to the AM system, such as the SLS, for building the mould. The SLS mould was produced and used mechanically to press the titanium-mesh plate to the required 3-dimensional profile of the missing cranium section. During the operation, the surgeon clears the scalp tissue of the defect area and fixates the perfectly pre-profiled plate onto the cranium using self-tapping screws. The scalp tissue is then replaced and sutured. At post operation recovery, results observed showed improved surgical results, reduced operation time and a reduced probability of complications.

8.2.2 *Hip Implant*

In one case, a 30-year-old female was diagnosed as having bilateral pseudohypo-chondroplasia with multiple epiphyseal dysplasia [13]. She had a poor range of movement and constant pain in her hips. A total hip arthroplasty with a custom designed femoral implant was recommended. Conventional X-ray (see Figure 8.8 (a)) showed very distorted femoral cavities which were wide proximally and extremely narrow in mid-shaft. In order to design a custom stem, CT scans and an AM model (see Figure 8.8(b)) were required, which showed a slot type femoral canal. An implant was designed from the CT scan and the model (see Figure 8.8(c)). The postoperative X-ray in Figure 8.8(d) shows that the implant produced a line-to-line fit with the cavity. A clinical follow-up one year after the operation showed that the hips were pain-free and functioning well.

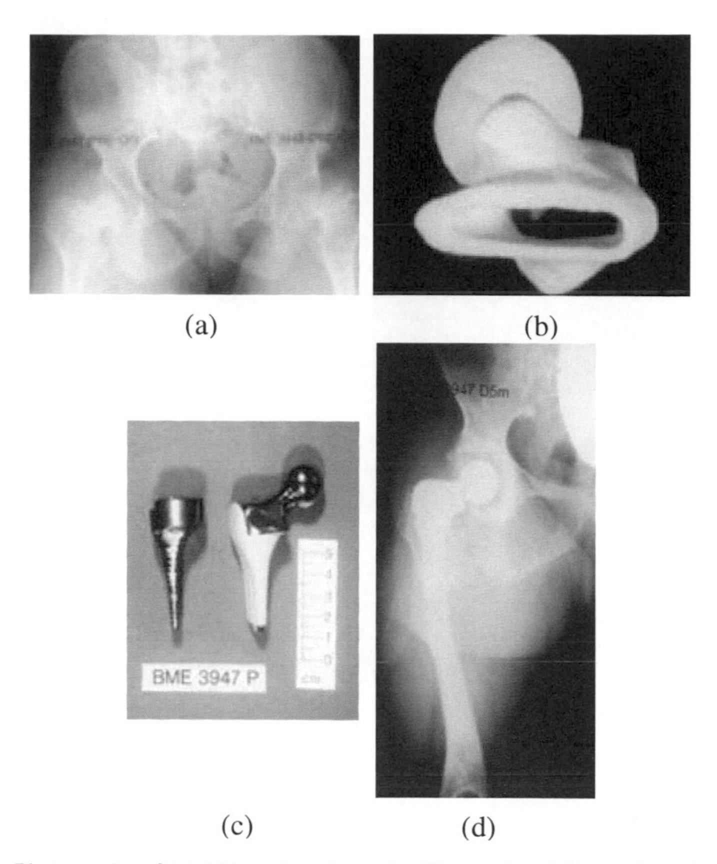

Fig. 8.8. Photographs of total hip arthroplasty. (a) Conventional X-ray showed very distorted femoral cavities which were wide proximally and extremely narrow in mid-shaft. (b) In order to design a custom stem, CT scans and an AM model were required, which showed a slot type femoral canal. (c) An implant was designed from the CT scan and the model. (d) Postoperative X-ray shows that the implant produced a line-to-line fit with the cavity. (Courtesy of Materialise)

In another case, a 46-year-old female was diagnosed with malunion of a femoral fracture [14]. She had previous multiple femoral fractures and an osteotomy. Conventional X-ray, as seen in Figure 8.9 (a), showed a misalignment in both A-P and M-L views, and there was an uncertainty on the cross section geometry of the femoral shaft at the level below the lesser trochanter. In order to design a custom implant and determine the

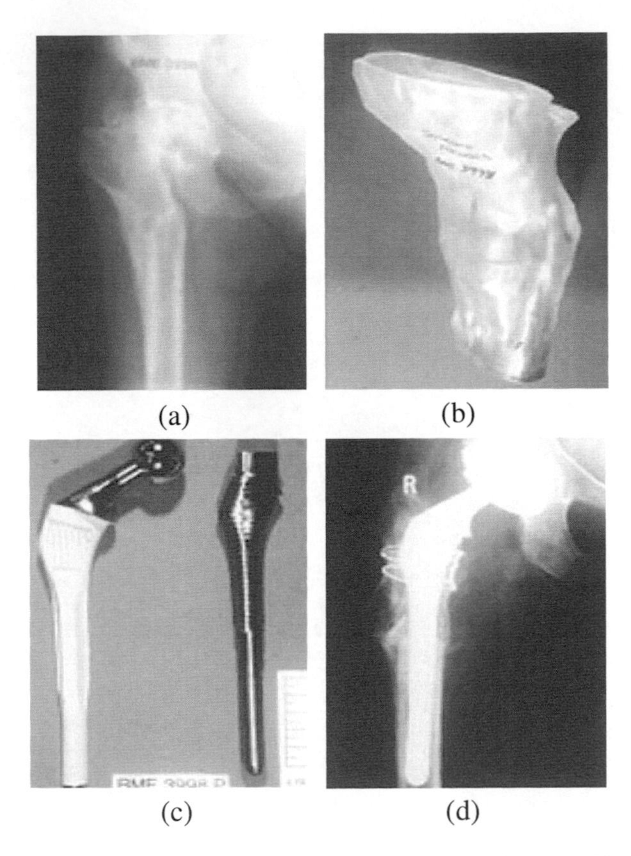

Fig. 8.9. Total hip arthroplasty for the case of femoral malunion. (a) X-ray showed a misalignment in both A-P and M-L views, and an uncertainty on the cross section geometry of the femoral shaft at the level below the lesser trochanter. (b) The model was made of a transparent resin material so that the internal anatomical geometry of the femoral cavity was clearly visible. (c) A subtrochanteric osteotomy was planned and a custom implant was designed with middle cutting flutes at the region of osteotomy to provide torsional stability. (d) Postoperative X-rays showed that the implant produced a good fit and stability. (Courtesy of Materialise)

form of surgery, CT scans and a medical model were used. The model (Figure 8.9 (b)) was made of a transparent resin material so that the internal anatomical geometry of the femoral cavity was clearly visible. A subtrochanteric osteotomy was planned and a custom implant (Figure 8.9 (c)) was designed with middle cutting flutes at the region of osteotomy to provide torsional stability. Postoperative X-rays, as seen in Figure 8.9(d), showed that the implant produced a good fit and stability.

It has been proven that AM models are very useful in terms of understanding bone deformity, designing implant and planning surgery.

8.2.3 *Knee Implants*

Engineers at DePuy Inc., a supplier of orthopaedic implants, have integrated CAD and AM into their design environment, using it to analyse the potential fit of implants in a specific patient and then modifying the implant design appropriately [8]. At DePuy, SLA plays a major role in the production process of all the company's products, standard and custom. The prototypes are also used as masters for casting patterns to launch a product or to do clinical releases of a product. For this application, there are several advantages over traditional casting tooling in that the lead times to manufacture the customised implants and the costs associated with these are significantly reduced.

8.2.4 *Inter-Vertebral Spacers*

Human spinal vertebras can disintegrate due to conditions such as osteoporosis or extreme forces acting on the spine. In the management of such situations, a spacer is usually required as part of the spinal fixation process. AM has been investigated for the production of such spacers as it is an ideal process to fabricate 3D structures with good interconnecting pores for the promotion of tissue in-growth. Other considerations for producing such an implant are that the material is biocompatible, and that mechanical compressive strength of the spacer is able to withstand spinal loads.

A process developed at the Nanyang Technological University for such a purpose uses a solid AM master pattern of the spacer to produce a soft mould. Stainless steel bearings coated with a formulated binder system are then cast into the soft mould under vacuum. Upon curing, the part is ejected from the mould. The part then undergoes debinding and sintering processes to produce the final part. The primary advantage of this process is its ability to use a solid AM pattern to produce a porous structure with controllable pore sizes and mechanical strengths [15, 16]. Figure 8.10 shows the inter-vertebral spacers produced using the AM system.

Fig. 8.10. Inter-vertebral spacers produced by Nanyang Technological University (NTU) using AM.

8.2.5 *Buccopharyngeal Stent*

A male child was diagnosed at birth with a persistent buccopharyngeal membrane [17]. The buccopharyngeal membrane forms a septum between the primitive's mouth and pharynx. Normally, it completely ruptures during embryo development but was not the case due to a genetic defect. Persistence of the buccopharyngeal membrane would have resulted in the partial fusion of his jaws, the inability in opening and thus speaking as he grows. Another problem was that the child was rapidly growing, and major anatomical changes were expected every six weeks up to the age of about four to five. There was also no readily available commercial buccopharyngeal stent as it was a rare disease, yet customisation of the stent was essential to 'morph' with the changing anatomy of growth and surgical procedures.

To solve this problem, the Kandang Kerbau Women's and Children's Hospital in Singapore worked collaboratively with NTU Additive Manufacturing Centre (NAMC) and the Rapid Prototyping Laboratory (RPL) at Nanyang Technological University to create a newly designed stent made with biocompatible materials. This material was soft, comfortable, yet rigid enough. The stent was designed to have excellent anti-migration properties when deployed at the pharynx region, yet easy to deploy and extract without causing any trauma to the patient (see Figure 8.11).

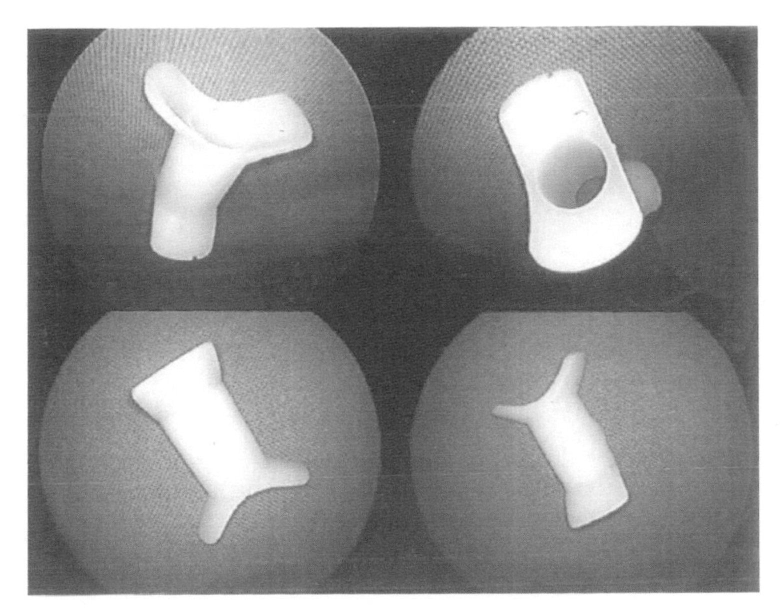

Fig. 8.11. Different views of the biocompatible buccopharyngeal stent.

To produce the stent, a master pattern was first fabricated using AM based on the patient's airway morphology. The stent master pattern was then used to create a solid silicone mould. The silicone mould was parted to remove the master pattern, thereby leaving a negative cavity of the implant. The silicone mould was then sprayed with a safety release agent and reassembled. The polyurethane-based resin was then prepared and vacuum cast into the silicone mould under 10^4 Torr in a controlled gas environment. The part was then ejected from the mould and postprocessing was carried out. Every six weeks, when it was time to produce a new stent, the CAD model for the previous stent was modified using a growth and surgical model semi-empirically developed with the surgeons. The process was repeated until the child reached the age of 5 years old when he was stable enough. With the newly developed stent, the patient was able to breathe, eat, and initiate vocalised sounds as his anatomical structure became more stabilised, thereby learning to talk.

8.2.6 *Customised Tracheobronchial Stents*

Stents for maintaining the patency of the respiratory channel have been investigated for production using AM techniques [18]. Customisation of these stents can be carried out to take into account compressive resistance with respect to stent wall thickness, as well as unique anatomical considerations. Measurements are taken of the actual forces required to open the airway channel to its original dimensions. The data is fed to the CAD system where modification of the stent design is carried out. Upon confirmation, the 3D data is fed to an AM system where the master pattern of the model is built. The master pattern then undergoes the silicone moulding vacuum casting process, reproducing the stent master pattern with a biocompatible material with all its strength, spring-back and anti-migration properties in place. Figure 8.12 shows four vacuum cast tracheobronchial stents in slightly differing sizes for an ideal intra-surgery custom fit. The stent is sterilised, packaged, and delivered to the operating theatre.

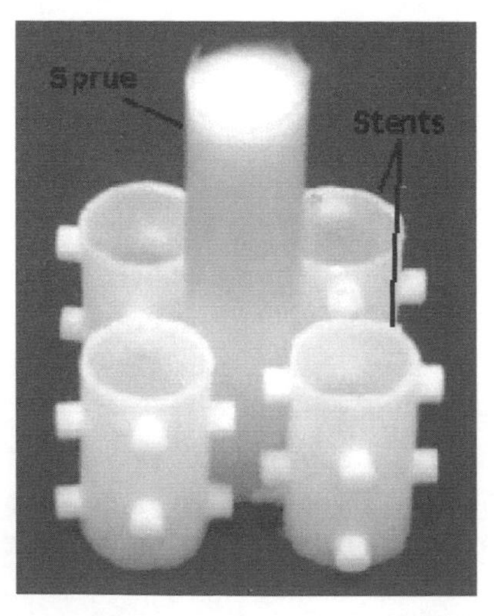

Fig. 8.12. Production of the customised stents in slightly differing sizes for an ideal custom fit.

8.2.7 *Obturator Prosthesis for Oncologic Patients*

Figure 8.13 shows the cavity in the mouth of a patient after resection of a tumour [19]. In order to protect the tissue weakened by irradiation and for the patient to be able to breathe and eat normally, this hole needs to be filled by an implant. A CT scan of the patient was made. The soft tissue around the cavity, clearly visible on the scans, was modelled using AM technology. This model served as a direct mould for the implant. The implant, called an obturator prosthesis, was cast from the mould in biocompatible silicone. No surgery was needed to implant the obturator prosthesis. As the silicone prosthesis is plastic deformable, it can be implanted very easily.

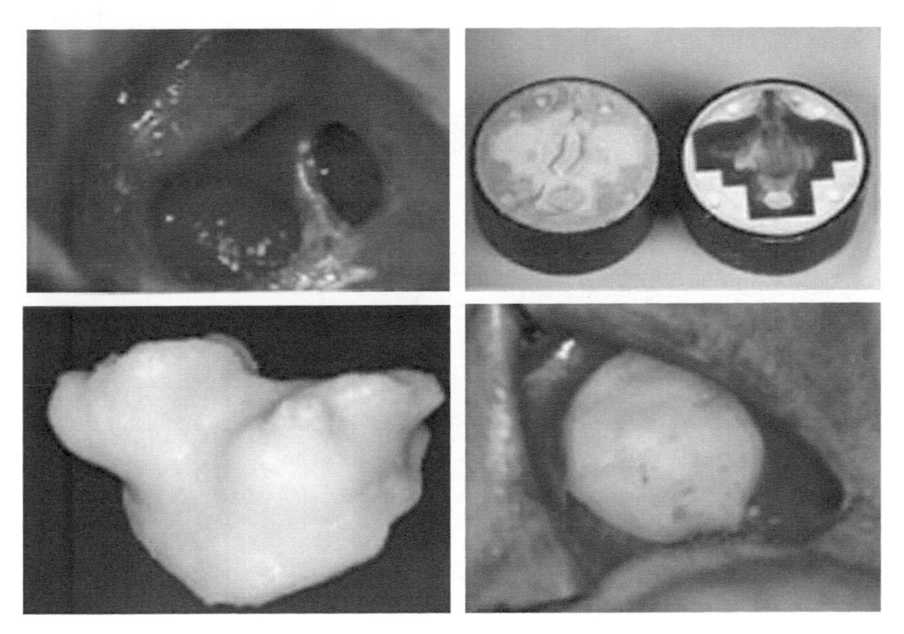

Fig. 8.13. Obturator prosthesis fits the cavity.

The prosthesis fitted the cavity much better than ever could have been achieved by using impression techniques. These traditional techniques produce a master of the obturator prosthesis by making an impression of the cavity in a plastic deformable material. The prostheses cast from such masters are always less accurate because of the presence of undercuts (the impression technique is not sensitive to local internal broadening of the cavity) and can severely damage the sensitive and vulnerable surrounding tissue. The soft prosthesis is fixed by means of magnets on a hard dental implant. This makes it possible to take it out for inspection and to replace it afterwards.

8.2.8 *Tissue Engineering Scaffolds*

Tissue engineering involves a combination of cells, scaffolds and suitable biochemical factors to improve or replace damaged or malfunctioning organs such as skin, liver, pancreas, heart valve leaflet, ligaments, cartilage and bone. A scaffold is a polymeric porous structure made of biodegradable materials such as PLA and PGA. They serve as supports to hold cells.

A typical process in tissue engineering is illustrated schematically in Figure 8.14 and is as follows:

(1) Examine the defects and determine if a scaffold is necessary.
(2) Isolate functional (undamaged) cells from donor tissue to be cultured.
(3) Select suitable materials to be prototyped.
(4) The patient is introduced to CT or MRI scanning to obtain the geometric data of the defects.
(5) Reconstruct CT data into a virtual model.
(6) Design a scaffold with suitable porous networks to fit the virtual model.
(7) Convert CAD model of scaffold to STL file.
(8) Create the scaffold using an AM machine.
(9) Transplant cells to scaffold.
(10) Implant scaffold to the defected receiver site with growth factors.
(11) Scaffold degrades gradually while cells grow and multiply.

One challenge in tissue engineering is improving the vitality of cells during transplantation. When transplanted to the scaffold, high cell density causes the inner cells to lack nutritional input from the exterior environment, causing some of them to die of malnutrition. Thus the microstructures of the scaffold are very important to the normal functions of cells.

The conventional method of fabricating a scaffold is to use organic solvent casting or particulate leaching [20]. However, this has three drawbacks: (1) the thickness is limited; (2) the sizes of the pores are not uniform; (3) the interconnectivity or distribution of the pores is irregular. This certainly will affect the number and quality of the cells that are seeding on the scaffold. To solve these problems, additive manufacturing was introduced [21–28]. Due to its advantage of fabricating intricate objects accurately and quickly, additive manufacturing has become more and more popular for fabricating tissue engineering scaffolds. Some researchers use additive manufacturing techniques to study the optimal microstructure of the scaffolds [29–35]. Based on cell type, interconnected networks of the scaffold can be predetermined in a computer CAD model. After conversion to STL file, a fine structure can be built layer by layer using additive manufacturing machines. The design of microstructure can be further advanced to make functionally graded scaffolds that closely mimic the structure of a native bone [36–41]. Some researchers apply AM techniques to study scaffold materials, for example, PEEK/HA composite [42–44], PVA/HA composite [43, 45, 46], PCL/HA composite [43, 47, 48] and PLLA/HA [49], which has been shown to have favourable potential. There are also researchers who work on theindirect fabrication of protein scaffolds such as collagen/gelatine scaffolds [50–54] and silk fibroin scaffolds [55–56]. In this approach, an intricate mould is made by AM first and then used to cast out the protein scaffold. Some researchers are interested in modelling the AM process for a better understanding of scaffold fabrication [57–58]. Other interesting works in this area include the invention of a mini-SLS to save materials [59] and an AM-made porous scaffold for cardiac tissue engineering [60].

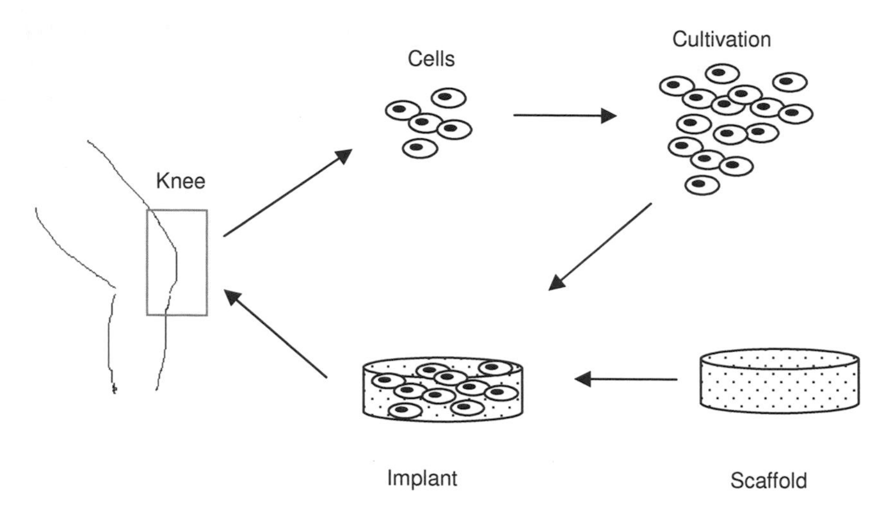

Fig. 8.14. An illustration of the tissue engineering process incorporating AM-made scaffolds.

8.3 Design and Production of Medical Devices

Additive manufacturing has impacted the biomedical field in several important ways. Besides biomodelling for surgical planning and medical implants, another obvious application is to design, develop and manufacture medical devices and instrumentation. Additive manufacturing technology is also being used to fabricate drug dosage forms with precise and complex time release characteristics. In addition, the market value of new designs of medical devices or instruments can be proved with the help of additive manufacturing.

8.3.1 *Biopsy Needle Housing*

Biomedical applications are extended beyond design and planning purposes. The prototypes can serve as a master for tooling, such as a urethane mould. At Baxter Healthcare, a disposable-medical-products company, designers rely on two AM processes, SLA and SGC, to create master models from which they develop metal castings [8]. The masters also serve as a basis for multiple sub-tooling processes. For example, after a master model has been generated via one of the AM machines, the engineers might build a urethane mould around it, cut open the mould, pull out the master, and then inject thermoset material into the mould to make prototype parts.

This process is useful in situations where multiple prototypes are necessary because the engineers can either reuse the rubber moulds or make many moulds using the same master. The prototypes are then delivered to customer focus groups and medical conferences for professional feedback. Design changes are then incorporated into the master CAD database. Once the design is finalised, the master database is used to drive the machining of the part. Using this method, Baxter Healthcare has made models of biopsy needle housing and many other medical products.

8.3.2 *Drug Delivery Device*

Drug delivery refers to the delivery of a pharmaceutical compound to humans or animals. The method of delivery can be classified into two ways, invasive and non-invasive. Most of the drugs adopt non-invasive ways by oral administration. Other medications, however, cannot be delivered by oral administration due to its ease of degradation, for example, proteins and peptide drugs. Generally they are delivered by the invasive method of injection.

Currently many research efforts are focusing on target delivery and sustained release formulation. Target delivery refers to delivery drugs at the target site only, for example, cancerous tissues, and not elsewhere along the route. Sustained release formulation refers to releasing the drug over a period in a controlled manner, thus achieving optimally therapeutic concentration.

Polymeric drug delivery devices play an important role in sustained release formulation. However, the current fabrication methods lack precision, which impairs the quality of the device, resulting in a decrease in the efficiency and effectiveness of drug delivery. The SLS process is explored to fabricate a polymeric matrix of drugs layer by layer using powdered materials [61-65]. The capability to build controlled released drug delivery device is also demonstrated in a study conducted by the Nanyang Technological University. In this research, a varying-porosity circular disc with a denser outer region that acts as a diffusion barrier region, and a more porous inner region that acts as a drug encapsulation region is fabricated using the SLS process (see Figure 8.15). Biodegradable powder materials are used. This research also concludes that control of SLS process parameters, such as laser power, laser scan speed and bed temperature will influence the porous microstructure of the polymeric matrix. Because the SLS process is not capable of processing two or more materials separately, research exploring the possibilities of using SLS to perform a dual material operation is therefore being carried out. In this research, two processes - space creation and secondary powder deposition - are integrated to form a foundation for future work in multimaterial polymeric drug delivery device fabrication [66, 67].

Fig. 8.15. Polymeric cylindrical DDD built with the SLS.

8.3.3 *Masks for Burn Victims*

Burn masks are plastic shields worn by patients with severe facial burns. They are used as treatment devices to prevent scar tissue from forming. Burn patients are required to wear the mask for at least 23 hours a day between tissue reconstructive surgery sessions.

The traditional methods for producing burn masks vary and are never simple. The most conventional way is to form a mould by applying a plaster material to the patient's face. The process can be extremely anxiety-provoking for the patient since their eyes and mouth are completely covered throughout the drying process. The weight of the plaster can shift the tissue on the face, making it virtually impossible to get an exact replication.

Additive manufacturing can be used to produce custom fit masks for burn victims. The process begins by digitising the patient using non-contact optical scanning. The scan data is used to produce an AM model of a mask that is a very precise representation of the patient's facial contours. The mask applies pressure to the face to slow the flow of blood to the healing skin. This, in turn, reduces the formation of scar tissue.

A five-year-old child and his family flew from their Colorado home to have a burn mask developed [68] (see Figure 8.16). Prior to their visit, physicians and occupational therapists had developed two burn masks using the conventional plaster method for the child, but encountered unsatisfactory results. The AM method enabled the child and his family to return to Colorado in less than 48 hours with a complete, more accurately

Fig. 8.16. A burn mask for a child developed by Total Contact. (Courtesy of Total Contact)

fitting mask. Today the child's soft tissue is healing better due to the accuracy of the burn mask, which creates smoother skin surfaces and causes less abnormal scarring.

8.3.4 *Functional Prototypes Help Prove Design Value*

Honeywell is one of the world's leading companies in militarised head-mounted display technology [69]. The display technology has also made minimally invasive surgeries easier. In actual surgeries, the functional prototype headset was worn by military surgeons, replacing the need for conventional cathode ray tube (CRT) monitors.

At the initial stage of developing the new miniature colour-display technology, a prototype was not used. However, the research scientists found that without a prototype they could not get meaningful feedback on their design. On the other hand, they could not justify the tooling costs for making a real product. They then turned to a design firm for additive manufacturing such a model. To allow the surgeons to provide truthful feedback, the prototype had to look and feel like the real product. A functional and durable material was required. After careful consideration, ABS was selected because the continuous copolymer phase gave the materials rigidity, hardness and heat resistance. In addition, it was easy to finish and approximate the injection-moulded plastics well using AM to build. For the AM machine, FDM was chosen for the electronic covers and case housings because the operating temperatures were to be high. For example, the electronics pack that supplies power to the displays was very small and can generate a great deal of heat in its prototype form. It contained a very high-density printed circuit board with the electronic equivalent of two television sets, plus video converters.

The final result was very pleasing. With the prototype, surgeons saw the benefits immediately and were quite certain that the newly developed display technology had proven its market value. AM has made a difference by being able to sell the concept to medical market leaders. The alternative could have resulted in a loss of funding for the programme due to communication gaps and a lack of understanding about the technology.

8.4 Forensic Science and Anthropology

Surgeons are not the only people interested in bones. Anthropologists and forensic specialists share the same interest, too. AM and related technologies have enabled these scientists to put a recognisable face on skeletons and share precious replicas of rare finds.

8.4.1 *The Mummy Returns*

Imagining how ancient people look different from the humans of today is made viable to us by AM. Viewing the faces of people some 3000 years ago through their skeletal finds is basically restoring history in real time.

In one study, an ancient Egyptian mummy from the collection of the Egyptian Museum (in Torino, Italy) was selected as a research example. This mummy was well-preserved and completely wrapped [70]. Without having to unravel the mummified wrappings, CT scans were obtained to establish geometric data. CT is the most important non-invasive imaging technique for obtaining fundamental data for 3D reconstructions of the skull and the body, especially with wrapped mummies. Based on the CT scan data, a 3D nylon model was built using the Selective Laser Sintering technology, which provided the frame for facial reconstruction of the mummy.

Anthropometric data, the conditions of the remaining dehydrated tissues, and the most accepted scientific and anthropological criteria were used to restore the physical appearance of the mummy. This was done by progressive layering plasticine onto the nylon model.

8.4.2 *"Officer Blue"*

One quiet day in Central Park, New York City, two horse policemen were making their rounds on horseback. Suddenly, a scuffle broke out and one cop rushed over to the scene. However, he was shot in the back by a sniper. The bullet had gone right through his protective vest, passed through his body, and lodged in his horse, Officer Blue. The policeman was dead, but the horse was still alive, with the bullet inside its spine. To determine the type and origin of the bullet, investigators had to examine it. However, without having to put down the horse, it was impossible for the bullet to be removed.

In this case, one of the AM technologies, the 3D printer, provided the solution. The horse was first brought to a CT scan to obtain the digital data of the bullet. Then, the CT scanned data was reconstructed and converted to an STL file, which was sent to the 3D printer to recreate an exact replica of the bullet. The 3D printer moulded the bullet layer-by-layer until a blue plaster prototype emerged. This bullet model was used as evidence in the court. In the end, the horse's life was spared.

8.4.3 *The Woman in the River*

In another criminal case, Laminated Object Manufacturing (LOM) was used [71] to perform forensic reconstruction of a murdered body. A dismembered body of a woman was uncovered in rural Wisconsin. However, it was impossible for the police to identify who she was because the skin of the face had been removed. Fortunately, the skull of the victim was preserved whole. Using CT scanning, a virtual model of the skull was constructed. After conversion to STL file, a model of the victim was created layer-by-layer using the LOM machine (Figure 8.17). Forensic anthropologists performed a facial reconstruction directly on the LOM model, and photographs of the reconstructed face were distributed for identification (see Figure 8.18). One response gave clues to search for the suspect. In the end, the murderer was arrested.

Fig. 8.17. (left) LOM model before decubing and (right) finished model.

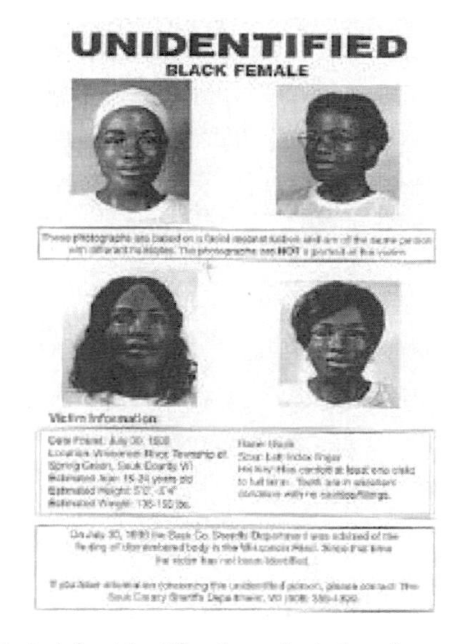

Fig. 8.18. The flyer distributed for identification, depicting the completed AM-assisted reconstructed face with a variety of hairstyles.

8.5 Visualisation of Bio-Molecules

8.5.1 *Bio-Molecular Models for Educational Purposes*

The stereo-structure of a molecule determines its physical and chemical properties. In biological sciences, the structures of proteins are intensively studied in order to understand biological activities. Researchers generally have a good spatial sense to imagine the structure using computerised virtual construction. However, to clearly explain it to students is very challenging, because words are always vague to some degree and each student has his own perception and understanding.

Additive manufacturing technology can be used to produce accurate, three-dimensional physical models of proteins and other molecular structures. From protein data banks where proteins of known structures are stored, the structure file of the protein of interest can be found. Based on xyz coordinates obtained from the structure file, a CAD model could be constructed. After conversion to STL file, a 3-D protein model can be rapidly prototyped (see Figure 8.19).

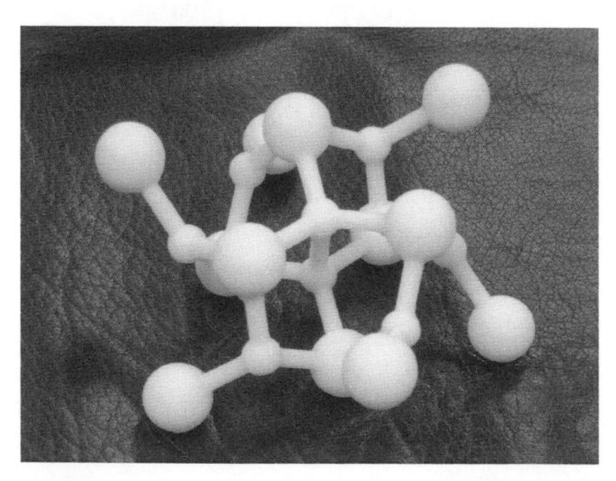

Fig. 8.19. AM model of a protein molecule made using Stratasys' PolyJet™ system.

Students are able to interact with the model and explore its shape and chemical properties. Colour schemes can be used to identify the helices and sheets.

8.5.2 *3DP for Modelling Protein–Protein Interactions Study*

Protein–protein interactions refer to the association and disassociation of protein molecules. They play a vital role for nearly every biological process in a living cell. A good understanding of the mechanisms of these interactions can help clarify the cause and development of disease and provide new therapeutic approaches.

One issue in the study of protein–protein interactions is to deduce the spatial arrangement of the complex (a protein carrying another one for a long time or a brief interaction just for modification); in other words, to predict the molecular structure of the protein complex using the known protein structures [72]. In principle, especially for a large complex, the interface of interaction is unique due to surface complementarity. This is where physical models come in and work.

Structures of the known proteins can be found in Protein Data Bank, thus xyz coordinates can be obtained to build CAD models. After conversion to STL file, accurate physical models are made using additive manufacturing machines. Then these models, the components of the complex, are manoeuvred by hand using translational and rotational orientations to quickly search for the most suitable and probable configuration. Based on the configuration identified by the models, a virtual complex is reconstructed in molecular graphics programs. After modification, the structure of the complex is considered to be finally determined.

In 2002, an investigative study proved that 3D-Rapid Prototyping is in general eligible for the modelling of protein–protein interaction [73].

8.6 Bionic Ear

A team of researchers from Princeton University led by Michael McAlpine made history when they successfully harnessed the power of AM to create a 'bionic ear' [74]. Figure 8.20 shows the bionic ear manufactured using a 3D printer. This bionic ear is capable of detecting frequencies a million times higher than the normal range of human hearing. The bionic ear is an evidence of how 3D printing can seamlessly integrate two materials, electronic and biological tissues, that traditionally do not blend well together due to the former being rigid and fragile, and the latter being soft and flexible.

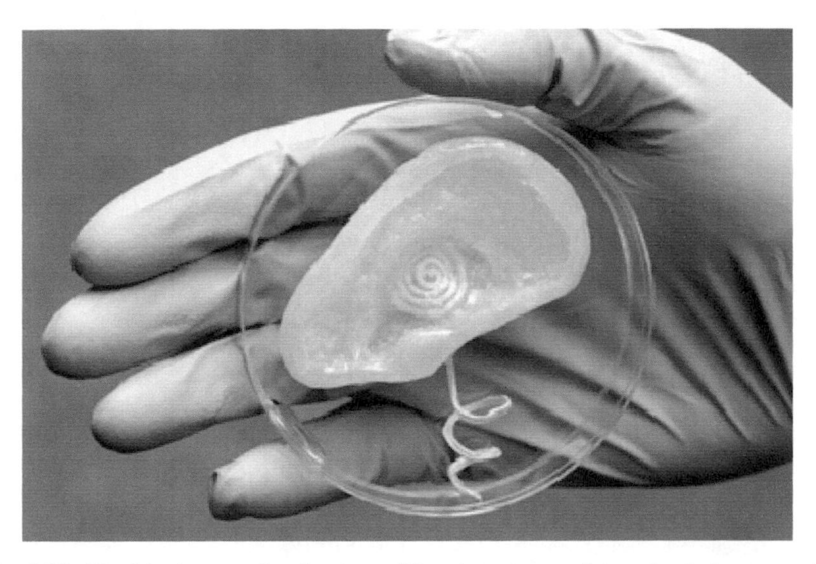

Fig. 8.20. The bionic ear printed using a 3D printer has an integral spiral antenna that resembles the ear's cochlear structure. (Courtesy of the American Chemical Society)

Using a computer model of an ear and an internal antenna coil connected to an external electrode, the 3D printer alternates among 3 'inks': a mix of bovine cartilage-forming cells suspended in a thick goo of hydrogel, a suspension of silver nanoparticles to form the coil and external cochlea-shaped electrodes, and silicone to encase the electronics (see Figure 8.21). The whole printing process takes about 4 hours, where the ear is then cultivated in a nutrient-rich liquid to allow for cell growth producing collagen and other molecules. The original surroundings are then replaced with cartilage [75].

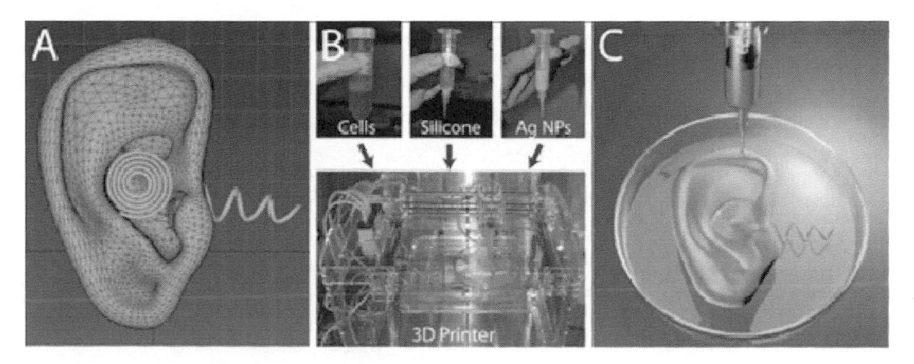

Fig. 8.21. The bionic ear computer model and the 3 'inks' required for 3D printing the bionic organ. (Courtesy of the American Chemical Society)

The ear is not able to replace the human ear but it was created to demonstrate the power of 3D printing technology in cybernetics, which combines biological tissue with electronics. Further research could lead to synthetic replacements for actual human functions and create additional senses to humans' 5 traditional senses. 3D printing may also address the main tissue engineering challenges of building organs with blood vessels.

8.7 Dentistry

As the world enters a new era of manufacturing revolution with 3D printing, the dental industry is showing an interest in 3D printing with a general shift towards digital dentistry aided by oral scanners. Using oral scanners combining CAD/CAM and 3D printing, dentists and lab owners are able to customise and speed up the production of crowns, bridges, stone models and a range of orthodontic appliances with quality and precision.

8.7.1 *Speeding Up the Production of Crowns and Bridges*

Dental technicians used to prepare every crown and bridge manually, which took a very long period of time. The patient was required to bite into a tray of gooey silicone at the dentist's office to make the models of the patient's teeth. After hardening, the impression was sent to the lab, where gypsum was poured on it to create the model. The entire process could take several weeks. AM is well positioned in dentistry by speeding up the entire process of making crowns and bridges, which are typically high-end products and require customisation to suit different patients' needs.

Highland Dental Laboratory, a full service dental lab based in Texas, USA utilises EnvisonTEC's Perfactory DDP4 3D printer to produce crowns and bridges. Previously, crowns were pressed or cast manually by hand, which was labour intensive and costly. In addition, a bulk of smaller and less complicated work was outsourced to cope with the demand. With the installation of the 3D printer, the lab was able to produce crowns of e.max, zirconia, full cast alloy and other materials in-house, thus giving the company a vastly significant improvement in quality control and job turnarounds [76]. Instantaneous data transferred from dentists straight to the lab help speed up the production of the model to the finished product.

8.7.2 *3D Printed Custom-Made Toothbrush*

The first toothbrush was invented in the 1780s, made from cattle bone and featured a swine bristle brush. Aside from basic modifications, the form of the humble toothbrush has remained unchanged for centuries but now, with AM technology, the way we brush our teeth will change forever. The Blizzident is a 3D printed toothbrush that is custom made from a scan taken by a dentist and tailor-fit for each individual customer [77]. The customised 3D mouthpiece looks like a set of dentures filled with hundreds of bristles (see Figure 8.22). It requires the customer to wear it in the mouth and uses biting and chewing motions to clean the teeth and gums in a matter of seconds. The Blizzident is not just a time-saving device but it also eliminates human errors of manual brushing. With plenty of soft bristles angled to clean the teeth, gums and even the tongue, the Blizzident will be able to brush all the traditionally difficult-to-reach places, giving the user a thoroughly cleaned mouth.

Fig. 8.22. The Blizzident toothbrush is 3D printed to custom fit a customer's mouth and teeth profile. (Courtesy of Blizzident)

To make the toothbrush, an impression of the teeth is scanned and uploaded to the company's server. The makers find the optimal placement of 400 bristles by performing simulations of biting and chewing motions and create a CAD model of the brush. The toothbrush is then 3D printed by high precision machines using stereolithography (SLA), which employs a vat of liquid ultraviolet curable photopolymer "resin" and an ultraviolet laser to build the parts layer-by-layer. The materials used to 3D print these devices are made from the same specially classified biocompatible printing materials used to form implantable medical devices. The company has also taken steps to ensure the materials and printers used are able to create totally smooth surfaces for perfect hygiene. Ordinary 3D printers produce rough surfaces where dirt can hide and get trapped easily. The Blizzident is a perfect example of what 3D printing can do in creating customised versions of everyday objects.

8.7.3 *Digital Dentistry and Orthodontics*

Traditional dentistry relies heavily on devices like stone moulds and braces to restore dental structure and health while digital dentistry relies on innovative technologies such as oral scanners and 3D printing. In 1997, a start-up company approached 3D Systems with a novel idea to leverage digital scanning and 3D printing to dramatically change the way teeth were straightened for orthodontic patients. This company believed that personal orthodontic treatment devices manufactured using the latest 3D printing technology could eliminate the reliance on metal or ceramic brackets and wires, thereby drastically reducing the aesthetic, discomfort and other limitations associated with braces. Align Technology is the company which invented Invisalign Systems. They worked closely with 3D Systems, who helped them develop a customised solution combining digital scanning and 3D printing technology [78]. Stereolithography technology is used to make the tools to manufacture numerous clear plastic aligners, which are worn sequentially by patients to move teeth into a desired final orientation. Being the early adopter of 3D printing for mass customisation has positioned Invisalign System as the market leader of invisible orthodontic system, generating over $500 million in revenue globally in 2012.

References

[1] Materialise NV. (2007). Retrieved from http://www.materialise.be/medical-rpmodels/case3_ENG.html

[2] Kaneko, T., Kobayashi, M., Tsuchiya, Y., Fujino, T., Itoh, M., Inomata, M., Uesugi, M., Kawashima, K., Tanijiri, T., & Hasegawa, N. (1992). Free surface 3-dimensional shape measurement system and its application to Mictotia ear reconstruction. In *The Inaugural Congress of the International Society for Simulation Surgery*.

[3] Chow, K. Y. (1996). *Development of a direct link between a laser digitiser and a rapid prototyping system.* Final year thesis, Nanyang Technological University, Singapore.

[4] Adachi, J., Hara, T., Kusu, N., & Chiyokura, H. (1993). Surgical simulation using rapid prototyping. In *Proceedings of the Fourth International Conference on Rapid Prototyping*, pp. 135-142.

[5] Koyayashi, M., Fujino, T., Chiyokura, H., & Kurihara, T. (1992). Preoperative preparation of a hydroxyapatite prosthesis for bone defects using a laser-curable resin model. In *The Inaugural Congress of the International Society for Simulation Surgery*.

[6] *Rapid prototyping helps separate conjoined twins.* (2006). Cyon Research Corporation. Retrieved from http://www.newslettersonline.com/user/user.fas/s=63/fp=3/tp=47?T=open_article,484858&P=article

[7] Materialise. Retrieved from http://www.materialise.be/medical-rpmodels/ case22_ENG.html

[8] Mahoney, D. P. (1995). Rapid prototyping in medicine. In *Computer Graphics World,* **18**(2): 42-48.

[9] Swaelens, B., & Kruth, J. P. (1993). Medical applications in rapid prototyping techniques. In *Proceedings of the Fourth International Conference on Rapid Prototyping*, pp. 107-120.

[10] Jacobs, A., & Hammer, B., Niegel, G., Lambrecht, T., Schiel, H., Hunziker, M., & Steinbrich, W. (1993). First experience in the use of stereolithography in medicine. In *Proceedings of the Fourth International Conference on Rapid Prototyping*, pp. 121-134.

[11] Faber, J., Berto, P. M., & Quaresma, M. (2006). Rapid prototyping as a tool for diagnosis and treatment planning for maxillary canine impaction. In *American Journal of Orthodontics and Dentofacial Orthopedics,* **129**(4):583-9.

[12] D'Urso, P. S., Williamson, O. D., & Thompson, R. G. (2005). Biomodeling as an aid to spinal instrumentation. In *Spine*, **30**(24):2841-5.

[13] Materialise NV. (2007). Retrieved from http://www.materialise.be/medical-rpmodels/case6_ENG.html

[14] Materialise NV. (2007). Retrieved from http://www.materialise.be/medical-rpmodels/case7_ENG.html

[15] Lim, C. S., Chandrasekeran, M., & Tan, Y. K. (2001). Rapid tooling of powdered metal parts. In *International Journal of Powder Metallurgy*, **37**(2): 63-66.

[16] Lim, C. S., & Chanrdrasekeran, M. (1999). A process to rapid tool porous metal implants. In *Internal Report, Nanyang Technological University, Singapore*, February: 1-6.

[17] Tan, S. S., Tan, H. K., Lim, C. S., & Chiang, W. M. (2006). A novel stent for the treatment of persistent buccopharyngeal membrane. In *International Journal of Pediatric Otorhinolaryngology.* **70**(9):1645-1649.

[18] Lim, C. S., Eng, P., Lin, S. C., Chua, C. K., & Lee, Y. T. (2002). Rapid prototyping and tooling of custom-made tracheobronchial stents. In *International Journal of Advanced Manufacturing Technology*, **20**(1): pp. 44-49.

[19] Materialise NV. (2007). Retrieved from http://www.materialise.com/medical-rpmodels/case2_ENG.html

[20] Yang, S., & et al. (2001). *The design of scaffolds for use in tissue engineering. Part I. Traditional factors.* Tissue Engineering, **7**(6): pp. 679-689.

[21] Yang, S., & et al. (2002). *The design of scaffolds for use in tissue engineering. Part II. Rapid prototyping techniques.* Tissue Engineering, **8**(1): pp. 1-11.

[22] Leong, K. F., Cheah, C. M., & Chua, C. K. (2003). *Solid freeform fabrication of three-dimensional scaffolds for engineering replacement tissues and organs.* Biomaterials, **24**(13): pp. 2363-2378.

[23] Yeong, W. Y., & et al. (2004). *Rapid prototyping in tissue engineering: Challenges and potential.* Trends in Biotechnology, **22**(12): pp. 643-652.

[24] Bártolo, P. J., & et al. (2009). *Biomanufacturing for tissue engineering: Present and future trends.* Virtual and Physical Prototyping, **4**(4): pp. 203-216.

[25] Derby, B. (2012). *Printing and prototyping of tissues and scaffolds.* Science, **338**(6109): pp. 921-926.

[26] Chua, C. K., Yeong, W. Y., & Leong, K. F. (2005). *Rapid prototyping in tissue engineering: A state-of-the-art report. in Virtual Modelling and Rapid Manufacturing - Advanced Research in Virtual and Rapid Prototyping.*

[27] Chua, C. K., Liu, M. J. J., & Chou, S. M. (2012). Additive manufacturing-assisted scaffold-based tissue engineering. in Innovative Developments in Virtual and Physical Prototyping. In *Proceedings of the 5th International Conference on Advanced Research and Rapid Prototyping.*

[28] Boland, T., & et al. (2007). *Rapid, prototyping of artificial tissues and medical devices.* Advanced Materials and Processes, **165**(4): pp. 51-53.

[29] Chua, C. K., & et al. (2003). Development of a tissue engineering scaffold structure library for rapid prototyping. Part 1: Investigation and classification. In *International Journal of Advanced Manufacturing Technology*, **21**(4): pp. 291-301.

[30] Chua, C. K., & et al. (2003). Development of a tissue engineering scaffold structure library for rapid prototyping. Part 2: Parametric library and assembly program. In *International Journal of Advanced Manufacturing Technology,* **21**(4): pp. 302-312.

[31] Chua, C. K., & et al. (2003). *Novel method for producing polyhedra scaffolds in tissue engineering. in Virtual Modelling and Rapid Manufacturing - Advanced Research in Virtual and Rapid Prototyping.*

[32] Cheah, C. M., & et al. (2004). *Automatic Algorithm for Generating Complex Polyhedral Scaffold Structures for Tissue Engineering.* Tissue Engineering, **10**(3-4): pp. 595-610.

[33] Naing, M. W., & et al. (2005). *Fabrication of customised scaffolds using computer-aided design and rapid prototyping techniques.* Rapid Prototyping Journal, **11**(4): pp. 249-259.

[34] Cai, S., Xi, J., & Chua, C. K. (2012). *A novel bone scaffold design approach based on shape function and all-hexahedral mesh refinement,* pp. 45-55.

[35] Ang, K. C., & et al. (2006). *Investigation of the mechanical properties and porosity relationships in fused deposition modelling-fabricated porous structures.* Rapid Prototyping Journal, **12**(2): pp. 100-105.

[36] Chua, C. K., & et al. (2010) *Process flow for designing functionally graded tissue engineering scaffolds.*

[37] Chua, C. K., & et al. (2011). *Selective laser sintering of functionally graded tissue scaffolds.* MRS Bulletin, **36**(12): pp. 1006-1014.

[38] Sudarmadji, N., & et al. (2011). *Investigation of the mechanical properties and porosity relationships in selective laser-sintered polyhedral for functionally graded scaffolds.* Acta Biomaterialia, **7**(2): pp. 530-537.

[39] Sudarmadji, N., Chua, C. K., & Leong, K. F. (2012). *The development of computer-aided system for tissue scaffolds (CASTS) system for functionally graded tissue-engineering scaffolds. Methods in molecular biology (Clifton, N.J.),* **868**: pp. 111-123.

[40] Chua, C. K., Sudarmadji, N., & Leong, K. F. (2007). Functionally graded scaffolds: The challenges in design and fabrication processes. In *Proceedings of the 3rd International Conference on Advanced Research in Virtual and Rapid Prototyping: Virtual and Rapid Manufacturing Advanced Research Virtual and Rapid Prototyping.*

[41] Leong, K. F., & et al. (2008). Engineering functionally graded tissue engineering scaffolds. In *Journal of the Mechanical Behavior of Biomedical Materials,* **1**(2): pp. 140-152.

[42] Tan, K. H., & et al. (2003). *Scaffold development using selective laser sintering of polyetheretherketone-hydroxyapatite biocomposite blends.* Biomaterials, **24**(18): pp. 3115-3123.

[43] Tan, K. H., & et al. (2005). *Selective laser sintering of biocompatible polymers for applications in tissue engineering.* Bio-Medical Materials and Engineering, **15**(1-2): pp. 113-124.

[44] Tan, K. H., & et al. (2005). Fabrication and characterization of three-dimensional poly(ether-ether-ketone)/-hydroxyapatite biocomposite scaffolds using laser sintering. In *Proceedings of the Institution of Mechanical Engineers, Part H: Journal of Engineering in Medicine,* **219**(3): pp. 183-194.

[45] Chua, C. K., & et al. (2004). Development of tissue scaffolds using selective laser sintering of polyvinyl alcohol/hydroxyapatite biocomposite for craniofacial and

joint defects. In *Journal of Materials Science: Materials in Medicine*, **15**(10): pp. 1113-1121.

[46] Wiria, F. E., & et al. (2008). Improved biocomposite development of poly(vinyl alcohol) and hydroxyapatite for tissue engineering scaffold fabrication using selective laser sintering. In *Journal of Materials Science: Materials in Medicine*, **19**(3): pp. 989-996.

[47] Ang, K. C., & et al. (2007). Compressive properties and degradability of poly(ε-caprolatone)/ hydroxyapatite composites under accelerated hydrolytic degradation. In *Journal of Biomedical Materials Research - Part A*, **80**(3): pp. 655-660.

[48] Wiria, F. E., & et al. (2007). *Poly-ε-caprolactone/hydroxyapatite for tissue engineering scaffold fabrication via selective laser sintering*. Acta Biomaterialia, **3**(1): pp. 1-12.

[49] Simpson, R. L., & et al. (2008). Development of a 95/5 poly(L-lactide-co-glycolide)/hydroxylapatite and β-tricalcium phosphate scaffold as bone replacement material via selective laser sintering. In *Journal of Biomedical Materials Research - Part B Applied Biomaterials,* **84**(1): pp. 17-25.

[50] Tan, J. Y., Chua, C. K., & Leong, K. F. (2010). *Indirect fabrication of gelatin scaffolds using rapid prototyping technology.* Virtual and Physical Prototyping, **5**(1): pp. 45-53.

[51] Tan, J. Y., Chua, C. K., & Leong, K. F. (2012). *Fabrication of channeled scaffolds with ordered array of micro-pores through microsphere leaching and indirect Rapid Prototyping technique*. Biomedical Microdevices, pp. 1-14.

[52] Tan, J. Y., Chua, C. K., & Leong, K. F. (2013*) Fabrication of channeled scaffolds with ordered array of micro-pores through microsphere leaching and indirect Rapid Prototyping technique.* Biomedical Microdevices, **15**(1): pp. 83-96.

[53] Yeong, W. Y., & et al. (2006). *Indirect fabrication of collagen scaffold based on inkjet printing technique.* Rapid Prototyping Journal, **12**(4): pp. 229-237.

[54] Yeong, W. Y., & et al. (2007). *Comparison of drying methods in the fabrication of collagen scaffold via indirect rapid prototyping*. Journal of Biomedical Materials Research - Part B Applied Biomaterials, **82**(1): pp. 260-266.

[55] Liu, M. J. J., & et al. (2013). *The development of silk fibroin scaffolds using an indirect rapid prototyping approach: Morphological analysis and cell growth monitoring by spectral-domain optical coherence tomography.* Medical Engineering and Physics, **35**(2): pp. 253-262.

[56] Tay, B. C. M., & et al. (2013). *Monitoring cell proliferation in silk fibroin scaffolds using spectroscopic optical coherence tomography*. Microwave and Optical Technology Letters, **55**(11): pp. 2587-2594.

[57] Ramanath, H. S., & et al. (2008). *Melt flow behaviour of poly-ε-caprolactone in fused deposition modelling.* Journal of Materials Science: Materials in Medicine, **19**(7): pp. 2541-2550.

[58] Wiria, F. E., Leong, K. F., & Chua, C. K. (2010). *Modeling of powder particle heat transfer process in selective laser sintering for fabricating tissue engineering scaffolds.* Rapid Prototyping Journal, **16**(6): pp. 400-410.

[59] Wiria, F. E., & et al. (2010). *Selective laser sintering adaptation tools for cost effective fabrication of biomedical prototypes.* Rapid Prototyping Journal, **16**(2): pp. 90-99.

[60] Yeong, W. Y., & et al. (2010). *Porous polycaprolactone scaffold for cardiac tissue engineering fabricated by selective laser sintering.* Acta Biomaterialia, **6**(6): pp. 2028-2034.

[61] Leong, K. F., & et al. (2001). Fabrication of porous polymeric matrix drug delivery devices using the selective laser sintering technique. In *Proceedings of the Institution of Mechanical Engineers, Part H: Journal of Engineering in Medicine*, **215**(2): pp. 191-201.

[62] Low, K. H., & et al. (2001). *Characterization of SLS parts for drug delivery devices.* Rapid Prototyping Journal, **7**(5): pp. 262-267.

[63] Cheah, C. M., & et al. (2002). Characterization of microfeatures in selective laser sintered drug delivery devices. In *Proceedings of the Institution of Mechanical Engineers, Part H: Journal of Engineering in Medicine*, **216**(6): pp. 369-383.

[64] Leong, K. F., & et al. (2006). *Building porous biopolymeric microstructures for controlled drug delivery devices using selective laser sintering.* International Journal of Advanced Manufacturing Technology, **31**(5-6): pp. 483-489.

[65] Leong, K. F., & et al. (2007). *Characterization of a poly-ε-caprolactone polymeric drug delivery device built by selective laser sintering.* Bio-Medical Materials and Engineering, **17**(3): pp. 147-157.

[66] Liew, C. L., & et al. (2001). Dual material rapid prototyping techniques for the development of biomedical devices. Part 1: Space creation. In *International Journal of Advanced Manufacturing Technology*, **18**(10): pp. 717-723.

[67] Liew, C. L., & et al. (2002). Dual material rapid prototyping techniques for the development of biomedical devices. Part 2: Secondary powder deposition. In *International Journal of Advanced Manufacturing Technology,* **19**(9): pp. 679-687.

[68] Total Contact. (2007). Retrieved from http://www.totalcontact.com

[69] Materialise NV. (2007). Retrieved from http://www.cadinfo.net/editorial/ honeywell.htm

[70] Cesarani, F., Martina, M. C., Grilletto, R., R. Boano, Roveri, A. M., Capussotto, V., Giuliano, A., Celia, M., & Gandini, G. (2004). Facial reconstruction of a

wrapped Egyptian mummy using MDCT. In *American Journal of Roentgenology*, **183**(3): pp.755-758.

[71] Crockett, R. S., & Zick, R. (2000). Forensic applications of solid freeform fabrication. In *Proceedings for the Solid Freeform Fabrication Symposium*, pp. 549-554.

[72] Shimizu, T. S., Le Novère, N., Levin, M. D., Beavil, A. J., Sutton, B. J., & Bray, D. (2000). *Molecular model of a lattice of signalling proteins involved in bacterial chemotaxis*. Nature Cell Biology, **2**(11): pp. 792 – 796.

[73] Laub, M., Chatzinikolaidou, M., Rumpf, H., & Jennissen, H. P. (2002). *Modelling of protein-protein interactions of bone morphogenetic protein-2 (BMP-2) by 3D-rapid prototyping*. Materialwissenschaft und Werkstofftechnik, **33**: pp. 729-737.

[74] Mannoor, M. S., & et al. (2013). *3D Printed Bionic Ears.* Nano Letters, 13(6), pp. 2634-2639.

[75] Young, S. *Cyborg Parts.* Retrieved from http://www.technologyreview.com/demo/517991/cyborg-parts/

[76] EnvisionTEC. Retrieved from http://envisiontec.com/case_studies/highland-dental-laboratory-produces-crowns-bridges-3d-printer/

[77] Collins, K. (Oct 2013). *Blizzident 3D-printed toothbrush cleans your gnashers in six seconds.* Wired.co.uk. Retrieved from http://www.wired.co.uk/news/archive/2013-10/01/blizzident

[78] Grynol, B. (2013). *Disruptive Manufacturing: The effects of 3D printing*. Deloitte. Retrieved from http://www.deloitte.com/assets/DcomCanada/Local%20Assets/Documents/Insights/Innovative_Thinking/2013/ca_en_insights_disruptive_manufacturing_102813.pdf

Problems

1. List several possible applications for AM in medical and biomedical engineering.

2. Name several advantages and disadvantages of applying AM to the field of medicine and biomedical engineering.

3. List some possible materials for use with AM in relation to in-vivo applications.

4. Discuss how AM can create value for surgical procedures relating to the separation of conjoined twins joined at the head.

5. How can AM models be useful to surgeons before and during the operating procedure to remove a tumor from the cranium?

6. Why and how is AM important when producing hip implants for non-standard sized patients requiring hip replacement?

7. Respiratory stents, such as the buccopharyngeal stent, have been used on babies with congenital buccopharyngeal defects. Explain how AM can be used to support the child with such a stent until he is old enough for major reconstructive surgery.

8. Discuss how AM can be used to support organ replacement by tissue engineering. Discuss what materials should be considered and why.

9. Explain the challenges of building an AM system for tissue engineering applications.

10. In what ways can an AM model assist in the design and laboratory testing of a substitution replacement mechanical heart valve? Discuss the AM selection considerations for such an application.

11. How can AM prove useful in forensic science applications? Compare the use of current techniques versus AM for such applications.

12. Being able to visualize scientific concepts in three-dimensions can be quite a challenge when it comes to protein–protein interaction research. How can AM create value in research or teaching laboratories for such an application?

13. Describe how a 3D printer is used to manufacture a bionic ear. What are the advantages of using 3D printing in this application?

14. Give two examples of the applications of AM in dentistry.

Chapter 9

BENCHMARKING AND THE FUTURE OF ADDITIVE MANUFACTURING

9.1 Using Bureau Services

The best way to experiment with AM is with a service bureau that owns and operates one or several systems. AM equipment, especially the industrial grade types, is still fairly expensive (US$50,000 or more, although some concept modellers cost only a few thousand or hundred dollars) and the cost of operator training, materials and installation of the equipment can easily double this cost. The volume of prototyping work in a company will probably not justify the acquisition of such a system. Thus, it may be more economical to engage a service bureau. Some factors to consider when engaging a service bureau include:

(1) Type of materials needed for the prototype,
(2) Size of the prototype,
(3) Accuracy,
(4) Surface finish,
(5) Experience of service bureau,
(6) Location of service bureau for communication and coordination of jobs,
(7) Provision of secondary processes such as machining or sandblasting,
(8) Cost.

Typically, service bureaus base their charges on the following items:

(1) Modelling and preparation of model for building (if necessary),
(2) Execution of program to convert computer models into .STL files,

(3) "Slicing" time in converting .STL file into cross sectional data,
(4) The amount of machining time required to make the part,
(5) The setup time in preparing the apparatus,
(6) The actual amount of material used to make the part,
(7) Post fabrication assembly such as gluing together several smaller pieces,
(8) Secondary processes such as machining or sand blasting.

Needless to say, the company can save costs on the first two items if it has its own CAD facility and CAD-AM interface. Fees from service bureaus can range from a few hundred dollars for a simple part to thousands of dollars for large and complex prototypes. The Wohler's Report [1], published annually, lists service bureaus worldwide.

9.2 Technical Evaluation Through Benchmarking

The execution of a benchmark test is a traditional practice, necessary for all kinds of highly productive and expensive equipment such as CAD/CAM workstation, CNC machining centre, etc. Wherever a relatively broad spread of possibilities is offered for specific users' requirements, the execution of a benchmark test is absolutely necessary. The dynamic development and increased range of commercially available AM systems on offer (currently more than 30 different types of equipment worldwide, partly in different types of equipment, partly in different sizes), mean objective decision making is essential. In analysing the benchmark test piece, some tests have to be conducted including visual inspection and dimension measurement.

9.2.1 *Reasons for Benchmarking*

Generally, benchmarking serves the following purposes:

(1) It is a valuable tool for evaluating strengths and weaknesses of the systems tested. Vendors have to produce the benchmark models in response to requests from potential buyers. In doing so, they will

not have the choice to demonstrate what they want, but have to show what is requested. Consequently, vendors cannot hide the limitations of the system. It is also a rigorous and therefore more revealing means of testing so that the potential buyer can verify the claims of the vendor.

(2) Since the benchmark model is specifically designed by the potential buyer, it can be custom-made to its own requirements and needs. For example, in the case of a company that makes parts which frequently have very thin walls, then the ability of the system to produce accurately built thin walls can be tested, measured and verified.

(3) Benchmarking has also become a means of helping various departments within a company comprehend what the AM system can do for them. This is vital in the context of concurrent engineering, whereby designers, analysts, manufacturing engineers work on the product concurrently. Today, a AM system's application areas extend beyond design models, to functional models and manufacturing models.

(4) Sometimes, a benchmark test may also help to identify applications for an AM system which had not previously been considered. Although this is not really a primary motivational factor for benchmarking, it is nevertheless a side benefit.

9.2.2 *Benchmarking Methodology*

There are four steps in the proposed benchmarking methodology:

(1) Deciding on the benchmarking model type.
(2) Deciding on the measurements.
(3) Recording time and measurements, tabulating and plotting the results.
(4) Analysing and comparing results.

9.2.2.1 *Deciding on the benchmarking model type*

In general, additive manufacturing benchmarking models can be categorised according to the following types:

(1) *Typical company products.* This is probably the most common type since the company needs to confirm how well its products can be prototyped and whether its requirements can be fulfilled. The company is also in the best position to comment on the results since it has intimate product knowledge. Examples are turbine blades, jewellery, cellular phones, etc.

(2) *Part classification.* According to M B Wall [2], a classification scheme based on general part structures is applicable for additive manufacturing parts. This classification scheme, based on part structure, has ten part classes, as seen in Table 9.1.

Table 9.1. Part structure classification scheme of 10 part classes

Part Class Number	Part Structure
Part Class 1	Compact parts
Part Class 2	Hollow parts
Part Class 3	Box type
Part Class 4	Tubes
Part Class 5	Blades
Part Class 6	Ribs, profiles
Part Class 7	Cover type
Part Class 8	Flat parts
Part Class 9	Irregular parts
Part Class 10	Mechanisms

Figure 9.1 shows some examples of such a classification scheme. The part sizes and part structures are mostly related to shrinkage, distortion and curling effects.

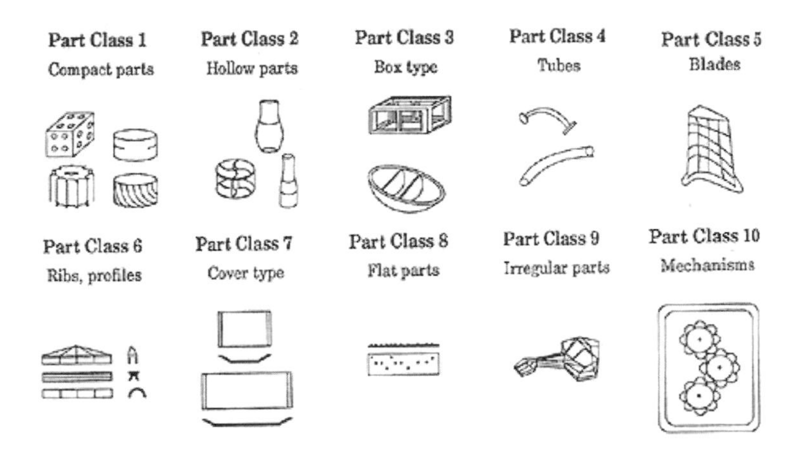

Fig. 9.1. Part structure classification scheme for AM systems.

(3) *Complex and hybrid parts.* Alternatively, determined complex parts can be designed, aiming to test the performance of available systems in specific aspects. Furthermore, this category can include a hybrid combining types 1 and 2 above.

9.2.2.2 *Deciding on the measurements*

In deciding the measurements, it must be stressed that, as far as possible, the benchmark model should:

(1) be relatively simple and designed with low expenditure,
(2) not utilise too much material, and
(3) allow simple measuring devices to determine the measurements.

In general, two types of measurements can be taken, namely main (large) measurements and detailed (small) measurements.

9.2.2.3 *Recording time and measurements, tabulating and plotting the results*

The measurement results are based on the deviations of the part built from the CAD model. These deviations of both the main and detailed measurements are tabulated for each of the AM systems. Subsequently, the results can be plotted to graphically present the performance of the systems in a single diagram. Thus, for the main measurements, one can visually compare the systems' performance using the main measurement diagram, and similarly, using the detailed measurement diagram.

The time results are based on three components – data preparation, building time and postprocessing. The total time is based on the addition of the three components. A table of all four-time results can be tabulated with each AM system alongside another.

9.2.2.4 *Analysing and comparing results*

In a general analysis, the deviations of all systems may or may not be in an acceptable range. The evaluator of the systems is usually decided by this acceptable range. The time component by itself gives one an idea of the length required for a task and directly affects the cost factor. Therefore, the time data can become useful for a full economic justification and cost analysis.

In comparison, one can determine, based on the time and measurement results, the strengths and weaknesses of the systems. In arriving at a conclusion, the evaluator must only consider the benchmarking results as *a component* of the overall evaluation study. The benchmarking results should never be taken as the only deciding factor of an evaluation study.

Finally, the above approach ignores the human aspects. For example, the skills, expertise and experience of the vendors' operators are not accounted for in the benchmark.

9.2.3 *Case Study*

9.2.3.1 *The button tree display*

With the compliments of Thomson Multimedia in Singapore, the results of a benchmarking study involving five machines are made available here. The test piece is a button tree display that is mounted in between the front cabinet of a hi-fi set and printed circuit board (PCB), as shown in Figure 9.2. The button tree display has a frame of length 128 mm and width 27 mm. It consists of three round buttons joined to the frame by 0.6 mm hinges. The buttons have "legs" that contact the tact switch on the PCB when it is depressed. There are also locating pins for location and light emitting diode (LED) holders on the frame. Catches are made from the side of the frame so that the button tree display can hold down firmly to the PCB. The five different test pieces are made from the principals of SGC, LOM (under Cubic Technologies), SOUP (under CMET), SCS and FDM (under Stratasys). A photograph of each test piece is shown in Figures 9.3, 9.4, 9.5, 9.6 and 9.7 respectively.

The measurements taken are linear, radial and angular dimensions. Coordinate measuring machine (CMM), profile projector and vernier calliper are used for the measurements. When choosing the type of dimensions to measure, a few criteria are considered:

(1) the overall dimensions are to be included.
(2) the important dimensions that will affect the operation of the button tree.
(3) there must be a variety of dimensions.
(4) there must be sufficient main and detailed dimensions to plot the graph, and give an indication of which method is superior.

Fig. 9.2. Button tree display is mounted on the front cabinet of a hi-fi.

Fig. 9.3. Benchmark test piece made from Solid Ground Curing (SGC).

Fig. 9.4. Benchmark test piece made from Laminated Object Manufacturing (LOM).

Fig. 9.5. Benchmark test piece made from Solid Object Ultraviolet Plotter (SOUP).

Fig. 9.6. Benchmark test piece made from Solid Creation Systems (SCS).

Fig. 9.7. Benchmark test piece made from Fused Deposition Modelling (FDM).

Figures 9.8, 9.9 and 9.10 show the technical drawings for the button tree display. The dimensions of the test piece are divided into three parts. The first part includes dimensions taken from the frame, locating pins and LED holder. The second part has the dimensions taken from the button set, and the dimensions of the catch fall into the third part. Related to the different parts, the results are sub-divided into main measurements (> 10 mm) and detailed measurement (= 10 mm). For both the readings, the deviations from the actual reading are computed and tabulated. The deviations for both the measurements are plotted and compared.

9.2.3.2 *Results of the measurements*

The graphs plotted facilitate a detailed evaluation concerning the five different techniques, namely the SGC, LOM, SOUP, SCS and FDM, in terms of main and detailed measurements and their deviations from the designed dimensions.

From visual inspections, none of the thin rib or wall is missing. This shows that all the five processes are capable of making wall thickness as thin as 0.5 mm. However, one of the locating pins of the LOM test pieces is tilted and the catches of the SGC and FDM test pieces are missing. This is due to the catches being too weak and coming off when the supports were removed. SLA objects are built on supports rather than directly on the elevator platform. Supports are used to anchor the part firmly to the platform and prevent distortion during the building process. The SOUP, FDM and SCS test pieces came with supports, thus they had to be removed before any measurements were taken. The button sets of the five test pieces are slanted due to its weight and the weak hinges. Tables 9.2 and 9.3 list the measurements taken and their deviations from the nominal values.

The plots illustrated in Figures 9.11 and 9.12 show that SGC achieved better measurements or fewer deviations as compared to other processes in detailed measurements. For the main measurements, FDM attained fewer deviations. On the other hand, SOUP produced the higher

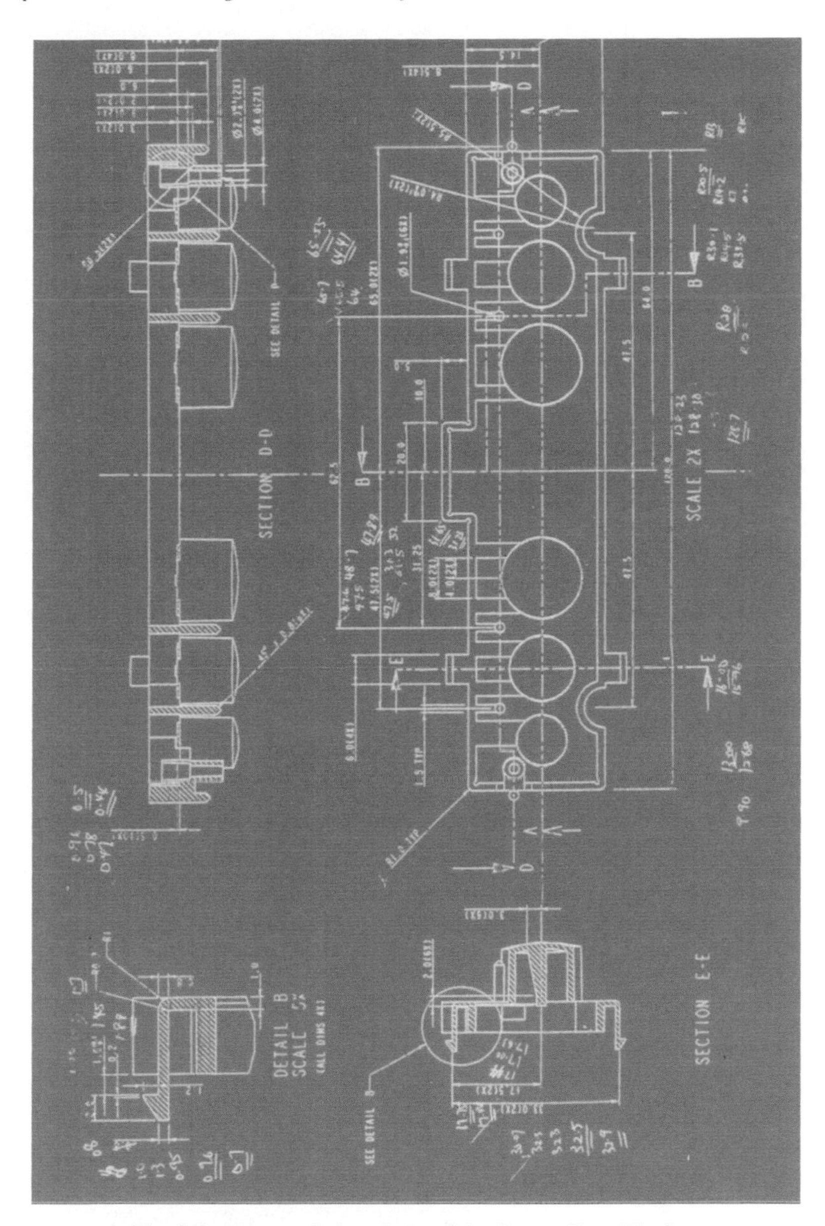

Fig. 9.8. Front and plan views of the Button Tree Display.

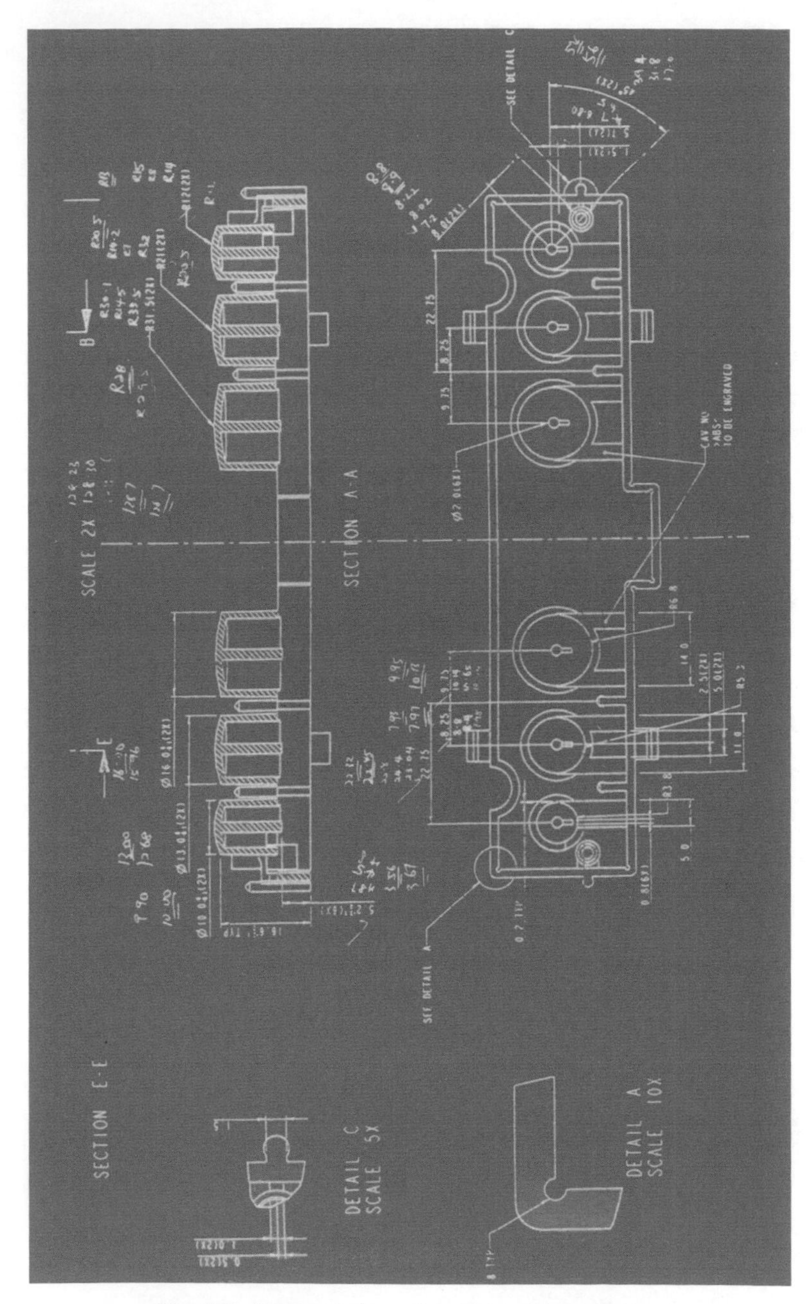

Fig. 9.9. Sectional view of the Button Tree Display.

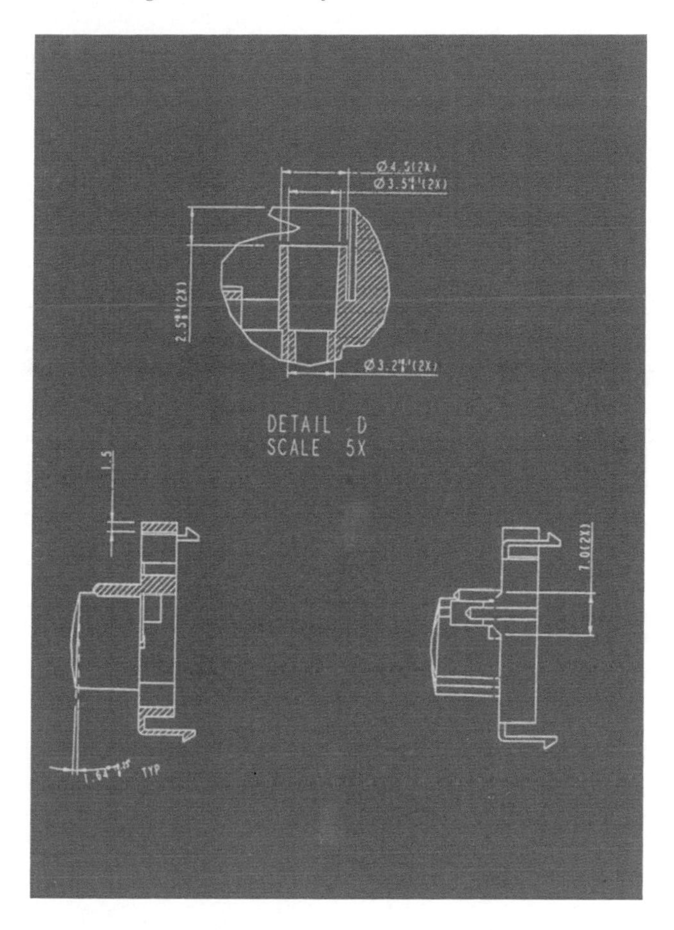

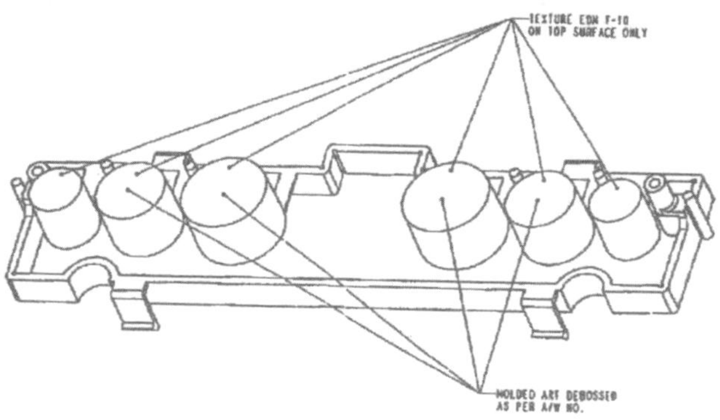

Fig. 9.10. (Top) Sectional and (bottom) isometric views of the Button Tree Display.

deviations for both the detailed and main measurements. From all the measurements taken, greater deviations are observed for the radii of the button set and the angles of the LED holder. These indicate that AM systems have limitations in producing good curve surfaces.

Table 9.2. Main measurements (>10 mm) of the five benchmark test pieces.

Main Measurements						
Drawing Dimensions (mm)		Measured Dimensions (mm) & Deviations (mm) in *italics*				
		SGC	SOUP	LOM	SCS	FDM
Frame, Location Pins and LED Hold	128.0	128.3 *+0.3*	129.1 *+1.1*	128.2 *+0.2*	128.7 *+0.7*	128.7 *+0.7*
	27.0	27.1 *+0.1*	27.3 *+0.3*	27.1 *+0.1*	27.1 *+0.1*	26.8 *-0.2*
	65.0	65.7 *+0.7*	65.5 *+0.5*	64 *-1.0*	64.4 *-0.6*	65.6 *+0.6*
	47.5	47.6 *+0.1*	47.5 *0.0*	48.7 *+1.2*	47.9 *+0.4*	48.7 *+1.2*
	31.25	31.3 *+0.05*	31.5 *+0.25*	32 *+0.75*	31.7 *+0.4*	31.2 *-0.05*
Button-Tree	R31.5	30.1 *-1.4*	14.5 *-17.0*	33.5 *+2.0*	28.0 *-3.5*	29.5 *-2.0*
	R21.0	14.2 *-6.8*	7.0 *-14.0*	32.0 *+11.0*	20.5 *-0.5*	20.5 *-0.5*
	R12.0	15.0 *+3.0*	8.0 *-4.0*	14.0 *+2.0*	13.0 *+1.0*	11.0 *-1.0*
	16.0	16.1 *+0.1*	16.1 *+0.1*	15.9 *-0.1*	16.0 *0.0*	16.0 *+0.0*
	13.0	13.0 *+0.0*	13.1 *+0.1*	12.9 *-0.1*	13.0 *0.0*	12.7 *-0.3*
	22.75	23.0 *+0.25*	23.8 *+1.05*	24.4 *+1.65*	22.45 *-0.3*	22.3 *-0.45*
Catch	33.0	32.3 *-0.7*	33.1 *+0.1*	32.3 *-0.7*	32.5 *-0.5*	32.9 *-0.1*
	17.5	17.4 *-0.1*	17.0 *-0.5*	17.6 *+0.1*	17.8 *+0.3*	17.4 *-0.1*

Table 9.3. Detailed measurements (= 10 mm) of the five benchmark test pieces.

Detailed Measurements						
Drawing Dimensions (mm)		Measured Dimensions (mm) & Deviations (mm) in *italics*				
		SGC	SOUP	LOM	SCS	FDM
Location Pins	8.0	7.2 *-0.8*	8.2 *+0.2*	8.0 *+0.0*	8.1 *+0.1*	8.8 *+0.8*
	5.7	4.7 *-1.0*	6.8 *-1.1*	6.5 *-0.8*	7.6 *-1.9*	7.4 *-1.7*
	45°	39.4 *-5.6*	31.8 *-13.2*	37 *-8.0*	56 *+11.0*	54 *+9.0*
	8.25	8.8 *+0.55*	8.0 *-0.25*	8.4 *+0.15*	8.0 *-0.25*	7.9 *-0.35*
Button-Tree	9.75	10.1 *+0.35*	10.2 *+0.45*	10.7 *+0.95*	10.1 *+0.35*	10.0 *+0.25*
	10.0	10.1 *+0.1*	10.1 *+0.1*	9.9 *-0.1*	10.0 *0.0*	9.9 *-0.1*
	5.2	4.9 *-0.3*	4.8 *-0.4*	4.5 *-0.7*	3.6 *-1.6*	3.7 *-1.5*
	0.5	0.5 *0.0*	1.0 *+0.5*	0.8 *+0.3*	0.5 *0.0*	0.4 *-0.1*
Catch	1.6	1.5 *-0.1*	1.1 *-0.5*	1.4 *-0.2*	1.7 *+0.1*	1.9 *+0.3*
	0.8	1.0 *+0.2*	1.0 *+0.2*	1.3 *+0.5*	1.0 *+0.2*	0.7 *-0.1*

The deviations could be due to the following errors in taking the measurements:

(1) The poor surface finishes of the test pieces result in inaccurate dimensions.
(2) The test pieces are not strong as they became deformed under pressure.
(3) The support of the SOUP test piece was not properly removed.

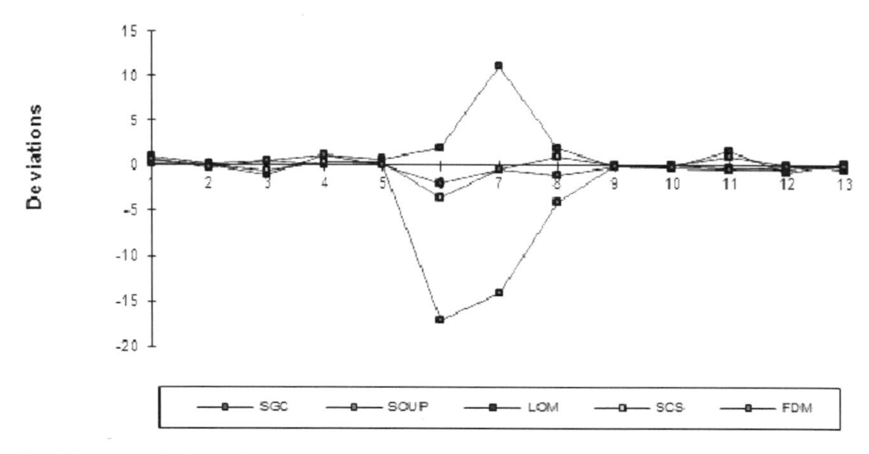

Fig. 9.11. Deviation of main measurements (>10 mm) from the nominal values for the five benchmark test pieces.

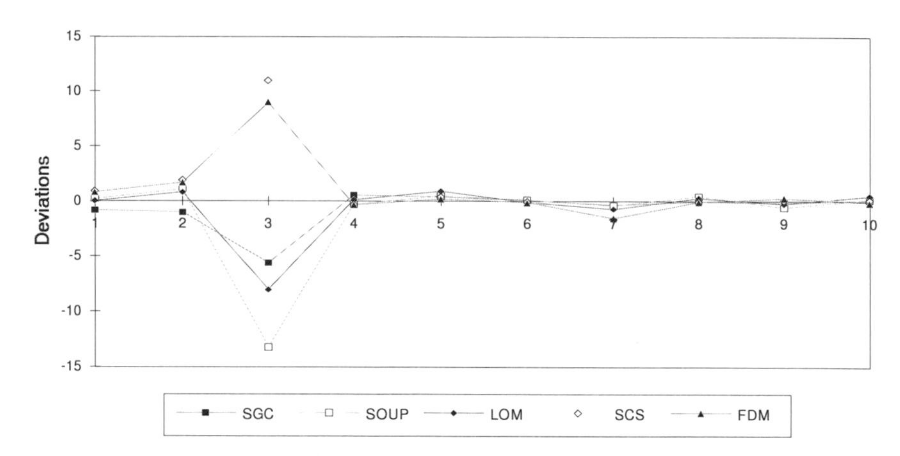

Fig. 9.12. Deviation of detailed measurements (= 10 mm) from the nominal values for the five benchmark test pieces.

9.2.4 *Other Benchmarking Case Studies*

Other than the benchmarking performed by Thomson Multimedia, there are also other benchmarking case studies done by various researchers. Tables 9.4(a)-(e) list a few from the year 2004 to 2013 [3-11].

Table 9.4(a). Benchmarking study published in 2013.

Benchmarking study published in 2013			
1.	M. Fahad and N. Hopkinson	Evaluation of parts produced by a novel additive manufacturing process	Applied Mechanics and Materials, Volume 315 (2013) 63-67
	This benchmark analyses the flatness of parts produced on High Speed Sintering (HSS), one of the AM processes being developed at Loughborough University. The designed benchmark part comprised various features such as cubes, holes, cylinders and cones on a flat base.		

Table 9.4(b). Benchmarking study published in 2012.

Benchmarking study published in 2012			
2.	E. Atzeni, L. Iuliano, P. Minetola and A. Salmi	Proposal of an innovative benchmark for accuracy evaluation of dental crown manufacturing	Computers in Biology and Medicine, Vol. 42, Issue 5, (2012) 548-555
	This innovative benchmark represents a dental arch with features relating to different types of prepared teeth. It includes tooth orientation and oblique surfaces similar to those of real prepared teeth. The evaluation procedure proves that the scan data can be used as a reference model for crown restoration design.		

Table 9.4(c). Benchmarking studies published in 2006.

Benchmarking studies published in 2006			
3.	K. Abdel Ghany and S.F. Moustafa	Comparison between the products of four RPM systems for metals	Rapid Prototyping Journal 12/2 (2006) 86-94
	This work evaluates and compares the quality of four identical benchmarks fabricated from different metallic powders by using four recently developed RPM systems for metals. The evaluation considers benchmark geometry, dimensional precision, material type, product strength and hardness, surface quality, building speed, materials, operation and running cost.		
4.	D. Dimitrov, W. van Wijck, K. Schreve and N. de Beer	Investigating the achievable accuracy of three dimensional printing	Rapid Prototyping Journal 12/1 (2006) 42-52
	This paper deals with current research towards the building of a full capability profile – accuracy, surface roughness, strength, elongation, build time and cost – of this important process.		
5.	M. Mahesh, Y.S. Wong, J.Y.H. Fuh and H.T. Loh	A Six-Sigma approach for Benchmarking of RP&M processes	International Journal of Advanced Manufacturing Technology (2006) 31: 374-387
	This paper presents a methodology of using six-sigma quality tools for the benchmarking of rapid prototyping & manufacturing (RP&M) processes. It involves the fabrication of a geometric benchmark part and a methodology to control and identify the best performance of the process to reduce variability in the fabricated parts.		

Table 9.4(d). Benchmarking studies published in 2005.

<table>
<tr><th colspan="4">Benchmarking studies published in 2005</th></tr>
<tr><td rowspan="2">6.</td><td>Todd Grimm</td><td>3D Printer Dimensional Accuracy Benchmark</td><td>T.A. Grimm & Associates, Inc.</td></tr>
<tr><td colspan="3">This benchmark analyses and quantifies the dimensional accuracy available from the Dimension® SST, InVision™ SR and ZPrinter® 310. This report illustrates the accuracy of each system with reverse engineering colour maps and comparative charts.</td></tr>
<tr><td rowspan="2">7.</td><td>Vito R. Gervasi, Adam Schneider and Joshua Rocholl</td><td>Geometry and procedure for benchmarking SFF and Hybrid fabrication process resolution</td><td>Rapid Prototyping Journal 11/1 (2005) 4-8</td></tr>
<tr><td colspan="3">This paper shares with the solid freeform fabrication community a new procedure and benchmark geometries for evaluating SFF proccss capabilities. The procedure evaluates the range capability of various SFF and SFF-based hybrid processes in producing rod and hole elements.</td></tr>
</table>

Table 9.4(e). Benchmarking studies published in 2004.

<table>
<tr><th colspan="4">Benchmarking studies published in 2004</th></tr>
<tr><td rowspan="2">8.</td><td>M. Mahesh, Y.S. Wong, J.Y.H. Fuh and H.T. Loh</td><td>Benchmarking for comparative evaluation of AM systems and processes</td><td>Rapid Prototyping Journal Vol 10, number 2, 2004 (pp. 123-135)</td></tr>
<tr><td colspan="3">This paper presents issues on AM benchmarking and aims to identify factors affecting the definition, fabrication, measurements and analysis of benchmark parts.</td></tr>
<tr><td rowspan="2">9.</td><td>K.W. Dalgarno and R.D Goodridge</td><td>Compression Testing of layer manufactured metal parts: the RAPTIA compression benchmark</td><td>Rapid Prototyping Journal, Vol 10, number 4, 2004 (pp 261-264)</td></tr>
<tr><td colspan="3">This paper reports the results of a compression test benchmarking study carried out to investigate the mechanical properties of layer manufactured metal components in order to assess their suitability in load-bearing applications. Compression tests were carried out on the DTM LaserForm St-100 material, ARCAM processed H13 tool steel, EOS DirectSteel (50μm), and ProMetal material.</td></tr>
</table>

9.3 Industrial Growth

9.3.1 *Industrial Growth of Additive Manufacturing*

The additive manufacturing industry, formerly known as rapid prototyping, has enjoyed tremendous growth since the first system was introduced in 1988. The rate of growth has also been significant. Right up to 1999, the industry was enjoying two-digit growth rates annually. In 2001, the AM industry continued to expand, although not anywhere near the same rate as before. More systems were installed, more materials for these systems used, and more applications for the technology were uncovered [12]. The rate of growth has tapered off significantly since 2000. The events and economic conditions of 2001 did not help to improve the situation. Then in 2003, the AM industry took a turn and revenues returned to levels of the past [13]. The growth of the AM industry once again leaped upwards in 2004 [14]. There has been no sign of slowing down in the AM industry growth since then.

Observations of four leading AM companies revealed a sales increase of approximately 15% in 2006, compared to 2005 [15]. 3D Systems still remains the leading company in the AMP industry, with a revenue of $135 million in 2006. But due to the fierce price competition it is facing from competitors like Stratasys, its revenue has slipped compared to the preceding year. Stratasys reported approximately $104 million in revenue for the year 2006, which was about a 26% increase from 2005 [16]. With the introduction of the only high-definition colour 3D printer in the market, Z Corporation (acquired by 3D Systems in April 2013) has enjoyed a 50% revenue growth [17]. EOS GmbH (Germany), the leading company for sintering systems, reported approximately $68 million in revenue during the fiscal year ending September 2006 [18].

Currently, the two giants in the AM industry, 3D Systems and Stratasys, are both undergoing rapid growth. According to the financials released by both companies, Stratasys ended the fourth quarter of 2013 with $155.1 million in revenue, which is more than a 60% increase compared to the previous year. The merger with Objet and the acquisition of

MakerBot further boosted Stratasys' market share and profits in 2013. Meanwhile, 3D Systems' fourth quarter revenue in 2013 reached a record of $154.8 million, with a 53% increase in gross profit. [19] The significant milestones of AM from 1998 to 2013 are captured in Table 9.5.

Table 9.5. Significant Milestones of Additive Manufacturing [20].

Time	Event
1988	3D Systems was founded by Charles Hull. The technique was named Stereolithography and a patent was obtained
1988	3D Systems developed SLA-250, the first version of 3D printers made available to the public
1988	Fused Deposition Modelling (FDM) was invented by Scott Crump
1989	Stratasys was founded by Scott Crump
1989	EOS was founded
1991	The Laminated Object Manufacturing (LOM) system was sold by Helisys
1992	The Selective Laser Sintering (SLS) system was sold by DTM to 3D Systems
1993	An inkjet-based machine, which can produce outstanding surface finish with relatively low speed, was founded by Solidscape
1993	3 Dimensional Printing (3DP) techniques were patented by Massachusetts Institute of Technology (MIT)
1995	Z Corporation obtained the license to use 3DP techniques from MIT to develop its own 3D printers
1996	"Genisys", the legacy FDM product, was introduced by Stratasys
1996	The term "3D Printer" was first used to refer to all the additive manufacturing machines
1997	EOS' stereolithography business was sold to 3D Systems
1997	The Electron Beam Melting (EBM) technique was used by Arcam AB to produce solid metal parts
2000	Concept Laser GmbH was founded to optimize the SLS process
2003	Selective Deposition Lamination (SDL) technology was invented by Dr. Conor MacCormack and Fintan MacCormack

Table 9.5. (*Continued*) Significant Milestones of Additive Manufacturing [20].

Time	Event
2005	The first high resolution 3D printer was released by Z Corporation
2008	Connex500™, which manufactures models with several different materials at the same time, was invented by Objet.
2010	The first 3D printed car, Urbee, was produced using Stratasys' 3D printers
2010	First fully bioprinted blood vessels were made by Organovo, Inc., a regenerative medicine company.
2011	3D food printer underwent development at Cornell University
2011	The travel speed of 3D printer was raised to 350 mm/second by Ultimaker, a Dutch 3D printer manufacturer
2011	First 3D printed bikini was manufactured by Shapeways and Continuum Fashion
2011	First 3D chocolate printer was invented by the University of Exeter, the University of Brunel and application developer Delcam in the UK
2011	First 3D printed aircraft was made by researchers at the University of Southampton
2011	RegenHU Ltd, which specialises in making bio-printers and producing 3D organomimetic models for tissue engineering, was founded
2012	ProJet 7000 stereolithography machine was released by 3D Systems
2012	3Z line of high-precision wax 3D printers and 3ZSupport were released by Solidscape
2013	Z Corporation was acquired by 3D Systems

9.4 Future Trends

As the whole AM industry moves forward, AM-driven activities will continue to grow. Compared to several years ago, more significant trends have appeared. Among them are the proliferation of low-cost 3D printers, the rise and fall of metal-based AM companies, AM in rapid manufacturing and mass customisation, and the use of AM in the biomedical engineering field. Continuous improvements in the area of speed and quality, as well as ease of use have also been observed. Some new initiatives in AM have also been explored and studied, such as the application of additive manufacturing to the food, construction, fashion, marine and offshore sectors.

9.4.1 *Low-Cost Office and Desktop 3D Printers*

Office and desktop 3D printers may have resulted from the further development of concept modellers, the breed of AM system (especially the development of FDM technology) geared towards producing prototypes for design reviews instead of physical testing or fully functional parts. This class of AM system is characterised by its higher speed, lower cost, and weaker accuracy and resolution compared to the higher-end class. It has been found that since the concept modeller is designed to operate in design offices, not at workshops, they feature clean and safe operations. These aspects have exceptional market appeal and have now become one of the most important aspects in choosing an AM system.

Over time, the problems with accuracy, surface finish, and material properties of 3D printers have been gradually addressed. Although current 3D printers are still of a different class from higher-end systems, the gap is narrowing. With the quality of 3D printers increasing, the system may capture more and more market share from high-end systems in the future [2]. An example of a high resolution desktop machine is Solidscape's 3Z series printers. They offer a layer thickness of only 0.0254 mm and are even used for creating investment casting patterns.

The growth of office and desktop 3D printers is also accompanied by the lowering of cost to purchase and maintain such systems. While in 2003 such a system may have cost around $50,000, in 2006, the price was already reduced to around $20,000. Examples of machines of this price class are the Stratasys Dimension and 3D Systems' LD 3D Printer.

In 2014, 3D Systems began offering personal 3D Printers such as Cube® 3 and ProJet® 1500, which are much more affordable and start from only $1500 [21]. These machines can produce prototypes with a minimum layer height of 0.1 mm [22]. Meanwhile, Stratasys has also revealed its own desktop 3D printer called Moji, which can produce finely resolved models in nine colours. Its cost is around $9,900 [23].

The increase in the number of office and desktop 3D printers produced and purchased is an indication that the market is enlarging to encompass smaller companies that have neither workshop-class facilities nor the funds to purchase and maintain high-end AM systems. With the reduction in the price of 3D printers, it will become more affordable for educational institutions below university level, such as polytechnics, colleges and schools, to purchase 3D printers.

9.4.2 *The Rise and Fall of Metal-Based AM System Companies*

As its name implies, metal-based AM systems offer output material of metal; as such, they have excellent potential to produce ready-to-use prototypes of metal parts. Also called direct metal technologies [24], these systems possess absolute advantage in delivering prototypes of metal parts with closer material properties with the final product than resin-based or plastic-based AM systems, while keeping the speed above regular CNC machining. Once the technology is perfected, it is not impossible for the systems to become rapid manufacturing systems, replacing traditional metal manufacturing systems.

As if tempted by the potential of metal-based AM systems, at the time of writing, at least eight companies [25] offer metal-based AM systems: 3D Systems, EOS, Concept Laser, MCP, Phenix Systems, Arcam, Optomec and Solidica. The technologies used by these companies mostly revolve around laser sintering and laser melting of a layer of powder. Certain companies like Optomec and Solidica use different techniques. In the case of Solidica, CNC milling has become one of its main processes other than its layered aluminum lamination. The details of the companies and their technologies have been discussed in earlier chapters.

Most of the current metal-based AM systems are not able to produce part quality surface finishes to the standards of CNC milling or investment casting. Nonetheless, the technology is rapidly growing and beginning to achieve the same quality as sand casting [25]. Techniques that introduce a hybrid between layer deposition and CNC milling like Solidica's may solve the surface finish problem, though it adds complexity to the process. A key advantage is that metal-based AM systems can also be used to make tooling parts and casting moulds.

The large number of companies producing metal-based AM systems may be an advantage for the consumers. As they compete with each other, the speed and quality of the machines, as well as the price will become competitive. However, as the survival of the fittest dictates, few will prevail in the end and the rest may disappear along the way. As such, it becomes quite risky for a customer to purchase one of the systems, as the company may cease to operate, resulting in the loss of the company's support and service to the customer.

9.4.3 *Rapid Manufacturing and Tooling*

Rapid manufacturing and tooling (RM&T) has evolved from the application of additive manufacturing. It refers to a process that uses AM technology to produce finished manufactured parts or templates, which are then used in manufacturing processes like moulding or casting, to directly build a limit volume of small prototypes. However, it is highly unlikely that this AM-driven process will ever reach the kind of production capacity of processes such as injection moulding [26]. With the enhancement of AM technologies, an increasing number of companies from aerospace, architecture, motor sports, medical, dental, and consumer products industries are turning to RM&T for customised and short-run production. The prospect of RM&T to grow and become the largest application of additive fabrication is optimistic [26].

Typically, the costs incurred using RM&T applications (around $1,000) are much lower than conventional high-end tooling cost (around $300,000). The tooling time is also notably shorter than conventional tooling time. This is mainly due to the fact that additive manufacturing processes are able to fabricate models directly and quickly without the need to go through various stages of the conventional manufacturing process (i.e. postprocessing). Moreover, it is more economical to build small intricate parts in low quantities using RM&T as there is compelling evidence that rapid manufacturing may be less expensive than traditional manufacturing approaches [27].

In addition, RM&T has the ability to make changes to the design at minimal cost even after the product has reached the production stage. Such a feature is highly desirable as it is usually very expensive to make modifications to the moulds (if the moulding process is the fabrication for the finished product) and the time needed to modify the mould is also likely to be lengthy. Thus RM&T is much preferred over the conventional tooling method in the early stages of the manufacturing cycle as modifications can be done almost instantaneously by simply editing the STL files, and the cost incurred is relatively low.

However, the downside to additive manufacturing processes is its lack of an appropriate range of materials for this application. Not all materials available are suitable for all products but nevertheless, current materials such as sintered nylon, epoxy resin and composite materials are often good enough for most low volume applications. Moreover, since RM&T has been successfully implemented in the aerospace industry, which imposes stringent quality demand, it is entirely possible that such applications can also be effectively implemented in other industries.

9.4.4 *AM in Mass Customisation*

Tightly related to the concept of rapid manufacturing, additive manufacturing technology may be crucial in turning the mass customisation of products into reality [28]. Mass customisation has been growing stronger and stronger in recent years, and it has been regarded as the future paradigm of manufacturing [29]. Multiple cases that indicate the shifting of market trends towards mass customisation have been recorded. One well-known case is Dell's mass customisation of personal computers [30]. Backed by the market trend and its distinct capability to customise, this will put AM in an even stronger position in the future.

The concept of mass customisation with AM does not stop at just high-end products. At the time of writing, an innovative business concept in the toy industry is being developed. Before Z Corp was acquired by 3D Systems, Z Corp and SolidWorks have been working together to bring the usage of AM to produce customised parts to kids. The project is named Cosmic Modelz, where kids can use SolidWorks' Cosmic Blobs software to create their own custom figures and toys. The data can then be sent to Z Corp to be produced for a price as low as $25 to $50 per model [31]. With the constant decreasing cost of AM, it is not impossible to see more and more of this kind of services in the future operated by smaller companies or even by individuals.

9.4.5 *Customised 3D Printing and DIY*

Over the last few years, the consumer and hobbyist 3D printing space has grown significantly. Competition and advancement of technology have seen prices for printers become greatly reduced, making it more affordable for end users to have 3D printers at home. Makerbot is the brand that has been trying to target the consumer market, and they are not alone. Many companies and new start-ups are trying to be the brand that brings consumer 3D printing into the mainstream, like what Apple did for home computing.

While some sceptics argue that AM technology is not efficient enough for mass production, many concede that AM is best suited for custom, one-of-a-kind products and the manufacturing of goods that have low but highly sporadic demand. In this case, customisation would be highly beneficial and extremely valuable to the unique characteristics of a customer. It would be very advantageous for a patient who needs a hip replacement or a growing child with a prosthetic hand. A father in the USA spent years searching for affordable ways to give his son, who was born without fingers, a working prosthetic hand. With the help of the Internet, he found a YouTube video of another user sharing his blueprints on how to create a prosthetic hand using AM [32]. An industrial-made prosthetic hand would have cost tens of thousands of dollars, but his own DIY 3D printed prosthetic hand merely cost him ten dollars for the materials.

As more companies make it their corporate mission to make 3D printing more accessible to the consumer market, more and more people will be able to create and print customised items from the comfort of their own homes.

9.4.6 *Biomedical Engineering Application*

When AM technologies were first introduced, they were mainly used in the automotive industry, where verification and evaluation of new design ideas are required before marketing new products. Today, the use of AM technologies has expanded beyond the automotive industry into a wide variety of industries and fields.

AM technologies have been used in a broad spectrum of applications in the field of biomedical engineering. A variety of AM systems has been used in the production of scale replicas of human bones and body organs to advance customised drug delivery devices and other areas of medical sciences, including anthropology, palaeontology and medical forensics [33].

Fused deposition modelling (FDM), selective laser sintering (SLS), ColorJet Printing (CJP), formerly known as Three-Dimensional Printing technology (3DP), and stereolithography (SLA) are the most common systems employed in fabrication of tissue engineering scaffolds.

Intensive studies on AM techniques related to biomedical applications have been conducted in recent years. Studies in various disciplines are done to evaluate AM techniques, compare AM techniques with conventional techniques, and improve the scaffold structures fabricated with various AM techniques, to customise implants and to explore new techniques/ materials [33-42].

Although some success in studies for tissue engineering using AM technologies are observed, this technique has yet to deliver significant progress in the clinical use. The utilisation of AM technologies will grow as more low-cost systems with improved features are introduced to the market, thus encouraging the wider use of AM technologies.

9.4.7 *Increased Use of AM in Construction*

While more studies were carried out, AM has been widely used by architects and designers in construction. In March 2013, a Canal House was designed and built by architects from a Dutch firm called DAS [43]. This shows the capability of additive manufacturing to produce prototypes on a large scale.

Furthermore, China has also produced buildings via AM technology in Shanghai in 2014. The buildings, made using 3D printers and developed by Suzhou Yingchuang Science and Trade Development Co., Ltd, look similar to buildings built using conventional methods. Even though each layer is around 3 centimetres tall, its hardness has been proven to be five times that of common construction materials [44].

However, the use of AM in construction is still under further study and development, especially in terms of the accuracy and materials used. The fire resistance and durability of AM-based buildings need to be further analysed before mass production can even be considered.

9.4.8 *Others*

Besides the seven prominent development trends discussed, three other trends have also been observed in the development of AM.

9.4.8.1 *Ease of use*

One of the development trends that has been going on continuously since the first AM system is to make additive manufacturing systems easier to use than before. All AM systems are supplied with user-friendly software that provides good visual feedback as well as useful automatic or semi-automatic functions to use with the 3D model and the machine.

Attempting to make postprocessing easier to carry out or to eliminate it altogether is another notable development trend. As systems are able to produce parts with high dimensional accuracy and better surface finish, the need for elaborate postprocessing is also diminished. Companies introduced various technologies to simplify, if not eliminate, postprocessing. Stratasys introduced water-soluble supports even on their low-cost Dimension system line. With SST (Soluble Support Technology), users simply immerse the prototype into a water-based solution and wash away the supports.

9.4.8.2 *Improvement of building speed and prototype quality*

One other development trend is the improvement of building speed and the prototype quality of additive manufacturing systems. As with the ease of use, this development has been continuously observed. Newer systems introduced are rapidly replacing older systems that are slow in building time and weak in accuracy. Some years ago, Stratasys introduced the FDM Titan upgrade, which offered an average of 54% increase in building speed [45]. In 2007, Stratasys launched FDM 200mc to replace the predecessor system, Prodigy Plus. When used with ABSplus material, FDM 200mc systems can fabricate parts achieving up to 67% stronger mechanical properties. ABSplus material is engineered to work optimally with FDM200MC [46]. Kira produced Katana, which has demonstrably built a camera mockup in just an hour [47]. Solidscape made huge improvements to its T76 Benchtop printer, which has three times the speed and twice the material strength of the previous model [48].

9.4.8.3 *New research areas and fields of AM*

Researchers and scientists are always keen in finding new areas that AM can step in to fill the gap. In 2014, researchers from the Federal Polytechnic School of Lausanne, Switzerland, the Imperial College, UK and the University of Waterloo, Canada started to carry out a study on the field of printed electronics and intelligent 3D printing. This research aims to produce intelligent multi-function components using AM [49].

In addition, experts in AM from Dshape, a pioneer in 3D printing reef units made out of non-toxic patented sandstone material, have also been utilizing AM in their marine and offshore study. AM enables them to make more intricate designs, which are similar to natural coral structures, and this has helped them to build an artificial reef area under the sea to protect the marine environment [50].

Meanwhile, to encourage the wider use of AM technologies, researchers have also looked into the use of AM in industries such as outer space, which NASA is currently doing, clothing, fashion, movies, food, weapons and many more. In the future, AM will form the backbone of the world's economy and it will be an integral part of industrial automation.

References

[1] *Wohlers Report 2013: Additive Manufacturing and 3D Printing State of the Industry Annual Worldwide Progress Report*. (2013). Wohlers Association, Inc.

[2] Wall, M. B. (1991). *Making sense of prototyping technologies for product design.* MSc Thesis, MIT, USA.

[3] Fahad, M., & Hopkinson, N. (2013). *Evaluation of parts produced by a novel additive manufacturing process.* Applied Mechanics and Materials, Vol. 315: pp. 63-67.

[4] Atzeni, E., Iuliano, L., Minetola, P., & Salmi, A. (2012). *Proposal of an innovative benchmark for accuracy evaluation of dental crown manufacturing.* Computers in Biology and Medicine, Vol. 42, Issue 5: pp. 548-555.

[5] Abdel Ghany, K. & Moustafa, S. F. (2006). *Comparison between the products of four RPM systems for metals.* Rapid Prototyping Journal, **12**(2): 86-94.

[6] Dimitrov, D., van Wijck, W., Schreve, K., & de Beer, N. (2006). *Investigating the achievable accuracy of three dimensional printing.* Rapid Prototyping Journal, **12**(1): 42-52.

[7] Mahesh, M., Wong, Y. S., Fuh, J. Y. H., & Loh, H. T. (2006). *A Six-Sigma approach for Benchmarking of RP&M processes.* International Journal of Advanced Manufacturing Technology, **31**(3-4): 374-387.

[8] Grimm, T. A. (2005). *3D Printer Dimensional Accuracy Benchmark*, T.A. Grimm & Associates, Inc.

[9] Gervasi, V. R., Schneider, A., & Rocholl, J. (2005). *Geometry and Procedure for benchmarking SFF and Hybrid fabrication process resolution*. Rapid Prototyping Journal, **11**(1): 4-8.

[10] Mahesh, M., Wong, Y. S., Fuh, J. Y. H., & Loh, H. T. (2004). *Benchmarking for comparative evaluation of RP systems and processes*. Rapid Prototyping Journal, **10**(2): 123-135.

[11] Dalgano, K. W., & Goodridge, R. D. (2004). *Compression Testing of layer manufactured metal parts: the RAPTIA compression benchmark*. Rapid Prototyping Journal, **10**(4): 261-264.

[12] Wohlers, T. (20000. *Wohlers Report 2000: Rapid Prototyping & Tooling State of the Industry*. Wohlers Association, Inc.

[13] Wohlers Associates. (2004). *Industry Briefing*. Retrieved from http://www.wohlersassociates.com/brief05-04.htm

[14] Wohlers, T. (2005). *Wohlers Report 2005: RP, RT, and RM State of the Industry, Time-Compression Technologies*. Retrieved from http://www. timecompress.com/ magazine/magazine_articles.cfm?article_id=344&issue_id=83&articles=344

[15] Castle Island Co. (2007). *The Rapid Prototyping Industry – Overview*. Retrieved from http://home.att.net/~castleisland/ind_10.htm

[16] Castle Island Co. (2007). *The Rapid Prototyping Industry - Major US-Based Vendors*. Retrieved from http://home.att.net/~castleisland/ind_21.htm

[17] Z Corporation. (2006). *Z Corporation Announces 50-Percent Revenue Growth in 2005*. Retrieved from http://www.zcorp.com/news/newsdetail.asp?ID=412& TYPE=1

[18] Castle Island Co. (2007). *The Rapid Prototyping Industry - Vendors Outside the US*. Retrieved from http://home.att.net/~castleisland/ind_22.htm

[19] *Stratasys and 3D Systems Release 2013 Financials & No Surprise, Both had a Pretty Good Year*. Retrieved from http://3dprintingindustry.com/2014/ 03/06/stratasys-3d-systems-release-2013-financials-surprise-pretty-good-year/

[20] *The History of 3D Printing*. Retrieved from http://www.3ders.org/3d-printing/3d-printing-history.html

[21] 3D Systems. *3D Systems' Personal 3D printers*. Retrieved from http://www.3dsystems.com/3d-printers/personal/overview

[22] 3D Systems. *ProJet® 1000 & 1500 Personal 3D Printers*. Retrieved from http://www.3dsystems.com/sites/www.3dsystems.com/files/projet-1000-1500-us.pdf

[23] *Price of Stratasys Moji 3D printer*. Retrieved from http://desktop-3d-printers.findthebest.com/q/28/12373/How-much-is-the-Stratasys-Mojo-3D-printer

[24] Grimm, T. (2005). *Direct Metal Technologies Tackle the Impossible, Time-Compression Technologies*. Retrieved from http://www.timecompress. com/magazine/magazine_articles.cfm?article_id=323&issue_id=78&articles=323

[25] Wohlers, T. (2003). *An Explosion of Metal-based RP Systems.* Wohlers Associates Inc. Retrieved from http://wohlersassociates.com/blog/2003/12/an-explosion-of-metal-based-rp-systems/

[26] Wohlers, T. (2001). *Wohlers Report 2001: Rapid Prototyping & Tooling State of the Industry.* Wohlers Association, Inc.

[27] Wuensche, R. (2000). *Using Direct Croning to Decrease the Cost of Small Batch Production. Engine Technology International 2000,* Annual Showcase Review for ACTech GmbH.

[28] Wohlers, T. & Grimm, T. (2006). *Rapid Prototyping in 2006, Time-Compression Technologies.* Retrieved from http://www.timecompress.com/magazine/magazine_articles.cfm?article_id=34&issue_id=45&articles=34

[29] Butcher, D. R. (2006). *Mass Customization: A Leading Paradigm In Future Manufacturing.* ThomasNet® IndustrialNewsRoom. Retrieved from http://news.thomasnet.com/IMT/archives/2006/02/innovative_manu.html

[30] Fern, E. J. MS, PMP (Time-to-Profit, Inc.). (2002). *Six Steps To The Future: How Mass Customization Is Changing Our World.* American Society for the Advancement of Project Management. Retrieved from http://www.asapm.org/asapmag/articles/SixSteps.pdf

[31] Wohlers, T. (2006). *Cosmic Modelz.* Wohlers Association, Inc. Retrieved from http://wohlersassociates.com/blog/ 2006/08/cosmic-modelz/

[32] Huffington Post. (Nov 2013). *Dad Uses 3D Printer to make his son a prosthetic hand.* Retrieved from http://www.huffingtonpost.com/2013/11/04/dad-prints-prosthetic-hand-leon-mccarthy_n_4214217.html

[33] Leong, K. F., Cheah, C. M., & Chua, C. K. (2003). *Solid Freeform fabrication of three-dimensional scaffolds for engineering replacement tissues and organs.* Biomaterials, **24**: 2363-2378.

[34] Yeong, W. Y., Chua, C. K., Leong, K. F., Chandrasekaran, M., & Lee, M. W. (2007). Comparison of Drying Methods in the Fabrication of Collagen Scaffold Via Indirect Rapid Prototyping. In *Journal of Biomedical Material Research Part B: Applied Biomaterials,* **82B**(1): 260-266.

[35] Yan, Y. N., Wu, R., Zhang, R., Xiong, Z., & Lin, F. (2003). *Biomaterial forming research using RP technology.* Rapid Prototyping Journal, **9**(3): 142-149.

[36] Wagner, M., Kiapur, N., Wiedmann-Al-Ahmad, M., Hübner, U., Al-Ahmad, A., Schön, R., Schmelzeisen, R., Mülhaupt, R., & Gellrich, N. C. (2007). Comparative in vitro study of the cell proliferation of ovine and human osteoblast-like cells on conventionally and rapid prototyping produced scaffolds tailored for application as potential bone replacement material. In *Journal of Biomedical Materials Research Part A,* **83A**(4): 1154 – 1164.

[37] Jiankang, H., Dichen, L., Bingheng, L., Zhen, W., & Tao, Z. (2006). *Custom fabrication of a composite hemi-knee joint based on rapid prototyping.* Rapid Prototyping Journal, **12**(4): 198-205.

[38] Singare, S., Dichen, L., Bingheng, L., Zhenyu, G., & Yaxiong, L. (2005). *Customized design and manufacturing of chin implant based on rapid prototyping*. Rapid Prototyping Journal, **11**(2): 113-118.

[39] Li, X., Li, D. C., Lu, B., Tang, Y., Wang, L., & Wang, Z. (2005). *Design and fabrication of CAP scaffolds by indirect solid free form fabrication*. Rapid Prototyping Journal, **11**(5): 312-318.

[40] Tan, L. L. (2007). *FEATURE: Plugging Bone the Painless Way*. Innovation magazine. Retrieved from http://www.innovationmagazine.com/ innovation/ volumes/v4n3/features1.shtml

[41] Hutmacher, D. W., Schantz, J. T., Lam, C. X. F., Tan, K. C., & Lim, T. C. (2007). State of the art and future directions of scaffold-based bone engineering from a biomaterials perspective. In *Journal of Tissue Engineering and Regenerative Medicine*, **1**(4): 245 – 260.

[42] Ringeisen, B. R., Othon, C. M., Barron, J. A., Young, D., & Spargo, B. J. (2006). *Jet-based methods to print living cells*. Journal of Biotechnology, **1**: 930–948.

[43] *The World's "First" 3D-Printed House Begins Construction*. (2014). Retrieved from http://3dprintingindustry.com/2014/01/22/worlds-first-3d-printed-house-begins-construction/

[44] *China's first buildings made with 3D printing installed Shanghai*. (2014). Retrieved from http://khon2.com/2014/04/18/chinas-first-buildings-made-with-3d-printing-installed-shanghai/

[45] *Stratasys Rapid Prototyping Upgrades Increase Speed and Capacity; FDM Titan gets 54% Speed-up; FDM Vantage gets 150% Build Volume Increase*. (2003). Retrieved from http://www.theautochannel.com/news/2003/11/ 26/173492.html

[46] Stratasys Inc. (2007). *Stratasys Introduces FDM 200mc Rapid Prototyping and Manufacturing System*. Retrieved from http://intl.stratasys.com/ media. aspx?id=873

[47] Kira Corporation. (2007). *What is RapidMockup System?* Retrieved from http://www.rapidmockup.com/eg/menu1_5_e.htm

[48] Solidscape®, Inc. (2004). *Solidscape®, Inc. Introduces World's First Benchtop 3D Printer...including 3X speed increase and 200% stronger material*. Retrieved from http://www.solid-scape.com/t6x_benchtop_release.html

[49] *MGI's Ceradrop Announces Three New Orders for Printed Electronics and Intelligent 3D Printing Applications*. Retrieved from https://3dprintingstocks.com /wp-content/uploads/2014/04/MGI-PR-Three-orders-for-MGI-CERADROP.pdf

[50] *Underwater City: 3D Printed Reef Restores Bahrain's Marine Life*. Retrieved from http://blogs.ptc.com/2013/08/01/underwater-city-3d-printed-reef-restores-bahrains-marine-life/#sthash.3AMs8zgm.dpuf

Problems

1. What are the considerations when choosing a service bureau?
2. What are the components that make up the total cost of a part built by a service bureau?
3. Is there a correlation between the type of AM system and the industry in which the prototypes are used? Why?
4. Name the reasons for benchmarking.
5. Describe in detail the AM benchmarking methodology.
6. How have the primary and secondary markets for AM performed over the years?
7. In terms of system sales, would concept modellers outstrip industrial grade AM systems? Why?
8. What are the likely trends in AM technology?
9. Describe some significant AM milestones in the last 25 years. In your opinion, why are they significant?

Appendix

LIST OF AM COMPANIES

3D-Micromac AG
Rosenbergstraße, 09126
Chemnitz
Germany
Tel: +49 (0) 371-400 43-0
Fax: +49 (0) 371-400 43-40
Email: info@3d-micromac.com
URL: www.3d-micromac.com

3D Systems Inc.
333 Three D Systems Circle
Rock Hill, SC 29730, USA
Tel: +1 (803) 326-3900
URL: www.3dsystems.com

Arcam AB (publ.)
Krokslätts Fabriker 27
SE-431 37 Mölndal
Sweden
Tel: +46-(0)31-710 32 00
Fax: +46-(0)31-710 32 01
Email: info@arcam.com
URL: www.arcam.com

CAM_LEM Inc.
1768 E. 25th St.
Cleveland, OH 44114, USA
Tel: 216-391-7750
Fax: 216-579-9225
Email: sales@camlem.com
URL: www.camlem.com

CMET Inc.
Sumitomo Fudosan Shin-Yokohama Bldg,
2-5-5, Shin-Yokohama
Kouhoko, Yokohama,
Kanagawa 222-0033, Japan
Tel: +81-45-478-5561
Fax: +81-45-478-5569
URL: www.cmet.co.jp

CONCEPT Laser GmbH
An der Zeil 8
96215 Lichtenfels
Germany
Tel: +49 (0) 9571 / 949-228
Fax: +49 (0) 9571 / 949-239
Email: info@concept-laser.de
URL: www.concept-laser.de

Cubic Technologies
2785 Pacific Coast Highway #295
Torrance, CA 90505, USA
Tel: (310) 619-9541
Email: info@cubictechnologies.com
URL: www.cubictechnologies.com

D-MEC Ltd.
Shiodome Sumitomo Building 1-9-2
Higashi-Shinbashi, Minato-ku
Tokyo, 105-0021, Japan
Tel: +81-3-6218-3582
Fax: +81-3-6218-3690
Email: tokyo@d-mec.co.jp
URL: www.d-mec.co.jp

The Ennex™ Companies
Los Angeles, California
Tel: (805) 451-4507
Email: contact@ennex.com
URL: www.ennex.com/

EnvisionTEC GmbH
Brüsseler Straβe 51
45968 Gladbeck
Germany
Tel: +49 2043 9875-0
URL: www.envisiontec.de

EOS GmbH
Electro Optical Systems
Robert-Stirling-Ring 1
D-82152 Krailling
Germany
Tel.: +49 89 893 36 - 0
Fax: +49 89 893 36 – 285
Email: info@eos.info
URL: www.eos.info

ExOne GmbH
Am Mittleren Moos 41
86167 Augsburg, Germany
Tel: +49 (0) 821 7476 0

Email: info@exone.com
URL: www.exone.com

Fraunhofer-Gesellschaft zur Förderung der angewandten Forschung e.V.
Postfach 20 07 33
80007 Munich, Germany
Tel: +49 89 1205-0
Fax: +49 89 1205-7531
URL: www.fraunhofer.de/

Kira Corporation
39-1 Nakagawanami
Tomiyoshi-Shinden Kira-cho, Nishio-city
Aichi Pref.444-0592, Japan
Tel: +81-563-32-0100
Fax: +81-563-32-3241
URL: www.kiracorp.co.jp

Mcor Technologies
Unit 1, IDA Business Park
Ardee Road, Dunleer,
Co. Louth, Ireland
Tel: +353 41 6862800
Tel (USA): +1 770 619 9972
Email: info@mcortechnologies.com
URL: www.mcortechnologies.com

MicroFabrica Inc.
7911 Haskell Ave.
Van Nuys, CA 91406, USA
Tel: 888-964-2763
Fax: 818-997-3322
Email: sales@microfabrica.com
URL: www.microfabrica.com

Optomec Inc.
3911 Singer N.E.
Albuquerque, NM 87109
Tel: (505) 761-8250
URL: www.optomec.com

Phenix Systems under 3D Systems
Parc Européen d'Entreprises, Rue Richard Wagner
63200 Riom, France
Tel: 33 (0) 4 73 33 45 85
Fax: 33 (0) 4 73 33 45 86
URL: www.phenix-systems.com

regenHU Ltd.
Z.i. du Vivier 22
1690 Villaz-St-Pierre
Switzerland
Tel: +41 26 653 72 20
Fax: +41 26 653 72 21
URL: www.regenhu.com

SLM Solutions GmbH
Roggenhorster Strasse 9c
D-23556 Lübeck
Germany
Tel: +49-451-16082-0
Fax: +49-451-16082-250
Email: info@slm-solutions.com
URL: www.stage.slm-solutions.com

Solidscape, Inc., a subsidiary of Stratasys Ltd
URL: www.solid-scape.com

Stratasys, Inc.
7665 Commerce Way

Eden Prairie, Minnesota
55344-2080 U.S.A.
Tel: + 1 800 937 3010
Fax: +1 952 937 0070
Email: info@stratasys.com
URL: www.stratasys.com

voxeljet AG
Paul-Lenz-Str. 1a
86316 Friedberg, Deutschland
Email: info@voxeljet.de
URL: www.voxeljet.de

COMPANION MEDIA PACK ATTACHMENT

Multimedia is a very effective tool in enhancing the learning experience. At the very least, it enables self-paced, self-controlled interactive learning using the media of visual graphics, sound, animation and text. To better introduce and illustrate the subject of *Additive Manufacturing* (AM), an executable multimedia program was created for this book. It serves as an important supplement learning tool to understand better the principles and processes of AM. The program can be retrieved from this book's Companion Media Pack. To download the Companion Media Pack, please use the following access token activation URL: http://www.worldscientific.com/r/9008-SUPP. You will be prompted to login/register an account. Upon successful login, you will be redirected to the book's page; click on the 'Supplementary' tab to locate the Companion Media Pack (see Figure I.)

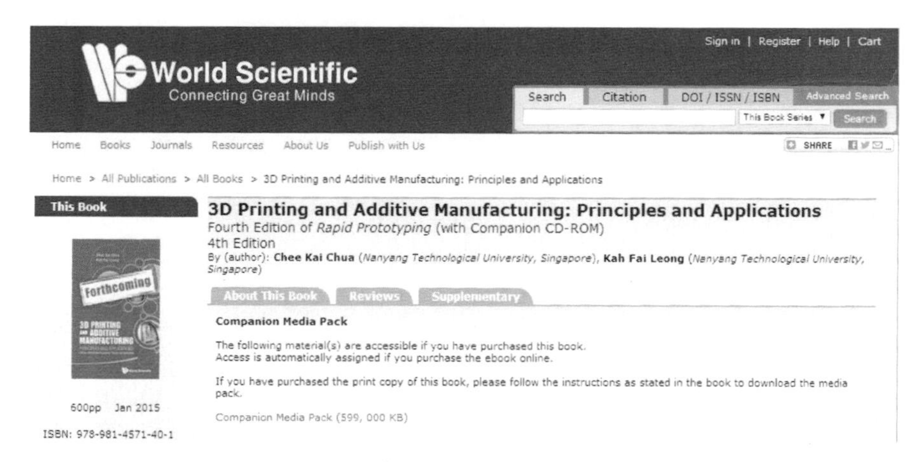

Fig. I. Screenshot of World Scientific's website to download the companion media pack. Site address: www.worldscientific.com/worldscibooks/10.1142/9008.

More than twenty different commercial AM systems are described in Chapters 3-5. However, only the six most matured techniques and videos on their processes are demonstrated in the multimedia pack. These six techniques and the length of their videos are listed below in Table I:

Table I: The six AM techniques and their corresponding video lengths.

Technique	**Movie Length/min**
Stereolithography Apparatus (SLA)	1:52
Polyjet	2:59
Selective Deposition Lamination (SDL)	1:14
Fused Deposition Modelling (FDM)	4:47
Selective Laser Sintering (SLS)	2:01
ColorJet Printing (CJP)	1:58

The working mechanisms of these six methods are interestingly different from one another. In addition, the multimedia pack also includes a basic introduction, the AM process chain, AM data formats, applications and benchmarking. While the book describes the principles and illustrates with diagrams the working principle, the companion multimedia pack goes a step further and shows the working mechanism in motion. Animation techniques through the use of Macromedia Flash enhance understanding through graphical illustration. The integration of various media (e.g. graphics, sound, animation and text) is done using Macromedia's Director.

Additional information on each of the techniques, such as product information, application areas and key advantages, makes the multimedia pack a complete computer-aided self-learning software about AM systems. Together with the book, the multimedia will also provide a directly useful aid to lecturers and trainers in the teaching of the subject on Additive Manufacturing.

COMPANION MEDIA PACK USER GUIDE

This user guide will provide the information needed to use the program smoothly.

System Requirements

The system requirements define the minimum configurations that your system needs in order to run the companion media pack smoothly. In some cases, not meeting this minimum requirement may also allow the program to run. However, it is recommended that the minimum requirements be met in order to fully benefit from this multimedia course package.

The recommended system requirements are:

- Intel Pentium PC 1.5 GHz or 100% compatible with 512MB of RAM
- Intel Integrated 3D Graphics card
- Windows XP or higher
- 4 GB of hard disk free space (if program is installed into the hard disk)
- Sound card and speakers

The preferred system requirements are:

- Intel Pentium PC 1.8 GHz or 100% compatible with 1GB of RAM

Installing AM Companion Media Pack

The program can be retrieved from this book's Companion Media Pack, downloadable from World Scientific's website under the Supplementary tab from http://www.worldscientific.com/worldscibooks/10.1142/9008.

Open the program by first unzipping the downloaded file, and then double-clicking the executable file start.exe to activate the courseware.

Getting Around the Program

The AM multimedia pack is a courseware developed for students, lecturers and anyone who is interested in learning more about AM. The user interface has been designed to be very simple, with a short animation introduction in order for first time users to understand how to move around the chapters and sub-topics in the program.

The next section will guide you through the various screens in the courseware and explain the functions of the icons on the screens in detail.

Main Introduction Screen

After the program has been started, a short introduction movie will be played. To skip this introduction movie, simply click on the "Skip" button. After playing the movie, you will be presented with the Main Introduction screen (see Figure I). When you click on the "Main Area" button shown in Figure I, it will direct you to the Main Menu of the courseware (see Figure II) where you can learn more of the key contents of the AM techniques, processes and the applications by clicking on any of the eight individual chapters desired.

Fig. I: Main Introduction page.

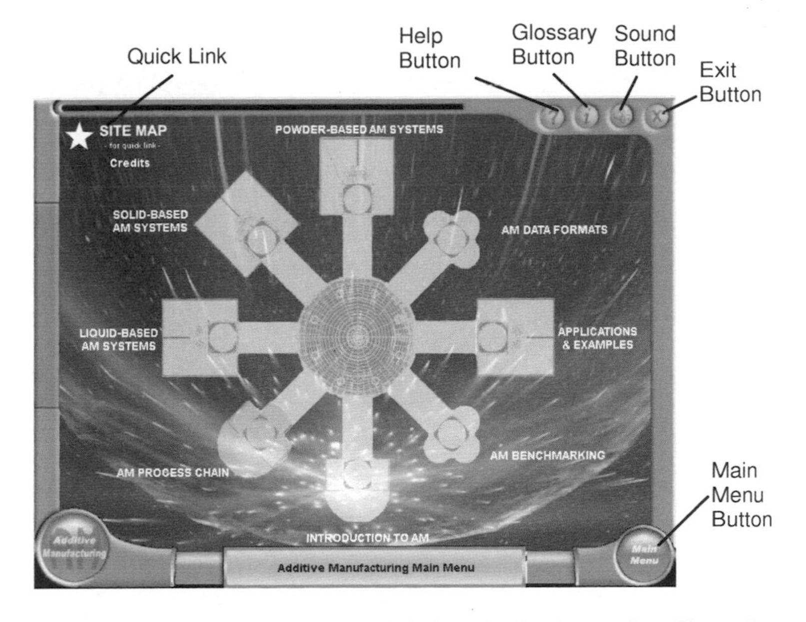

Fig. II. Main Menu (after you have clicked on the “Main Area” on Figure I).

Choosing a Chapter

To learn more about a particular chapter, move the cursor to the word over the chapter in the homepage or in the Main Menu. The words will light up to indicate that the particular chapter can be selected. Click once on the left key on the mouse when the word is lighted up to go into that chapter.

The eight chapters are: Introduction to AM; AM Process Chain; Liquid-based AM Systems; Solid-based AM Systems; Powder-based AM Systems; AM Data Formats; Applications and Examples; and Benchmarking and The Future of AM.

If you want to go to a specific sub-topic of any of the chapters, you can click on the quick link identified as "SITE MAP", which is located on the top left hand corner of the page. You will see the screen shown in Figure III once you click on the words. By moving the cursor over the title of each chapter, the sub-topics of the individual chapter available for selection will be shown on the right hand side of the screen. If you click on any of them, you will be guided directly into the page of the requested sub-topic.

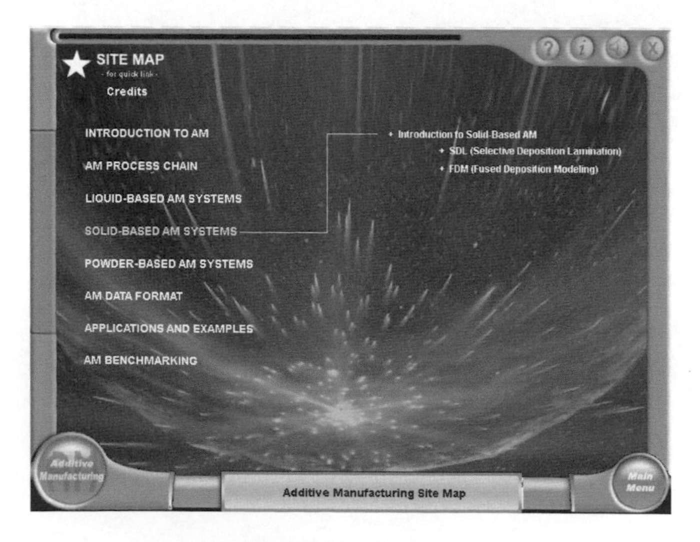

Fig. III. SITE MAP screenshot.

Graphical User Interface (GUI) Layout Design

The GUI Layout Design (the grey border in Figures I and II) will be the same throughout the whole courseware. More advanced interactive features of this program will be covered in a later section, "Exploring the Interactive Features".

Components of the GUI Layout Design

The GUI Layout Design comprises of the main text area and various icons, as shown in the screen shot (Figures I and II) above. The main text area is where the text information of a particular chapter and sub-topic will be displayed. It is also where the user will do most of the learning.

The icons are explained as follows:

(1) *Help button.* Click on the help button to assist you in navigating the specific chapter.
(2) *Glossary button.* Click on the glossary button icon, and definitions of some terms used in the content of the chapter will be provided for your understanding.
(3) *Sound button.* This is a toggle button. Click on the sound button icon to toggle the music on or off. If the music is playing, clicking on the icon will make the music stop playing. Click on the icon again and the music will start playing again. Note that the Sound button is not active when playing any of the movies.
(4) *Exit button.* Click on the exit button to end this program and return to Windows XP. When the exit button is clicked, a confirmation screen will appear. To confirm and exit the program, click on the "Yes" button. To cancel the exit program request, click on the "No" button.

Exploring the Interactive Features

The AM multimedia courseware is an interactive, fun and lively way to learn about AM and its workings. There are many interactive features that can be found in the program. Interactive features such as animations, graphics, tables and movies are scattered around the courseware to help enhance the learning of those topics and concepts. Do enjoy and have fun exploring!

Each chapter is designed with its own theme in order to provide you with a fresh look after you finish a chapter. The section of the chapter will highlight some of the interactive features (see Figure IV) that can be found within the courseware and how to activate them.

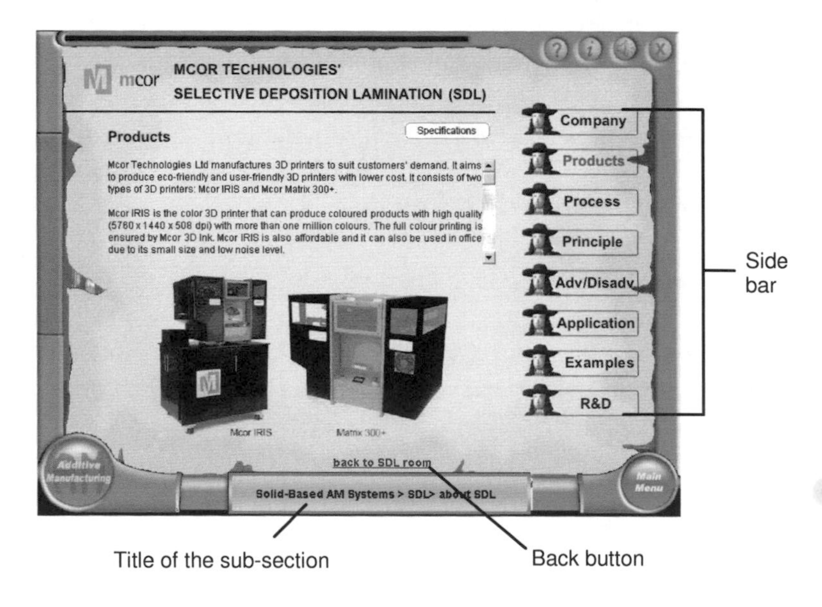

Fig. IV. Screenshot of the chapter on SDL.

Mouse Over a Hot Spot

There are many "hot spots" within the main text area in the learning screen. By moving the cursor over such "hot spots", hidden text, graphics or animations will appear (or move) on the screen.

For example, when the mouse cursor moves over "Data Preparation", the text explaining data preparation appears. There are many different "hot spots" in the entire courseware, so do explore and discover the hidden "hot spots" to help in the understanding of AM concepts in a fun and quick manner.

Buttons

In each of the six AM techniques presented (two for each of the Liquid-based, Solid-based and Powder-based AM Systems), there is a side bar for navigation within that section. Whenever the cursor is moved over any of the side bar buttons, they will be changed or highlighted in colour. Clicking once on the side bar button will bring you to the respective page. Figure IV shows the side bar at the "Company" page of the SDL technique.

Playing a Movie

There are six movies in this courseware to help explain the concepts of AM techniques (again, two for each of the Liquid-based, Solid-based and Powder-based AM Systems). Whenever a particular sub-topic or page contains a related movie, the icon and words "movie clip" will appear when you move over an active icon, as seen in Figure V. Clicking on the movie icon will launch the movie player and play the movie automatically. You can return to the program during or after the movie is played.

The movie player has the following controls:

(5) *Play/Pause button.* Play or pause the movie at the period of time.
(6) *Forward button.* Restart the movie by clicking on the button.
(7) *Sound scroll bar.* User is able to adjust the loudness of the movie.
(8) *Movie scroll bar.* User is able to fast forward or scroll back to the part of the movie which s/he had missed out.

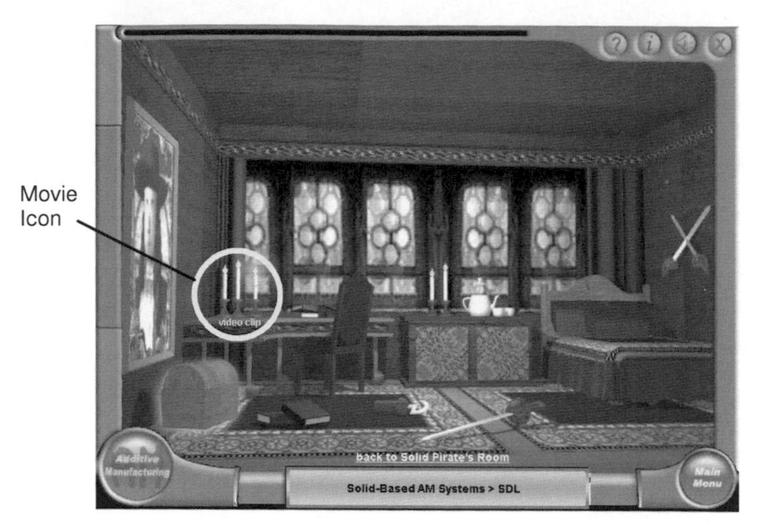

Fig. V. Screenshot with an active movie icon.

Self Check Quiz

There are self-quizzes in every part of the room for you to check if you have understood the content. To start the quiz, the user just has to click the "Start Quiz" button. After finishing the first round, you can choose to review the answer, retry the questions again or exit the quiz.

Fig. VI. Screenshot with active quiz icon.